New Advances in Postharvest Technology

Edited by İbrahim Kahramanoğlu

Published in London, United Kingdom

New Advances in Postharvest Technology
http://dx.doi.org/10.5772/intechopen.104277
Edited by İbrahim Kahramanoğlu

Contributors
Nariane Q. Vilhena, Lourdes Cervera-Chiner, Alejandra Salvador, Ana Moreno, N. Arunfred, V. Bini Marin, Anthony Tibaingana, Godswill Makombe, Tumo Kele, Mildred Osei-Kwarteng, Gustav K. Mahunu, Maurice Apaliya, Moomin Abu, Olapeju Olubusola Phorbee, Ganiyat Olatunde, Gift Oguzor, Lateef Sanni, Adeyinka Onabolu, Desmond Ikerionwu, Adeola Ojo, Apiradee Uthairatanakij, Natta Laohakunjit, Chalida Cholmaitri, John Golding, Pongphen Jitareerat, Oluyinka Benedicta Adewoyin, Ismael Aguilar-Ayala, Diana Herrera-Rojas, Annamaria Mincuzzi, Antonio Ippolito, Kian Sadeghi, Farhad Parnian-khajehdizaj, Mahdi Ganjkhanlou, Reza Faraji, Zahra Abdollahi, Subbaraman Sriram, Kodthalu Seetharamaiah Shivashankara, Darisi Venkata Sudhakar Rao, Pooja Shekar Patel, Dhia Ahmed Taain, Hamzah Abbas Hamzah, Abbas Mahdi Jasim, Bonga Lewis Ngcobo, Isa Bertling, Chalermchai Wongs-Aree, Hanh Thi Nguyen, Sompoch Noichinda, Ana Cristina Agulheiro-Santos, Sara Ricardo-Rodrigues, Marta Laranjo, Miguel Elias

First published in London, United Kingdom, 2023 by IntechOpen
IntechOpen is the global imprint of INTECHOPEN LIMITED, registered in England and Wales, registration number: 11086078, 5 Princes Gate Court, London, SW7 2QJ, United Kingdom

British Library Cataloguing-in-Publication Data
A catalogue record for this book is available from the British Library

Additional hard and PDF copies can be obtained from orders@intechopen.com

New Advances in Postharvest Technology
Edited by İbrahim Kahramanoğlu
p. cm.
Print ISBN 978-1-83768-541-7
Online ISBN 978-1-83768-542-4
eBook (PDF) ISBN 978-1-83768-543-1

For EU product safety concerns :
IN TECH d . o . o . , Prolaz Marije Krucifikse Kozulić 3 , 51000 Rijeka , Croatia ,
info@intechopen . com or visit our website at intechopen . com .

Meet the editor

İbrahim Kahramanoğlu is an associate professor at the Faculty of Agriculture, European University of Lefke, Northern Cyprus. He is an expert in horticultural production, postharvest biology and technology, and good agricultural practices. His main research interests are postharvest physiology and handling of fruits, natural and novel technologies for handling and storage, digital and precision farming (agri 4.0) for sustainability, and value-adding to horticultural crops. He has authored various books, book chapters, conference papers, and scientific publications related to his experience.

Contents

Preface

Producers worldwide devote significant resources to cultivating crops, with the laudable intention of feeding the world's population. A troubling reality, however, is that before they reach customers, almost one-third of these products are lost due to postharvest losses. The food supply and agricultural sustainability are seriously threatened by this worrying situation, necessitating immediate action to address this pressing problem. This edited book, *New Advances in Postharvest Technology*, was painstakingly created in answer to this challenge to share thorough and contemporary knowledge on the physiology of postharvest losses and the most recent technological developments in postharvest systems.

This book is essential because it places a focus on sustainable practices that can reduce postharvest losses while maintaining the inherent freshness and nutritional benefits of fresh fruits and vegetables. The need for wholesome food is increasing as a result of changing dietary habits and worldwide population expansion. Adopting eco-friendly treatments and creative handling techniques that not only reduce post-harvest losses but also support food security and improve the livelihoods of smallholder farmers become crucial.

Written by experts in postharvest technology from around the world, this book is thoughtfully organized into three sections, each of which focuses on a different aspect of environmentally friendly applications and current advancements in postharvest management techniques.

A variety of cutting-edge strategies are explored in Section 1, "Eco-Friendly Applications in Postharvest Handling", with the goal of reducing postharvest losses and improving the nutritional value of fresh fruit. This section's chapters examine the possibility of extending shelf life and maintaining quality using green technologies, enhanced fruit coatings, and innovative postharvest treatments. A closer look at the volatile organic chemicals these microorganisms create reveals their importance in the control of fruit postharvest illnesses. Additionally, a thorough investigation of the use of phyto-hormones, growth regulators, and calcium in both pre and postharvest phases sheds light on their favorable effects on fruit quality and storage life.

In Section 2, "Recent Developments in Handling Practices", handling practices that have revolutionized postharvest management are thoroughly discussed. This section's chapters explore a wide range of subjects, from creative ways to use post-harvest losses in feed production to an investigation of pre and postharvest elements that have a significant impact on the quality of harvested produce. The importance of suitable postharvest technologies for biofortified crops is also highlighted, with a focus on the part these technologies play in value addition, micronutrient retention, and improved utilization of biofortified crops. With a focus on the Indian context, a thorough review of food safety traceability systems indicates their crucial importance in preserving the integrity of postharvest systems. The section also provides a thorough summary of current advancements in the postharvest use of light-emitting diodes (LEDs) in horticulture, emphasizing how this technology has the potential to completely change the industry. Furthermore, the chapter on 1-MCP (1-methylcyclopropene) preharvest application provides insights into its effectiveness in enhancing postharvest fruit quality and extending shelf life.

The emphasis is narrowed to just a few fruits, specifically pomegranates and several *Prunus* species, in Section 3, "Special Considerations for Selected Fruits". The chapters in this section explore the postharvest fungal infections that damage pomegranates and the cutting-edge techniques used to control them. Moreover, the difficulties encountered in preserving the quality of *Prunus* spp. during postharvest handling are examined, along with potential futures that provide creative solutions for these difficulties. Additionally, the impact of pre and postharvest treatments on the protein patterns and storage behavior of date palm fruit is examined, offering important insights for improving postharvest storage conditions. Finally, a thought-provoking study on Ugandan smallholder maize farmers' storage practices highlights the need for better storage methods to ensure food security, a problem that is felt all over the world.

The authors' collective efforts have resulted in a thorough body of information that is both understandable and illuminating. Each chapter highlights the authors' expertise. This book makes a substantial contribution to the subject of postharvest technology and provides scholars, practitioners, and other interested parties with a useful tool for managing fresh fruit sustainably. I sincerely thank each one of the chapter authors who made a significant contribution to the creation of this book. This volume is a monument to the developments in postharvest technology thanks to their commitment and intellectual insights, which have enhanced the contents.

As the book's editor, I hope that it will act as a guiding light for future study and invention to address the problems caused by postharvest losses. I hope that this volume will reignite interest and collaboration among various stakeholders, resulting in more environmentally friendly practices and a decline in postharvest losses around the world.

İbrahim Kahramanoğlu
European University of Lefke,
Department of Horticulture,
Gemikonağı, Northern Cyprus, Turkey

Section 1

Eco-Friendly Applications in Postharvest Handling

Chapter 1

Green Technology for Reducing Postharvest Losses and Improving the Nutritional Quality of Fresh Horticultural Produce

Apiradee Uthairatanakij, Natta Laohakunjit, Pongphen Jitareerat, Chalida Cholmaitri and John Golding

Abstract

Fresh fruits and vegetables provide essential nutrition to the diet, and it is critical to maintain product quality and nutrition from harvest through to the consumer. Fresh fruit and vegetables are still 'alive' even after detached from the plants and continue to respire. Besides, the climacteric fruits ripen after harvest. Therefore, it is important to manage the ripening process and prevent decay to reduce postharvest losses. In addition, foodborne illnesses are a major public health concern, and postharvest practices to improve food safety are essential. While traditional postharvest technologies such as synthetic chemicals have been effective at controlling postharvest decay and maintaining fruit quality during storage, there is an urgent need to develop alternative 'green technologies' to maintain product quality through to the consumer. Many new innovative green postharvest technologies are being developed to delay ripening, reduce pathogenic microorganisms, maintain freshness, and improve nutrition. This chapter discusses some new innovative green postharvest technologies such as the application of edible coatings and films, light emitting diode (LED), ultrasound, UVC irradiation, and plasma technology, which have been shown to reduce postharvest losses and improve the nutritional quality of fresh produce.

Keywords: eco-friendly technology, food waste, food safety, nutritional quality, postharvest losses, sanitizing

1. Introduction

Fresh produce including fruits and vegetables are perishable crops. The major problems are quality degradation, microbial spoilage, and postharvest disease infection, resulting in quality and nutritional losses and short storage life. Postharvest handling is a crucial step to maintain the quality of fresh produces. However, applying chemical treatments to reduce microbial growth and inhibit postharvest diseases may cause chemical residue when it is overused, and chemical methods cannot apply to organic produce. Therefore, using postharvest green technology such as ultrasound, light emitting diode (LED), edible coatings, UVC irradiation, and plasma technology to reduce postharvest losses and improve the nutritional quality of fresh horticultural produce may be taken into consideration as an alternative treatment for applying after harvest.

2. Edible coatings and films

2.1 Edible coatings

Edible coatings are a thin layer (˂0.3 mm) of biodegradable materials wrapped or coated around the surface of fruits which act as a barrier for the exchange of moisture and gases between fruits and the surrounding environment which can then delay fruit respiration, ethylene production, and slow down microbial growth [1–3]. Edible coatings are commonly classified into three main groups. Lipid-based materials (such as fatty acids and waxes), protein-based materials (such as zein, casein, whey protein, soy protein, egg albumin, and gelatin), and polysaccharide-based materials (such as starch, cellulose, chitosan, alginate, and gums) or mixtures of them [4]. These edible coatings are considered 'green technologies' which are simple, safe, and eco-friendly and can be applied in liquid form by dipping, spraying, or brushing. The fundamental characteristics of edible coatings are that they must be food-safe, tasteless, odorless, and flexible [5]. Water loss from harvested fruit and vegetables is primarily through transpiration. Edible coatings have been widely reported to reduce weight loss in a range of fruit and vegetables such as sweet cherries, peaches, and plums [6–8]. Edible coatings primarily reduce quality losses by creating a semipermeable gas barrier around the product to restrict transpiration and regulate gaseous (O_2 and CO_2) exchange between the internal atmosphere of fruits and the external environment [9]. Therefore, edible coatings can enhance shelf life by decreasing weight loss, retarding physicochemical changes, and delaying fruit ripening. Hong et al. [10] reported that the use of 2% chitosan in coating guava fruits reduced weight loss, delayed changes in total soluble solids (TSS) and titratable acidity (TA), and maintained firmness during 12 days of storage at 11°C. Green pepper fruit coated with 2% chitosan exhibited a reduction of weight loss and extend postharvest life for 16 days at 12°C [11]. Similarly, Mandal et al. [12] found that 2% chitosan coatings on mango fruits reduced weight loss, remained green of color peel, and increased the shelf life compared to the untreated fruit at room temperature. Moreover, the application of 1% chitosan also delayed changes in weight loss, TSS, TA, and external color compared to the untreated fruit [13]. In addition, Ruelas-Chacon et al. [14] suggested that guar gum (1.5%) coating had great potential in reducing the respiration rate of tomatoes. Li et al. [15] reported that 1% peach gum polysaccharides coating on the surface of cherry tomatoes decreased weight loss, reduced respiration rate, maintained firmness, and extended the shelf life of cherry tomatoes. Similarly, sweet cherry coated with 5% alginate reduced respiration rate and higher in fruit firmness and TA compared to untreated fruit [16]. Ratra et al. [17] reported that aloe vera gel (50%) reduced weight loss, lower the respiration rate, and increased the shelf life of bananas. Moreover, other research confirmed that sweet cherry, pineapple, and apple benefited from aloe vera coating with significantly lower weight loss and delayed fruit ripening [18–20]. In addition, 'Alberta' peaches coated with methyl cellulose and sodium alginate reduced respiration rate by 62% and 68%, respectively, during storage at 15°C [21].

The mixture of high concentrations of chitosan and low concentrations of glycerol has been reported to decrease the respiration rate and delay the ripening of coated tomatoes [22]. Duong et al. [23] demonstrated that sodium alginate solution mixed with 40 mL L^{-1} calcium chloride ($CaCl_2$) significantly reduced the weight loss and respiration rate of the rose apple for 10 days at 13°C, while Mandal et al. [24] reported that 3% carboxymethyl cellulose blended with 2% chitosan and wax extended the shelf life of tomato for 24 days at 20–25°C by delaying the ripening process. These numerous examples clearly demonstrate the application of 'green' edible coatings to improve the shelf life of fresh produce and minimize postharvest losses.

2.2 Edible films

Edible films are usually prepared by dissolving in water, alcohol, or a mixture of solvents. A plasticizer is often added to the solution in order to improve flexibility and elasticity of film. Other additives, such as antimicrobial agents, colors, and flavor can be added to the solution to create specific film properties and functionality [25]. Polysaccharide-based edible films such as carrageenan and chitosan can be used as a strong barrier to non-polar aroma compounds, reducing aroma loss and oxidation [26]. Hydrocolloid-based edible films such as alginate and carboxymethylcellulose (CMC) have the potential to prevent moisture losses [27]. Another essential factor to contemplate for selecting edible film material is the ability to serve as an effective carrier for antimicrobials. Edible chitosan film combined with bioactive compounds and essential oils allowed for the reduction of *Escherichia coli* and *Listeria monocytogenes* and the enhancement of the overall quality of broccoli [28]. Pullulan-based edible films combined with antibrowning and antimicrobial agents prevented enzymatic browning, delayed tissue softening, decreased weight loss, reduced respiration rates, and inhibited the growth of microorganisms in fresh-cut apples [29]. Although edible coatings and films are not replacing conventional packaging completely, they can be used as alternative packaging. Edible films do not only decrease the postharvest losses but also reduce the environmental pollution in long term.

3. Light emitting diode (LED)

3.1 LEDs induce phytonutrients in fresh produces

Light is an important environmental factor that influences plant growth and development, as it offers both the source of energy for photosynthesis and the signal for a wide range of physiological and biochemical processes. LED technology can produce monochromatic light within a narrow wavelength between 400 nm and 700 nm, where LEDs of different wavelengths (such as red, blue, green, and white light) can trigger different responses in plants. The blue and red LED lights are the most effective wavelengths for plant photosynthesis, while yellow and green light have a negligible effect because the absorption spectra of the photosynthetic pigments mainly target blue (400–500 nm) and red (600–700 nm) light spectra [30]. While LEDs are used widely in protected cropping, LED lights are becoming widely studied for their postharvest applications to extend the shelf life and maintain the postharvest quality of fresh produce due to their wavelength specificity, long lifespan, low thermal energy, and non-toxicity [31]. The quality of light has been shown to affect the accumulation of phytonutrients and enhance the levels of phytonutrients in plants

LED light	Light conditions	Fresh produces	Phytonutrients	References
	470 nm, 40 W m^{-2}	Strawberry	Total phenolics and anthocyanins	[33, 34]
	525 nm, 20 W m^{-2}	Broccoli	Total phenolics	[35]
Blue	436 nm, 1.52 W	'Dongdori' cabbage	Total phenolics, chlorophyll, and vitamin C	[36]
	400–500 nm, 80 μmol m^{-2} s^{-1}	Purple pepper	Anthocyanins	[37]
	625 nm, 12 W	Table grapes	Total phenolics and flavonoids	[38]
	663 nm, 1 W	Cherry tomato	Lycopene and β-carotene	[39]
Red	638–nm and 665–nm, 210 μmol m^{-2} s^{-1}	Basil and parsley microgreens	Total phenolics, α-tocopherol, and vitamin C	[40]
Green	510 nm, 30 μmol m^{-2} s^{-1}	Baby leaf lettuce	α-Carotene and anthocyanins	[41]
	52 μmol m^{-2} s^{-1}	Chinese cabbage	Total polyphenols and total flavonoids	[42]
Blue:red	Blue (30%) and red (70%)	Eggplant	Phenolic acid, chlorogenic acid, and gallic acid	[43]
	2016 kJ m^{-2}, 16 h	Carrot sprouts	Carotenoids	[44]

Table 1.
Effect of LEDs in inducing the levels of phytonutrients on fresh produces.

[32]. Ambient light supplemented with blue, red, green, or blue: red LEDs enhanced the total phenolics, flavonoids, anthocyanins, lycopene, α-tocopherol, and other compounds in several fruits and vegetables (**Table 1**).

LEDs' role in the induction of bioactive compound production in plants seems to be associated with phenylalanine ammonia-lyase (PAL) enzyme, which is engaged in the initial step of the phenylpropanoid pathway [45]. It has previously been reported that LED increased accumulation of primary metabolites in plants and induced the suppression of the translocation of photosynthetic products. LEDs also affect the signal transduction pathways by inducing the production of secondary metabolites in plants [46]. LED lights have a significant effect on the accumulation of phytonutrients and therefore could be used as an alternative technique for enhancing the quantity and quality of the phytonutrient profiles linked to nutrition and human health. However, the use of single- or combination-spectral light ratios may vary effect depending on the plant species or cultivars [47]. Therefore, more investigation is required to establish the spectrum qualities that make the best choice for enhancing the phytonutrient properties of fresh produce.

3.2 LEDs induce disease resistance in fresh produces

Nowadays, LEDs are becoming increasingly popular as a practical tool for protecting fruits from pathogen attacks. The specific wavelengths of light, especially red and

LED light	Light conditions	Fresh produces	Effect on disease	References
	200 μmol m^{-2} s^{-1}	Lettuce	Induced resistance against gray mold by *Botrytis cinerea*	[48]
Blue	50–150 μmol m^{-2} s^{-1}	Tomato	Induced resistance against gray mold by *B. cinerea*	[49]
	40 μmol m^{-2} s^{-1}	Tangerine	Induced resistance against *P. digitatum*	[50]
	8 μmol m^{-2} s^{-1}	Mandarin	Induced resistance against *P. italicum*	[51]
Red	287 μ W cm^{-2}	Bell pepper	Induced resistance against *Phytophthora capsici*	[52]
	80 μmol m^{-2} s^{-1}	Grapevine leaves	Induced resistance against gray mold by *B. cinerea*	[53]

Table 2.
Effect of LEDs in the inducing disease resistance in fresh produces.

blue LEDs, can induce disease resistance in plants against a wide range of microorganisms (**Table 2**). The mechanism of LED inactivation against pathogens is direct damage to DNA and cell membrane caused by reactive oxygen species (ROS) produced by LED light [54]. Blue light enhanced the production of phytoalexin scoparone thus, preventing the postharvest decay caused by *Penicillium digitatum* and *P. italicum* in citrus fruits [55]. Moreover, it was also reported that blue light stimulated the formation of phospholipase A2 and octanal, which resulted in a reduction of *P. digitatum* and *P. italicum* activities [50]. Red light inhibited disease development by increasing the expression of defense-related genes and promoting the production of phytoalexin stilbenenic [56]. LED lights affect the preservation of fresh produce after harvest by a dual effect; 1) the inhibition of microbial growth and 2) stimulation of plant defense responses. Therefore, further investigation is required to truly understand the mechanism, which may potentially lead to the development of LEDs that are effective tools for reducing postharvest disease.

4. Ultrasound

Treatment of fruit and vegetables with ultrasound is a relatively new method of maintaining fruit and vegetable quality after harvest. Ultrasound is a type of vibrational energy in the frequency above 20 kHz, which is higher than the range of human hearing. Based on the frequencies, ultrasound waves are divided into three categories, including diagnostic ultrasound (1–500 MHz), high-frequency ultrasound (100 kHz–1 MHz), and low-frequency ultrasound (20–100 kHz). The applications of ultrasound are classified into low and high energy depending on the sound power (W), sound intensity (W m^{-2}), or sound energy density (W s m^{-3}). High-energy ultrasound with low frequencies (20–100 kHz) refers to as 'power ultrasound,' which is widely used in the food industry to preserve food quality and provide food safety

[57]. The principle of ultrasound on microbial inactivation is to produce a cavitation phenomenon through the liquid medium. During the sonication process, longitudinal ultrasound waves pass through liquid media and produce alternating regions of compression and rarefaction to promote the cavitation phenomenon. A certain amount of gas can be trapped in cavities and form cavitation bubbles in the regions of different pressure. During rarefaction, the negative pressure increases the bubble volume, while the positive pressure decreases the bubble volume as the sound wave changes to compression. During treatment, very localized high temperatures (5500 K) and pressures (up to 50–100 MPa) are created by shock waves, which also accelerate bubble collapse. When the bubble collapses, the breakdown of water molecules inside the bubble explosion produces free radicals such as hydrogen atoms [H^+] and hydroxyl radicals [OH^-]. Hydrogen peroxide [H_2O_2] is formed when OH radicals combine with water [H_2O], which is a cause of disruption of the cell wall, thus inhibiting the growth of microorganisms [58].

Microbial spoilage is a major cause of postharvest waste and is caused by fungi, bacteria, yeast, and molds. Fresh produces are normally protected from microbial attack by using synthetic chemicals, but the application of green technologies such as ultrasound also reduce postharvest losses. Cao et al. [59] reported that ultrasound treatment at 40 kHz and 250 W for 10 min could be applied to reduce mesophilic aerobic mold and yeast by 1.49 log CFU g^{-1} and 1.73 log CFU g^{-1}, respectively, and maintain the quality of strawberry at 5°C for 8 days. Birmpa et al. [60] reported that ultrasound treatment at 37 kHz and 30 W for 45 min was effective in reducing the growth of *E. coli*, *Listeria innocua*, *Salmonella enteritidis*, and *Staphylococcus aureus* inoculated on lettuce and strawberries. Similarly, Alexandre et al. [61] showed that the number of *L. innocua* inoculated on red bell pepper was reduced by 1.9 log CFU g^{-1} after being treated with ultrasound treatment at 35 kHz and 350 W for 2 min. Gani et al. [62] observed that the use of ultrasound treatment of strawberries at 33 kHz and 60 W for 40 min contributed to a reduction in the number of bacterial from 5.91 to 3.91 log CFU g^{-1} and yeast and mold from 4.80 to 3.58 log CFU g^{-1}. Supapvanich and Kijka [63] demonstrated that the ultrasound treatment at 40 kHz and 150 w for 10 min maintained visual appearance, inhibited decay incidence, and reduced weight loss of 'Kim Ju' guava fruits during storage at 28 ± 1°C for 6 days. Furthermore, Guerrero et al. [64] confirmed that ultrasound treatment damages the cell wall and membrane integrity of the microbial pathogen. Thus, the ultrasound treatment reduces postharvest microbial growth due to cellular disruption by shear force or increasing temperature and pressure during bubble collapse, which creates hydroxyl radicals and leads to severe cell wall damage [65]. The produced highly reactive radicals can destroy the cell wall and membrane of microorganisms [66], which helps in the prevention of postharvest disease in fresh produces.

While ultrasound treatment has been shown to reduce the levels of microbes in fruits and vegetables, its application in combination with other green technology treatments, such as sanitizers, organic acids, essential oils, and other antimicrobials, has been studied to enhance the efficiency of this technique. For example, Chen and Zhu [67] reported that the combination of ultrasound at 40 kHz and 100 W for 10 min with chlorine dioxide at 40 mg L^{-1} reduced initial microflora and maintained the quality of Japanese plum. Similarly, the population of *Salmonella and E. coli* O157: H7 contaminated on the surface of apples was reduced by 2 and 1.5 log CFU g^{-1} in combination with ultrasound at 170 kHz and chlorine dioxide at 40 mg L^{-1} for 10 min [68]. Sagong et al. [69] observed that the effect of 2% lactic acid, 2% citric acid, and 2% malic acid combined with ultrasound treatment at 40 kHz for 5 min was effective to reduce *E. coli, S. typhimurium*, and *L. monocytogenes* on organic fresh lettuce. Yang et al. [70] also reported that the combination of ultrasound at 40 kHz for 10 min with salicylic acid mM could reduce blue mold caused by *Penicillium expansum* in peach fruit. Millan-Sango et al. [71] demonstrated that the combined treatment of ultrasound at 26 kHz and 200 W for 5 min with thyme essential oil at 0.018% reduced *Salmonella enterica* in lettuce. These examples demonstrate the effectiveness of ultrasonic combination treatments to reduce food safety and fruit pathogens; however, more work is required to optimize these treatments to kill pathogens and have minimal effects on product quality.

5. Ultraviolet (UV) radiation

UV radiation is widely used as a microbial sterilizing treatment in a range of diverse applications such as the treatment of wastewater, process water, surface disinfection, and air disinfection in food processing, packing, pharmaceuticals, and hospitals [72, 73]. In the postharvest treatment of fresh produce, UV radiation has been used for more than four decades. The main uses of postharvest UV treatments in postharvest include the following: (1) its use as defect sorting, that is, separating defective fruit (discolored and wounds) from perfect fruit, (2) control of spoilage and pathogenic microorganisms (i.e., *Alternaria* sp., *B. cinerea*, *Colletotrichum* sp., *Lasiodiplodia theobromea, Monilinia* sp., *Fusarium* sp., *Rhizopus* sp., and *Penicillium* spp.), (3) enhancement of bioactive compounds, (4) delayed fruit ripening and senescence, and (5) control of human pathogens (i.e., *E. coli*, *Salmonella*, and *Listeria*) that can contaminate fresh produce [72, 73]. However, this review will focus on the application of UV-C in the control of plant pathogens that cause postharvest diseases and improving the quality of fresh produce. UV radiation is commonly considered non-ionizing radiation where the wavelength of UV radiation is 100–400 nm, and its light region is between X-ray and visible light spectrum. The electromagnetic spectrum of UV is commonly divided into three regions: UV-C is a short wavelength of 100–280 nm, UV-B is a medium wavelength of 280–315 nm, and UV-A is a long wavelength of 315–400 nm [73]. Among these different UV spectra, UV-C shows the highest antimicrobial effect [72]. The most efficient wavelength for damaging genetic materials (DNA) of microbial and all biological tissues and eliciting antimicrobial compounds is 254 nm [72, 74]. Microbial death caused by UV-C is not only by genetic material damage but also by the overproduction of ROS, which can oxidize membrane lipids and inactivate the activity of cellular enzymes. It has been shown that gram-negative bacteria are more sensitive than gram-positive bacteria, followed by yeasts and eukaryotic organisms [73].

5.1 UV-C treatment in the control of postharvest disease

Postharvest disease is one of the major economic losses of fresh produce where the most postharvest diseases are caused by fungi, yeasts, and bacteria. Synthetic fungicides are currently used to prevent disease in fresh fruit and vegetables, but consumers and regulators are seeking safe and effective alternative treatments to manage postharvest decay. Treatments such as UV-C are ideal alternatives to synthetic chemicals as it is a physical treatment and does not leave chemical residues [72]. The UV-C desired dose required controlling postharvest diseases, and improving produce quality is dependent on a range of factors such as the type of produce, maturity stage, cultivar, and harvest season, but generally ranges between 0.2 kJ m^{-2} and 20 kJ m^{-2} [72, 73].

The antimicrobial effect of UV-C treatment on the control of postharvest diseases is thought to occur through two different mechanisms: (1) a direct germicidal effect on biomolecules of plant pathogens such as nucleic acid, membrane, and proteins [75] and (2) *via* an indirect effect on plant pathogens through the induction of the plant defense mechanisms following UV-C treatment such as plant defense enzymes, plant pathogenesis-related proteins (PRs), and secondary metabolite accumulation (i.e., antimicrobial agents—phenolic compounds and phytoalexins) [73, 76]. The limitation of a direct effect of UV-C treatment against pathogens is the low penetration of this radiation treatment into plant tissue, where only pathogens present on the surface can be killed due to the poor penetration ability of UV-C into the tissue (only 50–300 nm) [72]. Thus, to get complete coverage fresh produce must be rotated during treatment to ensure that all surfaces are exposed to UV-C light.

In general, plant defense responses can be induced by biotic and abiotic elicitors such as pathogens, antagonistic microorganisms, chemical treatments, and physical treatments like UV-C irradiation. Several studies have shown that UV-C treatment can elicit a range of responses such as the accumulation of antimicrobial compounds, the increase in the activities of various enzymes associated with plant defense, and plant pathogenesis-related proteins (PRs) such as phenylalanine ammonia-lyase (PAL), chitinase (CHI), β-1,3 glucanase (GLU), and peroxidase (POD). PAL is the key enzyme that plays a role in the biosynthesis of phenolics and other secondary metabolites phytoalexins (phenol), and even salicylic acid, which is a crucial signal molecule involved in plant protection from pathogen attack [75, 77]. UV-C radiation has been shown to induce plant defense mechanisms in a range of different fresh produce such as strawberry [74], pear [75], mangosteen [76], mango [78], peach [79], tomato [80, 81], and banana [82], thereby extending the shelf life of these fruits.

5.2 UV-C treatment in postharvest nutritional values

UV-C has been shown to affect a range of postharvest quality attributes including changes in plant pigments, antioxidants, nutrition, firmness, flavor, and aroma of fresh produce. These physiological changes may be desirable or undesirable depending on the type of produce as these changes directly affect eating quality. UV-C treatment has been shown to enhance antioxidant systems for both antioxidant compounds and oxidative enzyme activities in various fresh produce pear [75], mangosteen [76], mangoes [78], broccoli [83, 84], leaf vegetables [85], garlic [86], pepino fruit [87], and tomato [88, 89]. While many studies have shown that UV-C radiation treatment can induce many bioactive compounds, but some studies have shown that UV-C treatment did not enhance polyphenols, β-carotene, ascorbic acid, chlorophyll contents in persimmon and cucumber [90], and total phenolic content, anthocyanin compounds, and antioxidant activity in grape [81].

5.3 Combined UV-C treatment with other postharvest technologies

UV-C treatment can combine with other postharvest treatments such as chemical, physical, and biological treatments to improve postharvest quality. For example, Sripong et al. [78] demonstrated that UV-C irradiation (6.16 kJ m^{-2}) combined with hot water treatment at 55°C for 5 min significantly reduced the incidence and severity of postharvest anthracnose disease development in mango fruit compared with UV-C or hot water treatment alone. The reduction of disease was related to an increase in plant defense-related enzyme activities and their related gene expressions. In addition, the combined treatments could retard fruit ripening by maintaining firmness, slowing color change, retarding the increase of total soluble solids, and the decrease of titratable acidity (TA). This result could be explained that the treatments may slow down the activity of the enzymes associated with plant cell wall degradation and the hydrolysis of starch to sugar. The acidity of fruit has been known that it is used as a respiratory substrate. Thus, low respiration rate is a cause of slowing down the decrease in TA.

5.4 The advantage and disadvantages of UV-C treatment

As a physical treatment to improve the storage life of fruit produce, UV-C treatment has numerous practical advantages that include the following: (1) fast and simple handling, (2) low cost and investment, (3) low maintenance, (4) ability to operate at low temperatures and not requiring additional water, (5) less space for treatment, (6) it is non-ionizing treatment, (7) it is non-thermal technology which does not induce heat in the tissue of fresh produce, (8) fewer regulatory restrictions as compared to other irradiation methods such as gamma irradiation, and (9) approved by food control agencies [72, 78]. However, the disadvantages of UV-C radiation include the following: (1) low penetration power, and the potential penetration into plant tissue is only 50–300 nm. Thus, only pathogens on the plant surface are inactivated after exposure to this radiation, (2) UV-C rays are dangerous a workplace health and safety (WHS) issue and can damage the eyes and skin [72]. Thus, during UV-C treatment application even on the laboratory scale and a commercial scale, the treatment must be conducted with complete protection to all users, (3) generation of ozone gas during the operation with UV radiation, particularly at a wavelength below 260 nm, will produce ozone gas. Ozone is hazardous to the human health and can cause the air pollution in the working atmosphere, and thus, air ventilation or treatment may be necessary. The U.S. Environmental Protection Agency (EPA) states that the national ambient air quality standard for ozone is 70 ppb for an 8-hour average [82]. However, activated carbon filters are being used for ozone removal in the atmosphere [83, 84]. In summary, UV-C technology is a potential physical treatment that has many benefits for the storage of fresh fruit and vegetables. UV-C treatment has been shown to induce defense systems within the fruit through the signaling molecules and antimicrobial compounds to protect against plant pathogens, improve the nutritional quality of some fresh produce, and delay senescence resulting in extending the shelf life of fresh fruit and vegetables.

6. Plasma technology

6.1 Types of plasma and plasma generation

Plasma technology is a novel approach to preserving the quality and inactivating spoilage microorganisms and pathogens of food processes and maintaining the postharvest quality of fresh produce. It is an accepted technology for industry because of its relatively low input cost [85]. Plasma is commonly known as the fourth state of matter, after gases, liquids, and solids. Plasma is quite like a gas than a solid or liquid, but its property differs from solid, liquid, and gas [86]. Two major types of plasma are differentiated based on thermodynamics: thermal plasma and non-thermal plasma (referred to cold plasma, temperature < 60°C) [86–88]. Thus, cold plasma (CP) is selected to apply to various foods including fruit and vegetables to avoid the loss of nutrition and sensory contributions. CP can be generated by many techniques including corona discharge (CD), dielectric barrier discharge (DBD), gliding arc discharge (GAD), microplasma, radio frequency (RF) discharge, plasma spray, plasma needle, and atmospheric pressure plasma jet [86, 88]. In all plasma techniques, the feed gas is required to energize into plasma [88]. CP is commonly generated by energizing matter with an electric current (energy), and the solid state of matter changes from liquid to gas and finally to a plasma state, respectively. When the initial gas is treated with sufficiently high energy, the structure of molecules and intra-atomics is broken resulting in the formation of free electrons, neutrals, and other reactive ion species. The initial feed-in gas for plasma generation can be either a single gas or a mixture of various gases, i.e., carbon dioxide, oxygen, nitrogen, argon, neon, helium, or even air [86, 87]. In addition, moisture can be mixed with the feed-in gases [88]. The treatment plasma state consists of a mix of excited atoms and molecules (ionization), positive and negative charged particles, UV photons, ozone, ROS (such as superoxide [O_2^-], hydroxyl [$OH^{\cdot}\cdot$], hydroperoxyl [HO_2]), hydrogen peroxide [H_2O_2]), and reactive nitrogen species (RNS such as nitric oxide [NO], nitrogen dioxide [NO_2], dinitrogen pentoxide [N_2O_5], nitrate anions [NO_3^-], and nitrite anions [NO_2^-]) [86–88].

Sufficient concentrations of the plasma compositions in the environment (atmosphere or liquid plasma) have been shown to inactivate microbes such as bacteria, fungi, spores, and viruses without hazardous chemical residues [86, 89]. CP can be subdivided into non-thermal atmospheric plasma (NTAP) and plasma-activated water (PAW). PAW is conducted by applying plasma discharge in water (plasma generated directly in the water) or on the water surface (plasma generated over the water surface) [90, 91]. An increase in many chemically reactive molecules (reactive oxygen and nitrogen species, RONS) created in PAW is correlated with high oxidation-reduction potential (ORP), electrical conductivity (EC), and low acidity pH levels [92] that are considered to have high antimicrobial effects. However, the use of some oxidizing chemical solutions instead of water for plasma generation is called the plasma-activated solution (PAS) and can be used to increase the antimicrobial mechanism of the treatment, as indicated by the increases in ORP and EC values [92]. ORP, EC, and low pH values provide synergistic inactivation effects along with RONS. In addition to anti-microbial effects, CP can significantly slow the loss of fresh produce quality deterioration *via* the slowdown ripening and senescence processes, thereby maintaining the nutrition and organoleptically appearance [91].

6.2 Mode of action of cold plasma

There are large numbers of studies examining the mechanisms of CP treatment against microbes. It is generally thought that the mode of action of CP involves the interaction of the active chemical species, radicals, and reactive molecules generated by CP with microbial cell membranes and cellular functions through lipid peroxidation, protein denaturation, and genetically material degradation leading to cell damage, pore formation, cell lysis, and cytoplasm shrinkage [85]. However, the long lifespan of RONS such as hydrogen peroxide, nitrate, nitrite, and ozone may help maintaining the inactivation effect on microbial for a longer period than short-lifespan species such as hydroxyl radicals, superoxide, singlet oxygen, nitric oxide, peroxynitrite, and peroxynitric [92].

6.3 Cold plasma for food-borne pathogens

There are many studies on application of cold plasma treatment for inhibiting food-borne pathogens contamination, postharvest disease infection, and maintaining the quality of fruit and vegetables. For example, the inactivation effects of cold plasma on food-borne pathogens have been reported in a large number of produce varieties: *Salmonella typhimurium* in radish sprouts [93], *Salmonella stanley* and *E. coli* O157: H7 in apples [94], *E. coli* O157:H7 *Salmonella* sp., *L. monocytogenes* in apples, cantaloupe, and lettuce [95], etc.

As a case study, the antimicrobial efficiency of PAS treatment on cell structure of the food-borne pathogen and the contaminated fresh lettuce was investigated and showed that a 3% (v/v) of H_2O_2 solution treated with DBD plasma and incubated with *Staphylococcus aureus* for 10–30 min resulted in a decrease in the PAS-treated *S. aureus* population, which was correlated with the increase in PAS incubation time. Although the cell wall of gram-positive *S. aureus* is thick, it is composed of a peptidoglycan (PG) layer, which is less sensitive to ROS. It is thought that the high ROS formation by PAS attacks the PG layer, membrane, and DNA leading to cell death. Higher leakage of intracellular components DNA, proteins, and K^+ of PAS-treated *S. aureus* were also found in longer incubation time than in short incubation time. *In vitro* tests where soaking lettuce samples with PAS for 10 min have been shown to reduce the total bacteria counts by 1.25 log $CFUg^{-1}$, and this treatment also retained the nutrient quality of lettuce, as indicated by the stability of green color and vitamin C and chlorophyll contents when compared with the control (deionized water) [96].

6.4 Cold plasma for postharvest quality and disease control

Postharvest decays caused by microbial contamination and infection are the major factors of postharvest deterioration. The effects of plasma treatments against various postharvest diseases in a range of different fresh produce have been reported. For example, the application of CP treatment on postharvest fungal pathogens and spoilage fungi has been reviewed in *Aspergillus niger* in date palm [97], *P. italicum* in Satsuma mandarins [98], *P. digitatum* in lemon [99], *Alternaria alternata, A. niger,* and *P. italicum* in wash water from cherries, grapes, and strawberries [100].

PAW treatment for 45 min has been shown to inhibit postharvest decay of kumquat fruit caused by *Penicillium italicum* and maintained the fruit firmness, while color, vitamin C, total flavonoid, and carotenoids were not affected by PAW treatment [101]. The results of decontamination of strawberries inside closed packages with different gas mixtures (O_2, CO_2, and N_2), and then treated with DBD plasma at the ambient temperature had a similar antimicrobial effect to decontaminate total aerobic mesophiles and yeasts and molds by 3.0 log reductions. Strawberry fruit treated with plasma at high oxygen gas mixtures (65% O_2 + 16% N_2 + 19% CO_2) had lower fruit respiration and higher firmness and brightness than plasma treatment at a high nitrogen gas mixture (90% N_2 + 10% O_2) [102]. The quality of the mushroom (*Agaricus bisporus*) after soaking with PAW for 1, 10, and 15 min showed that PAW treatment reduced bacterial and fungal contamination by 1.5 and 0.5 log and also delayed the mushroom softening. In addition, mushroom color, pH, and antioxidant properties were not affected [103].

The effects of plasma treatment on either postharvest disease inhibition or quality maintenance in tropical fruit have been reported. For example, the plasma-activated solution (PAS) obtained by treating with a dielectric barrier discharge (DBD) in the phosphate buffer showed a decrease in spore germination and spore viability of *Colletotrichum asianum,* a causal agent of mango anthracnose disease. Indeed, the incidence of *Colletotrichum* in PAS-treated mango fruit was 48% lower than in non-treated fruit. SEM microstructure of the treated *Colletotrichum* spores showed that the subcellular structure was damaged by plasma treatment [104].

Browning of fresh and minimally processed fresh produce is an undesirable characteristic that is involved with two major browning enzymes: polyphenol oxidase (PPO) and peroxidase (POD). In postharvest trials with treated banana fruit, the PPO and POD activities of the atmospheric plasma-treated banana slices were 70% and 100%, respectively, lower than untreated fruit. In addition, the levels of phytonutrient compounds (total phenol and flavonoid), antioxidant activity, and vitamin B6 also increased following treatment. However, the plasma treatment produced porous structures in the fruit leading to a decrease in hardness [105]. A similar observation resulted in other studies with banana slices has shown that CP treatment inactivated the enzymatic browning activities (PPO and POD) resulting in low quinones and less browning. However, plasma treatment also induced ROS levels in cells and morphological surface changes (rougher, fissures, and cracks) with the increase in treatment time [106].

The shelf life of fruit and vegetables is usually limited by plant pathogens and senescence as indicated by their respiration rate. Cold storage and modified atmosphere packaging (MAP) are commercial preservation technologies, which are commercially used to prolong the shelf life of fresh produce [107]. The combined effect of cold storage, MAP, and plasma treatments on the safety and quality of cherry tomatoes has been studied. Cherry tomatoes treated with PAW and packed in the polypropylene (with perforated-oriented polypropylene film to generate the equilibrium MAP condition) and then stored were found to have reduced the microbial load on the surface of the tomato, whereas MAP and low storage temperature could prevent water loss and maintain the total soluble solids content resulting in shelf life extension [108]. However, there is less information about the effect of cold plasma on the activation of the natural defense mechanisms in fresh produce, and more work is required in this area.

6.5 Advantages and disadvantages of plasma technology

Cold plasma is heat-free technology that is suitable for fresh produce as it prevents postharvest losses by inactivating plant pathogens, reducing microbial contamination, slowing the loss of freshness and firmness, and delaying nutritional losses: vitamins, colors (i.e., chlorophyll, carotenoids, and anthocyanin), various phytochemical compounds, antioxidant properties, without no harmful synthetic chemical residues [86]. However, the efficiency of cold plasma technology for decontamination and quality maintenance is dependent on the variety of gas, gas flow rate, and process time of treatment. Undesirable conditions related to these factors may provide negative effects such as loss of color, nutrition, and bioactive compounds. For example, the decline in brightness (L^* value) of strawberry fruit treated with atmospheric cold plasma under a high nitrogen environment is probably due to superficial bleaching [102]. The total color difference (ΔE^*) value of the carrot slices treated with atmospheric pressure cold plasma treatment is unacceptable and may be caused by high oxidation of the surface carotene and the loss of moisture on the surface. A major disadvantage of the use of CP is the high investment in a state-of-the-art cold plasma system [86].

7. Conclusion

Postharvest losses represent economic losses in fresh food supply chains but can be drastically reduced and prevented by using appropriate postharvest technologies. Reducing postharvest losses have the potential to enhance food availability and minimize waste. Green postharvest technologies such as those presented above including light emitting diodes, edible coatings, and UV can be applied to fresh produce to enhance nutritional values and also maintain postharvest quality. In addition, the applications of ultrasound, UV-C, and plasma technology could reduce postharvest diseases, eliminate microbial contamination, and improve food safety, resulting in reducing food loss and food waste. However, there are still many technical and commercial challenges that need to be overcome to maintain the quality and extend the shelf life of fresh produce for modern trade. Each green technology has desirable limitations, and thus, using the hurdle approach to improve its effects on the optimization of the combined treatments may be a focus of future research for industrial application. For example, the application of ultrasound combined with UVC may have synergistic effects on microbial reduction. The combined technologies including edible coatings/films together with UVC, LED with UVC-LED, or hot water treatments with plant extracts or essential oils can be improved bioactive compounds and antioxidant activity. More research is needed to provide efficiency and cost-effectiveness for industry.

Acknowledgements

This is in part a contribution from the Horticulture Innovation 'Citrus Postharvest Program' (CT19003).

Conflict of interest

The authors declare no conflict of interest.

Author details

Apiradee Uthairatanakij[1*], Natta Laohakunjit[1], Pongphen Jitareerat[1], Chalida Cholmaitri[1] and John Golding[2]

1 School of Bioresources and Technology, King Mongkut's University of Technology Thonburi, Bangkok, Thailand

2 NSW Department of Primary Industries, Gosford, NSW, Australia

*Address all correspondence to: apiradee.uth@kmutt.ac.th

References

[1] Lin D, Zhao Y. Innovations in the development and application of edible coatings for fresh and minimally processed fruits and vegetables. Comprehensive Reviews in Food Science and Food Safety. 2007;**6**(3):60-75. DOI: 10.1111/j.1541-4337.2007.00018.x

[2] Kahramanoğlu I, Chen C, Gan Z, Chen J, Wan C. The effects of edible coatings on the postharvest quality of citrus fruits as affected by granulation. Journal of Food Quality. 2020;**2020**:1-8. DOI: 10.1155/2020/8819233

[3] Wan C, Kahramanoğlu I, Okatan V. Application of plant natural products for the management of postharvest diseases in fruits. Folia Horticulturae. 2021;**33**(1):203-215. DOI: 10.2478/fhort-2021-0016

[4] Pascall MA, Lin SJ. The application of edible polymeric films and coatings in the food industry. Journal of Food Processing & Technology. 2013;**4**:e116. DOI: 10.4172/2157-7110.1000e116

[5] Prasad N, Batra E. Edible coating (the future of packaging): Cheapest and alternative source to extend the post-harvest changes: A review. Asian Journal of Biochemical and Pharmaceutical Research. 2005;**3**(5):2231-2560

[6] Yaman O, Bayoundurlc L. Effect of an edible coating and cold storage on shelf life and quality of cherries. LWT–Food Science and Technology. 2002;**35**:146-150. DOI: 10.1006/fstl.2001.0827

[7] Hazrati S, Beyraghdar KA, Habibzadeh F, Tahmasebi-Sarvestani Z, Sadeghi AR. Evaluation of Aloe vera gel as an alternative edible coating for peach fruits during cold storage period. Gesunde Pflanzen. 2017;**69**(3):131-137. DOI: 10.1007/s10343-017-0397-5

[8] Kumar P, Sethi S, Sharma RR, Srivastav M, Singh D, Varghese E. Edible coatings influence the cold–storage life and quality of 'Santa Rosa' plum (*Prunus salicina* Lindell). Journal of Food Science and Technology. 2018;**55**(6):2344-2350. DOI: 10.1007/s13197-018-3130-1

[9] Ncama K, Mditshwa A, Tesfay SZ, Mbili NC, Magwaza LS. Topical procedures adopted in testing and application of plant-based extracts as bio-fungicides in controlling postharvest decay of fresh produce. Crop Protection. 2019;**115**:142-151. DOI: 10.1016/j.cropro.2018.09.016

[10] Hong K, Xie J, Zhang L, Sun D, Gong D. Effects of chitosan coating on postharvest life and quality of guava (*Psidium guajava* L.) fruit during cold storage. Scientia Horticulturae. 2012;**144**:172-178. DOI: 10.1016/j.scienta.2012.07.002

[11] Gholamipour Fard K, Kamari S, Ghasemnezhad M, Ghazvini RF. Effect of chitosan coating on weight loss and postharvest quality of green pepper (*Capsicum annum* L.) fruits. Acta Horticulturae. 2010;**877**:821-826. DOI: 10.17660/actahortic.2010.877.109

[12] Mandal D, Sailo L, Hazarika T, Shukla AC. Effect of edible coating on shelf life and quality of local mango cv. Rangkuai of Mizoram. Research on Crops. 2018;**19**:419-424. DOI: 10.31830/2348-7542.2018.0001.10

[13] Hossain MS, Iqbal A. Effect of shrimp chitosan coating on postharvest quality of banana (*Musa sapientum* L.)

fruits. International Food Research Journal. 2016;**23**:277-283

[14] Ruelas-Chacon X, Contreras-Esquivel JC, Montañez J, Aguilera-Carbo AF, Reyes-Vega ML, Peralta-Rodriguez RD, et al. Guar gum as an edible coating for enhancing shelf–life and improving postharvest quality of Roma tomato (*Solanum lycopersicum* L.). Journal of Food Quality. 2017;**2017**:1-9. DOI: 10.1155/2017/8608304

[15] Li C, Tao J, Zhang H. Peach gum polysaccharides–based edible coatings extend shelf life of cherry tomatoes. 3 Biotech. 2017;**7**(3):168. DOI: 10.1007/s13205-017-0845-z

[16] Díaz-Mula HM, Serrano M, Valero D. Alginate coatings preserve fruit quality and bioactive compounds during storage of sweet cherry fruit. Food and Bioprocess Technology. 2011;**5**(8):2990-2997. DOI: 10.1007/s11947-011-0599-2

[17] Ratra S, Kaur L, Thukral B. Effect of Aloe vera and wheat grass juice as an edible coating to prolong the shelf life of bananas. International Journal of Engineering Research and Technology. 2016;**3**:2648-2655

[18] Martínez-Romero D, Alburquerque N, Valverde JM, Guillén F, Castillo S, Valero D, et al. Postharvest sweet cherry quality and safety maintenance by Aloe vera treatment: A new edible coating. Postharvest Biology and Technology. 2006;**39**(1):93-100. DOI: 10.1016/j.postharvbio.2005.09.006

[19] Adetunji CO, Fawole OB, Arowọra KA, Nwaubani SI, Ajayi ES, Oloke JK, et al. Effects of edible coatings from Aloe vera gel on quality and postharvest physiology of *Ananas comosus* (L.) fruit during ambient storage. Global Journal of Science Frontier Research Bio-Tech & Genetics. 2012;**12**(5):38-43

[20] Ergun M, Satici F. Use of Aloe vera gel as biopreservative for 'Granny Smith' and 'Red Chief' apples. Journal of Animal and Plant Sciences. 2012;**22**:363-368

[21] Maftoonazad N, Ramaswamy HS, Marcotte M. Shelf-life extension of peaches through sodium alginate and methyl cellulose edible coatings. International Journal of Food Science and Technology. 2008;**43**(6):951-957. DOI: 10.1111/j.1365-2621.2006.01444.x

[22] Paul SK, Sarkar S, Sethi LN, Ghosh SK. Development of chitosan based optimized edible coating for tomato (*Solanum lycopersicum*) and its characterization. Journal of Food Science and Technology. 2018;**55**(7):2446-2456. DOI: 10.1007/s13197-018-3162-6

[23] Duong NTC, Uthairatanakij A, Laohakunjit N, Jitareerat P, Kaisangsri N. An innovative single step of cross-linked alginate-based edible coating for maintaining postharvest quality and reducing chilling injury in rose apple cv. 'Tabtimchan' (*Syzygium samarangenese*). Scientia Horticulturae. 2022;**292**:110648. DOI: 10.1016/j.scienta.2021.110648

[24] Mandal D, Lalrinawmi E, Hazarika T, Shukla AC. Bilayer edible coating with chitosan and wax extended shelf life and maintained quality of tomatoes at ambient storage. Research on Crops. 2017;**18**:662-667. DOI: 10.5958/2348-7542.2017.00113.9

[25] Janjarasskul T, Krochta JM. Edible packaging materials. Annual Review of Food Science and Technology. 2010;**1**(1):415-448. DOI: 10.1146/annurev.food.080708.100836

[26] Hambleton A, Debeaufort F, Bonotte A, Voilley A. Influence of

alginate emulsion-based films structure on its barrier properties and on the protection of microencapsulated aroma compounds. Food Hydrocolloids. 2009;**23**(8):2116-2124. DOI: 10.1016/j.foodhyd.2009.04.001

[27] Dragich A, Krochta JM. Whey protein solution coating for fat-update reduction in deep fried chicken breast strips. Journal of Food Science. 2009;**73**(3):M378-M383. DOI: 10.1111/j.1750-3841.2009.01408.x

[28] Alvarez MV, Ponce AG, Moreira MR. Antimicrobial efficiency of chitosan coating enriched with bioactive compounds to improve the safety of fresh cut broccoli. Food Science and Technology. 2013;**50**:78-87. DOI: 1016/j.lwt.2012.06.021

[29] Wu S, Chen J. Using pullulan-based edible coatings to extend shelf–life of fresh-cut 'Fuji' apples. International Journal of Biological Macromolecules. 2013;**55**:254-257. DOI: 10.1016/j.ijbiomac.2013.01.012

[30] Wang L, Li Y, Xin G, Wei M, Yang Q, Mi Q. Effects of red and blue light quality on nitrogen levels, activities and gene expression of key enzymes involved in nitrogen metabolism from leaves of tomato seedlings. Acta Horticulturae Sinica. 2017;**44**:768-776. DOI: 10.16420/j.issn.0513-353x.2017-0002

[31] Massa GD, Kim HH, Wheeler RM, Mitchell CA. Plant productivity in response to led lighting. HortScience. 2008;**43**:1951-1956. DOI: 10.21273/HORTSCI.43.7.1951

[32] Yeow LC, Chew BL, Sreeramanan S. Elevation of secondary metabolites production through light-emitting diodes (LEDs) illumination in protocorm–like bodies (PLBs) of Dendrobium hybrid orchid rich in phytochemicals with therapeutic effects. Biotechnology Reports. 2020;**27**:e00497. DOI: 10.1016/j.btre.2020.e00497

[33] Kim BS, Lee HO, Kim JY, Kwon KH, Cha HS, Kim JH. An effect of light emitting diode (LED) irradiation treatment on the amplification of functional components of immature strawberry. Horticulture, Environment, and Biotechnology. 2011;**52**(1):35-39. DOI: 10.1007/s13580-011-0189-2

[34] Xu F, Cao S, Shi L, Chen W, Su X, Yang Z. Blue light irradiation affects anthocyanin content and enzyme activities involved in postharvest strawberry fruit. Journal of Agricultural and Food Chemistry. 2014;**62**(20):4778-4783. DOI: 10.1021/jf501120u

[35] Zhan LJ, Hu JQ, Li Y, Pang LG. Combination of light exposure and low temperature in preserving quality and extending shelf-life of fresh-cut broccoli (*Brassica oleracea* L.). Postharvest Biology and Technology. 2012;**72**:76-81. DOI: 10.1016/j.postharvbio.2012.05.001

[36] Lee YJ, Ha JY, Oh JE, Cho MS. The effect of LED irradiation on the quality of cabbage stored at a low temperature. Food Science and Biotechnology. 2014;**23**(4):1087-1093. DOI: 10.1007/s10068-014-0149-6

[37] Liu Y, Schouten R, Tikuno Y, Liu X, Visser R, Tan F, et al. Blue light increases anthocyanin content and delays fruit ripening in purple pepper fruit. Postharvest Biology and Technology. 2022;**192**:112024. DOI: 10.1016/j.postharvbio.2022.112024

[38] Nassarawa SS, Abdelshafy AM, Xu Y, Li L, Luo Z. Effect of light-emitting diodes (LEDs) on the quality of fruits and vegetables during postharvest period: A review. Food and Bioprocess

Technology. 2021;**14**:388-414. DOI: 10.1007/s11947-020-02534-6

[39] Lo C, Manurung R, Rachmi ER. Enhancement of lycopene and β-carotene production in cherry tomato fruits (*Solanum lycopersicum* L. var. cerasiforme) by using red and blue light treatment. International Journal of Technical Research and Application. 2014;**2**:7-10

[40] Samuolienė G, Brazaitytė A, Viršilė A, Jankauskienė J, Sakalauskienė S, Duchovskis P. Red light-dose or wavelength-dependent photoresponse of antioxidants in herb microgreens. PLoS One. 2016;**11**(9):e0163405. DOI: 10.1371/journal.pone.0163405

[41] Samuolienė G, Brazaitytė A, Sirtautas R, Viršilė A, Sakalauskaitė J, Sakalauskienė S, et al. LED illumination affects bioactive compounds in romaine baby leaf lettuce. Journal of the Science of Food and Agriculture. 2013;**93**(13):3286-3291. DOI: 10.1002/jsfa.6173

[42] Lee MK, Arasu MV, Park S, Byeon DH, Chung SO, Park SU, et al. LED lights enhance metabolites and antioxidants in Chinese cabbage and kale. Brazilian Archives of Biology and Technology. 2016;**59**:e16150546. DOI: 10.1590/1678-4324-2016150546

[43] Jarerat A, Techavuthiporn C, Chanthana C, Nimitkeatkai H. Enhancement of antioxidant activity and bioactive compounds in eggplants using postharvest LEDs irradiation. Horticulturae. 2022;**8**:134. DOI: 10.3390/horticulturae8020134

[44] Martínez Zamora L, Castillejo N, Gómez P, Artés-Hernández F. Amelioration effect of LED lighting in the bioactive compounds synthesis during carrot sprouting. Agronomy. 2021;**11**:304. DOI: 10.3390/agronomy11020304

[45] Wang CY, Fu CC, Liu YC. Effects of using light–emitting diodes on the cultivation of Spirulina platensis. Biochemical Engineering Journal. 2007;**37**:21-25. DOI: 10.1016/j.bej.2007.03.004

[46] Heo JW, Kang DH, Bang HS, Hong SG, Chun C, Kang KK. Early growth, pigmentation, protein content, and phenylalanine ammonia–lyase activity of red curled lettuces grown under different lighting conditions. Korean Journal of Horticultural Science & Technology. 2012;**30**:6-12

[47] Kami C, Lorrain S, Hornitschek P, Fankhauser C. Light–regulated plant growth and development. Plant Development. 2010;**91**:29-66. DOI: 10.1016/s0070-2153(10)91002-8

[48] Kook HS, Park SH, Jang YJ, Lee GW, Kim JS, Kim HM, et al. Blue LED (light–emitting diodes)–mediated growth promotion and control of botrytis disease in lettuce. Acta Agriculturae Scandinavica Section B–Soil and Plant Science. 2013;**63**:271-277. DOI: 10.1080/09064710.2012.756118

[49] Kim K, Kook HS, Jang YJ, Lee WH, Kamala-Kannan S, Chae JC, et al. The effect of blue–light emitting diodes on antioxidant properties and resistance to Botrytis cinerea in tomato. Journal of Plant Pathology & Microbiology. 2013;**4**:203. DOI: 10.4172/2157-7471.1000203

[50] Alferez F, Liao HL, Burns JK. Blue light alters infection by *Penicillium digitatum* in tangerines. Postharvest Biology and Technology. 2012;**63**:11-15. DOI: 10.1016/j.postharvbio.2011.08.001

[51] Yamaga I, Takahashi T, Ishii K, Kato M, Kobayashi Y. Suppression of blue mold symptom development in Satsuma mandarin fruits treated by low–intensity blue LED irradiation. Food Science and Technology Research. 2015;**21**(3):347-351. DOI: 10.3136/fstr.21.347

[52] Babadoost M, Islam SZ. Red–light treatment induces resistance in bell pepper to *Phytophthora capsici*. Acta Horticulturae. 2022;**1337**:89-92. DOI: 10.17660/ActaHortic.2022.1337.12

[53] Ahn SY, Kim SA, Yun HK. Inhibition of Botrytis cinerea and accumulation of stilbene compounds by light–emitting diodes of grapevine leaves and differential expression of defense–related genes. European Journal of Plant Pathology. 2015;**143**:753-765

[54] Ghate VS, Ng KS, Zhou WB, Yang H, Khoo GH, Yoon WB, et al. Antibacterial effect of light emitting diodes of visible wavelengths on selected foodborne pathogens at different illumination temperatures. International Journal of Food Microbiology. 2013;**166**:399-406. DOI: 10.1016/j.ijfoodmicro.2013.07.018

[55] Yamaga I, Nakamura S. Blue LED irradiation induces scoparone production in wounded Satsuma mandarin 'Aoshima unshu' and reduces fruit decay during long–term storage. The Horticulture Journal. 2018;**87**(4):474-480. DOI: 10.2503/hortj.OKD–147

[56] Jeandet P, Douillt-Breuil AC, Bessis R, Debord S, Sbaghi M, Adrian M. Phytoalexins from the vitaceae: Biosynthesis, phytoalexin gene expression in transgenic plants, antifungal activity, and metabolism. Journal of Agricultural and Food Chemistry. 2002;**50**:2731-2741. DOI: 10.1021/jf011429s

[57] Arefi-Oskoui S, Khataee A, Safarpour M, Orooji Y, Vatanpour V. A review on the applications of ultrasonic technology in membrane bioreactors. Ultrasonics Sonochemistry. 2019;**58**:104633. DOI: 10.1016/j.ultsonch.2019.104633

[58] Gao S, Lewis GD, Ashokkumar M, Hemar Y. Inactivation of microorganisms by low-frequency high-power ultrasound: 1. Effect of growth phase and capsule properties of the bacteria. Ultrasonics Sonochemistry. 2014;**21**(1):446-453. DOI: 10.1016/j.ultsonch.2013.06.006

[59] Cao S, Hu Z, Pang B, Wang H, Xie H, Wu F. Effect of ultrasound treatment on fruit decay and quality maintenance in strawberry after harvest. Food Control. 2010;**21**(4):529, 532. DOI: 10.1016/j.foodcont.2009.08.002

[60] Birmpa A, Sfika V, Vantarakis A. Ultraviolet light and ultrasound as non-thermal treatments for the inactivation of microorganisms in fresh ready-to-eat foods. International Journal of Food Microbiology. 2013;**167**(1):96-102. DOI: 10.1016/j.ijfoodmicro.2013.06.005

[61] Alexandre EMC, Brandao TRS, Silva CM. Impact of non–thermal technologies and sanitizer solutions on microbial load reduction and quality factor retention of frozen red bell peppers. Innovative Food Science and Emerging Technologies. 2013;**17**:199-205. DOI: 10.1016/j.ifset.2012.11.009

[62] Gani A, Baba WN, Ahmad M, Sha U, Khan AA, Wani IA, et al. Effect of ultrasound treatment on physico-chemical, nutraceutical and microbial quality of strawberry. LWT–Food Science and Technology. 2016;**66**:496-502. DOI: 10.1016/j.lwt.2015.10.067

[63] Supapvanich S, Kijka C. Efficiency of ultrasonic treatment on postharvest quality and bioactive compounds of

'Kim Ju' guava fruit during short-term storage at room temperature. Current Applied Science and Technology. 2020;**21**(2):208-217

[64] Guerrero S, Tognon M, Alzamora SM. Response of *Saccharomyces cerevisiae* to the combined action of ultrasound and low weight chitosan. Food Control. 2005;**16**(2):131-139. DOI: 10.1016/j.foodcont.2004.01.003

[65] Chemat F, Rombaut N, Meullemiestre A, Turk M, Périno-Issartier S, Fabiano-Tixier AS, et al. Review of green food processing techniques. Preservation, transformation, and extraction. Innovative Food Science & Emerging Technologies/ Innovative Food Science and Emerging Technologies. 2017;**41**:357-377. DOI: 10.1016/j.ifset.2017.04.016ff. ffhal-01552142f

[66] Lobo V, Patil A, Phatak A, Chandra N. Free radicals, antioxidants and functional foods: Impact on human health. Pharmacognosy Reviews. 2010;**4**(8):118-126. DOI: 10.4103/0973-7847.70902

[67] Chen Z, Zhu CH. Combined effects of aqueous chlorine dioxide and ultrasonic treatments on postharvest storage quality of plum fruit (*Prunus salicina* L.). Postharvest Biology and Technology. 2011;**61**(2-3):117-123. DOI: 10.1016/j.postharvbio.2011.03.006

[68] Huang ST, Yang RC, Lee PN, Yang SH, Liao SK, Chen TY, et al. Anti-tumor and anti-angiogenic effects of *Phyllanthus urinaria* in mice bearing Lewis lung carcinoma. International Immunopharmacology. 2006;**6**(6):870-879. DOI: 10.1016/j.intimp.2005.12.010

[69] Sagong HG, Lee SY, Chang PS, Heu S, Ryu S, Choi YJ, et al. Combined effect of ultrasound and organic acids to reduce *Escherichia coli* O157:H7, *Salmonella typhimurium*, and *Listeria monocytogenes* on organic fresh lettuce. International Journal of Food Microbiology. 2011;**145**(1):287-292. DOI: 10.1016/j.ijfoodmicro.2011.01.010

[70] Yang Z, Cao S, Cai Y, Zheng Y. Combination of salicylic acid and ultrasound to control postharvest blue mold caused by *Penicillium expansum* in peach fruit. Innovative Food Science & Emerging Technologies. 2011;**12**(3):310-314. DOI: 10.1016/j.ifset.2011.04.010

[71] Millan-Sango D, Garroni E, Farrugia C, JFM VI, Valdramidis VP. Determination of the efficacy of ultrasound combined with essential oils on the de–contamination of salmonella inoculated lettuce leaves. LWT–Food Science and Technology. 2016;**73**:80-87. DOI: 10.1016/j.lwt.2016.05.039

[72] Civello PM, Vicente AR, Martínez GA. UV-C technology to control postharvest diseases of fruits and vegetables. Recent advances in alternative postharvest technologies to control fungal diseases in fruits and vegetables. Transworld Research Network. 2006;**37**(661):71-102

[73] Darré M, Vicente AR, Cisneros-Zevallos L, Artés-Hernández F. Postharvest ultraviolet radiation in fruit and vegetables: Applications and factors modulating its efficacy on bioactive compounds and microbial growth. Food. 2022;**11**(5):653. DOI: 10.3390/foods11050653

[74] Nigro F, Ippolito A, Lattanzio V, Di Venere D, Salerno M. Effect of ultraviolet-C light on postharvest decay of strawberry. Journal of Plant Pathology. 2000;**82**(1):29-37

[75] Li J, Zhang Q, Cui Y, Yan J, Cao J, Zhao Y, et al. Use of UV-C treatment

to inhibit the microbial growth and maintain the quality of yali pear. Journal of Food Science. 2010;**75**(7):M503-M507. DOI: 10.1111/j.1750-3841.2010.01776.x

[76] Sripong K, Jitareerat P, Uthairatanakij A. UV irradiation induces resistance against fruit rot disease and improves the quality of harvested mangosteen. Postharvest Biology and Technology. 2019;**149**:187-194. DOI: 10.1016/j.postharvbio.2018.12.001

[77] Almagro L, Gómez Ros LV, Belchi-Navarro S, Bru R, Ros Barceló A, Pedreño MA. Class III peroxidases in plant defence reactions. Journal of Experimental Botany. 2009;**60**(2):377-390. DOI: 10.1093/jxb/ern277

[78] Sripong K, Jitareerat P, Tsuyumu S, Uthairatanakij A, Srilaong V, Wongs-Aree C, et al. Combined treatment with hot water and UV–C elicits disease resistance against anthracnose and improves the quality of harvested mangoes. Crop Protection. 2015;**77**:1-8. DOI: 10.1016/j.cropro.2015.07.004

[79] El Ghaouth A, Wilson CL, Callahan AM. Induction of chitinase, β-1, 3-glucanase, and phenylalanine ammonia lyase in peach fruit by UV-C treatment. Phytopathology. 2003;**93**(3):349-355. DOI: 10.1094/PHYTO.2003.93.3.349

[80] Charles MT, Tano K, Asselin A, Arul J. Physiological basis of UV-C induced resistance to Botrytis cinerea in tomato fruit. V. Constitutive defence enzymes and inducible pathogenesis-related proteins. Postharvest Biology and Technology. 2009;**51**(3):414-424. DOI: 10.1016/j.postharvbio.2008.08.016

[81] Charles MT, Mercier J, Makhlouf J, Arul J. Physiological basis of UV-C-induced resistance to Botrytis cinerea in tomato fruit: I. Role of pre- and post-challenge accumulation of the phytoalexin-rishitin. Postharvest Biology and Technology. 2008;**47**(1):10-20. DOI: 10.1016/j.postharvbio.2007.05.013

[82] Schwartz J. The year of ozone. American Journal of Respiratory and Critical Care Medicine. 2016;**193**(10):1077-1079. DOI: 10.1164/rccm.201512-2533ED

[83] Lee P, Davidson J. Evaluation of activated carbon filters for removal of ozone at the PPB level. American Industrial Hygiene Association Journal. 1999;**60**(5):589-600. DOI: 10.1080/00028899908984478

[84] Fisk W, Spears M, Sullivan D, Mendell M. Ozone removal by filters containing activated carbon: A pilot study. In: Proceedings of the Healthy Building; 01 September 2009; Syracuse, NY. 2009. pp. 1-4

[85] Kim B, Yun H, Jung S, Jung Y, Jung H, Choe W, et al. Effect of atmospheric pressure plasma on inactivation of pathogens inoculated onto bacon using two different gas compositions. Food Microbiology. 2011;**28**(1):9-13. DOI: 10.1016/j.fm.2010.07.022

[86] Mani A, Krishna KR, Mirza A. Postharvest applications of cold plasma technology: A review. Agricultural Reviews. 2021;**R2221**:1-14. DOI: 10.18805/ag.R-2221

[87] Hernández-Torres CJ, Reyes-Acosta YK, Chávez-González ML, Dávila-Medina MD, Verma DK, Martínez-Hernández JL, et al. Recent trends and technological development in plasma as an emerging and promising technology for food biosystems. Saudi Journal of Biological Sciences. 2021;**29**(4):1957-1980. DOI: 10.1016/j.sjbs.2021.12.023

[88] Bora J, Khan T, Mahnot NK. Cold plasma treatment concerning quality

and safety of food: A review. Current Research in Nutrition and Food Science Journal. 2022;**10**(2):427-446. DOI: 10.12944/CRNFSJ.10.2.3

[89] Siddique SS, Hardy GSJ, Bayliss KL. Cold plasma: A potential new method to manage postharvest diseases caused by fungal plant pathogens. Plant Pathology. 2018;**67**(5):1011-1021. DOI: 10.1111/ppa.12825

[90] Wang Q, Salvi D. Recent progress in the application of plasma–activated water (PAW) for food decontamination. Current Opinion in Food Science. 2021;**42**:51-60. DOI: 10.1016/j.cofs.2021.04.012

[91] Misra NN, Tiwari BK, Raghavarao KSMS, Cullen PJ. Nonthermal plasma inactivation of food–borne pathogens. Food Engineering Reviews. 2011;**3**(3):159-170. DOI: 10.1007/s12393-011-9041-9

[92] Thirumdas R, Kothakot A, Annapurec U, Siliverud K, Blundelle R, Gattf R, et al. Plasma activated water (PAW): Chemistry, physico-chemical properties, applications in food and agriculture. Trends in Food Science and Technology. 2018;**77**:21-31. DOI: 10.1016/j.tifs.2018.05.007

[93] Oh YJ, Song AY, Min SC. Inhibition of *salmonella typhimurium* on radish sprouts using nitrogen-cold plasma. International Journal of Food Microbiology. 2017;**249**:66-71. DOI: 10.1016/j.ijfoodmicro.2017.03.005

[94] Niemira BA, Sites J. Cold plasma inactivates *Salmonella stanley* and *Escherichia coli* O157: H7 inoculated on golden delicious apples. Journal of Food Protection. 2008;**71**(7):1357-1365. DOI: 10.4315/0362-028X-71.7.1357

[95] Critzer F, Kelly-Wintenberg K, South S, Golden D. Atmospheric plasma inactivation of foodborne pathogens on fresh produce surfaces. Journal of Food Protection. 2007;**70**(10):2290. DOI: 10.4315/0362-028x-70.10.2290

[96] Zhao J, Qian J, Zhuang H, Luo J, Huang M, Yan W, et al. Effect of plasma-activated solution treatment on cell biology of *Staphylococcus aureus* and quality of fresh lettuces. Food. 2021;**10**(12):2976. DOI: 10.3390/foods10122976

[97] Ouf SA, Basher AH, Mohamed AAH. Inhibitory effect of double atmospheric pressure argon cold plasma on spores and mycotoxin production of *Aspergillus niger* contaminating date palm fruits. Journal of the Science of Food and Agriculture. 2015;**95**(15):3204-3210. DOI: 10.1002/jsfa.7060

[98] Won MY, Lee SJ, Min SC. Mandarin preservation by microwave–powered cold plasma treatment. Innovative Food Science & Emerging Technologies. 2017;**39**:25-32. DOI: 10.1016/j.ifset.2016.10.021

[99] Hayashi N, Yagyu Y, Yonesu A, Shiratani M. Sterilization characteristics of the surfaces of agricultural products using active oxygen species generated by atmospheric plasma and UV light. Japanese Journal of Applied Physics. 2014;**53**(5S1):05FR03. DOI: 10.7567/jjap.53.05fr03

[100] Ouf SA, Mohamed AAH, El–Sayed WS. Fungal decontamination of fleshy fruit water washes by double atmospheric pressure cold plasma. CLEAN–Soil, Air, Water. 2015;**44**(2):134-142. DOI: 10.1002/clen.201400575

[101] Guo J, Qin D, Li W, Wu F, Li L, Liu X. Inactivation of *Penicillium italicum* on kumquat via plasma–activated water and its effects on quality

attributes. International Journal of Food Microbiology. 2021;**343**:109090. DOI: 10.1016/j.ijfoodmicro.2021.109090

[102] Misra NN, Moiseev T, Patil S, Pankaj SK, Bourke P, Mosnier JP, Keener KM, Cullen PJ. Cold plasma in modified atmospheres for post–harvest treatment of strawberries. Food and Bioprocess Technology. 2014;7(10):3045-3054. DOI: 10.1007/s11947-014-1356-0

[103] Xu Y, Tian Y, Ma R, Liu Q, Zhang J. Effect of plasma activated water on the postharvest quality of button mushrooms, *Agaricus bisporus*. Food Chemistry. 2016;**197**:436-444. DOI: 10.1016/j.foodchem.2015.10.144

[104] Wu Y, Cheng JH, Sun DW. Subcellular damages of *Colletotrichum asianum* and inhibition of mango anthracnose by dielectric barrier discharge plasma. Food Chemistry. 2022;**381**:132197. DOI: 10.1016/j.foodchem.2022.132197

[105] Pour AK, Khorram S, Ehsani A, Ostadrahimi A, Ghasempour Z. Atmospheric cold plasma effect on quality attributes of banana slices: Its potential use in blanching process. Innovative Food Science and Emerging Technologies. 2022;**76**:102945. DOI: 10.1016/j.ifset.2022.102945

[106] Gu Y, Shi W, Liu R, Xing Y, Yu X, Jiang H. Cold plasma enzyme inactivation on dielectric properties and freshness quality in bananas. Innovative Food Science and Emerging Technologies. 2021;**69**:102649. DOI: 10.1016/j.ifset.2021.102649

[107] Sandhya. Modified atmosphere packaging of fresh produce: Current status and future needs. LWT–Food Science and Technology. 2010;**43**(3):381-392. DOI: 10.1016/j.lwt.2009.05.018

[108] Bremenkamp I, Ramos AV, Lu P, Patange A, Bourke P, Sousa-Gallagher MJ. Combined effect of plasma treatment and equilibrium modified atmosphere packaging on safety and quality of cherry tomatoes. Future Foods. 2021;**3**:100011. DOI: 10.1016/j.fufo.2021.100011

Chapter 2

Improved Postharvest Techniques for Fruit Coatings

Chalermchai Wongs-Aree, Hanh Thi Nguyen and Sompoch Noichinda

Abstract

Fruits, particularly tropical fruits, have a high moisture content, distinct morphological characteristics, and physiological changes, all of which contribute to their high rate of perishability. Nonetheless, their organoleptic and nutritional qualities make them one of the most important horticultural products. Fruit coating, which imitates natural packaging, is a postharvest solution that is practical and cost-effective for a variety of applications, including on-shelf display, transportation, and storage in support of the supply chain of fruits and vegetables. Gas and moisture permeability, microbiological resistance, and esthetic enhancement are the coating functions. Using modified materials and procedures, edible coatings for fresh and freshly cut fruits are currently being developed. Edible coatings infused with essential oils or volatiles may help to prevent disease resistance while also providing consumers with a fragrant preference. When considering how to advance fruit coating technology when agricultural wastes are the primary source of new coating materials, composite coatings, nanoparticles, encapsulation, and multiple-layer coatings all hold a great deal of promise. Future research may center on the optimal material for particular fruits during the logistics phase.

Keywords: edible coating, modified materials, modified techniques, quality maintenance, hypoxia, tropical fruits

1. Introduction

Fruits, especially tropical fruits, are highly perishable and quickly lose their quality after being picked. Fruit postharvest loss in tropical areas could be up to 50%. The high level of physiological changes that occur in fruit is mostly attributable to the high level of moisture content [1]. The amount of water that evaporates from fruit is a significant factor in determining both the quality of the fruit and the economic quantity losses [2]. As a result, packaging that includes coating has the potential to successfully limit the loss of water from the fruit that has been stored. On the other hand, standard packaging might not be appropriate for certain fruits or environments. For instance, the sharp spines of a durian can easily puncture the plastic film, or the headspace in the package can accumulate condensed water and cause the fruit to rot. Edible coatings are a simple kind of fruit packaging that is applied to individual fruits in order to facilitate easier postharvest handling [3]. However, for fruit coating to be successful, it is necessary to consider not just different coating processes but also the characteristics of the fruit itself.

2. Background related to fruit-based coatings

The extracellular barrier of cuticles protects the plant's surface and interior from environmental stresses. The cuticle, composed of polysaccharides and a lipid matrix termed "cutin" [4], can dramatically transform the epidermis's outer cell wall. Many factors are essential determinants of shelf life and storage capacity, including the cuticle's moderating effect on water transpiration, fruit dehydration, and vulnerability to rots, pests, and diseases. Fruit cuticles have been proven to be highly responsive to environmental factors, and their quality changes even after harvest. When a fruit matures, the cuticle grows thicker [5]. Cutin polymers can be formed from long-chain fatty acids (C_{20}-C_{34}) but are typically made up of esterified and oxygenated C_{16} and C_{18} fatty acids [6]. It may also contain trace amounts of glycerol, phenylpropanoids, primary and secondary alkanes, alcohols, aldehydes, ketones, etc. [7]. Fatty acids are essential for the biosynthesis of cutin and wax, which occurs primarily in chloroplasts [8]. The components of wax extracted from the fruit surface of guavas comprise fatty acids and primary alcohols as the major components, followed by sterols, *n*-alkanes, and aldehydes. Interestingly, the wax also contains triterpenoids, a natural pesticide against insects and pathogenic fungi [9]. Thus, understanding the components and function of the cuticle should be useful in improving the coating materials used in fresh fruit markets.

Fresh fruits are perishable due to their quick deterioration due to dehydration, softening, discoloration, microbiological decay, disorder, and loss of nutrients [1]. These properties result from the metabolic processes occurring within the fruit, which are accelerated mainly by an improper relative humidity and gas composition in the storage atmosphere. High water loss and rapid metabolic changes associated with ripening and senescence are the primary causes of fresh fruit deterioration after harvest [10–13]. Artificial coatings mimic the protective properties of the fruit's natural cuticle and can postpone ripening and senescence by acting as a barrier to water and gases. Currently, edible coatings that are safe for food or handling provide a sustainable and effective solution for preserving the high quality of fruit throughout the postharvest value chain [3]. It is generally accepted that edible coatings as a type of MAP (modified atmosphere packaging) add value to fruits by shielding them from contamination, improving their esthetics, and preventing the loss of flavorful volatiles during preparation and storage (**Figure 1**). For the success of a fruit coating business, a thorough understanding of fruit characteristics, nature, respiration, and transpiration behavior is essential.

2.1 Fruit types

Three distinct fruit categories can be identified based on their floral structure: simple fruit, aggregation fruit, and multiple fruit. Some fruits, like the papaya, the rose apple, and the chili, have an internal cavity despite having a variety of morphologies. In addition, some fruits produce an inner supplementary tissue called aril (mangosteen, durian). Fruits that are either an aggregate (sugar apple, strawberry) or multiple fruit (pineapple, jackfruit) are made up of many fruitlets that have fused

Figure 1.
Fruit coating on mature green "Nam Dok Mai" mango during air drying.

together. Variations in fruitlet maturation might cause uncertain ripening of the entire fruit [14]. Fruit coating could improve the ripening quality of the uncertain ripening by maintaining moisture and gas diffusion in the whole fruit.

2.2 Fruit structure/fruit parts

Dermal tissues (peel), cortical tissues (pulp), vascular bundles (veins), seeds, and intercellular spaces make up the anatomy of most fruits. The peel consists primarily of parenchyma, of which some are transformed into guard cells (**Figure 2A**) and the lenticular opening channel (**Figure 2B**). Fruit stomata and lenticels are responsible for gas and water vapor exchanges, which could be an issue for fruit coating. In rambutan (*Nephilium lappaceum* L.), the epidermal tissues, spinterns, include up to 100–200 apertures/mm^2 of stomata [10] that are constantly open [15]. Thus, stomata on spinterns that connect up to 20 groups of vascular bundles in the mesocarp are primarily responsible for moisture loss from rambutan fruit (**Figure 2A**). Replacement of water lost by spinterns with water from the skin [16]. The aril is still edible, despite water loss from the pericarp, producing skin withering, spintern drying, and pericarp browning (**Figure 3**).

Vascular bundles (phloem and xylem tissues) are gathered at the fruit's stylar end (peduncle/pedicel) that connects the fruit to the mother plant. In tangerine fruit, the vascular bundles start from the stylar end through the albedo mesocarp flavedo under the peel (**Figure 4**). The maturing fruit's peel is typically covered with the cuticle. Consequently, the cut peduncle is where the fruit quickly loses its moisture to the air after being harvested. Fruit edible flesh could be derived from the ovary wall, some accessory organs (receptacle and petal), or the unique tissue of "aril" covering the seed, which comprises a majority of parenchymal cells carrying high water content, stored chemicals, and biological metabolisms. Furthermore, in some fruits, there is a fruit cavity (papaya, rose apple) or seed cavity (mango, apple) that is a gas container in the fruit. The fruit's intercellular spaces and cavity (**Figure 2B**) play crucial roles in the fruit's ability to transpire and respire. The gaseous atmosphere in the intercellular

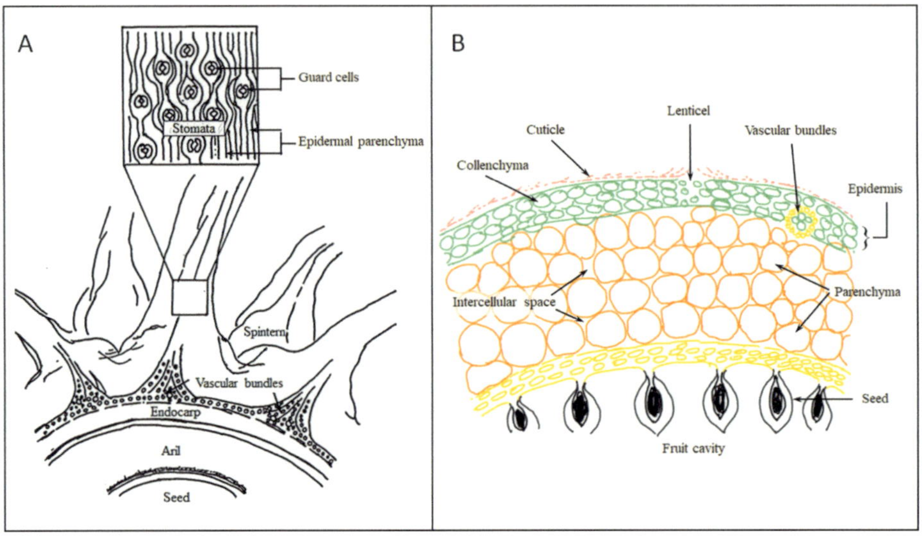

Figure 2.
Anatomical structure of cross-sectioned rambutan fruit showing spintern stomata and vascular bundle networks (A) and cross-sectioned papaya fruit showing lenticel and pericarp tissues (B).

Figure 3.
Fruit pericarps and arils of ordinary fruit (left) and dehydrated fruit (right).

spaces and cavity can be modified depending on the respiration rate and metabolism of the cells. The portions of intercellular spaces in several tropical fruits are shown in **Table 1**. Moreover, in fruit cavities, such as papaya, the different varieties exhibit varying volumes of cavities [17]. Intercellular spaces are important for cellular modification under hypoxic conditions [18]. Thus, the higher portion of intercellular air spaces is more tolerant to hypoxic conditions caused by fruit coatings. For example, apples and rose apples are tolerant to hypoxia from fruit coatings, in contrast to tangerines and guavas, which contain tiny intercellular spaces. Excessive CO_2 from respiration could accumulate in the room to reduce cellular toxicity, or O_2 demand can be taken from the portions to delay hypoxic cellular conditions. Coating material types and concentrations must be carefully considered. An off-flavor may result if a wax coating effectively prevents water loss from fruits with few intercellular spaces, such as tangerines. Thus, the modulation of the proper coating formula for each fruit depends on the fruit characteristics of each cultivar.

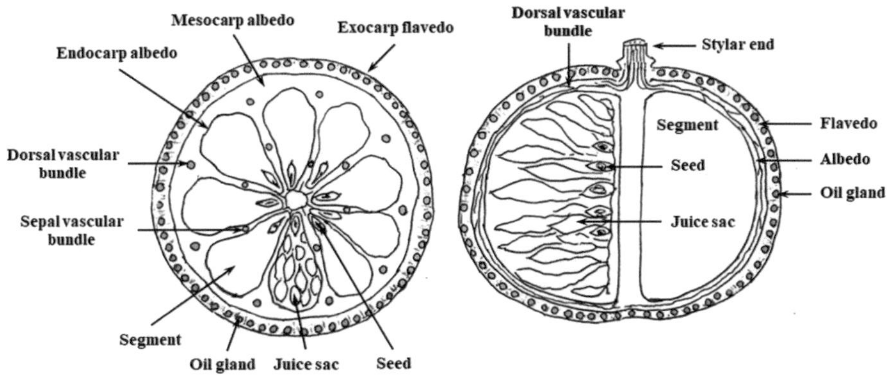

Figure 4.
Fruit structure of tangerine fruit when cross-sectioned (left) and longitudinally sectioned (right).

Commodity	Total intercellular air spaces (%)
'Gala' apple	1.25
'Klom Sali' guava	0.24
'Phet' rose apple	3.32
'Khaew Wan' tangerine	0.42

Table 1.
Percentages of intercellular spaces in some tropical fruits (own data).

2.3 Maturity, respiration, and ethylene production characters

Fruits are classified as climacteric or non-climacteric based on their respiratory and ethylene production patterns during maturation. The success of fruit coatings depends on the fruit's maturation, transpiration, and respiration rates. After fruit setting and during fruit growth due to cell division, the respiration rate is at its highest; it then gradually decreases to its lowest point during the early stage of fruit maturation. During ripening, respiration and ethylene production rates in climacteric fruits rise sharply, peak, and drop off, while both rates are not apparent in non-climacteric fruits. Storage temperatures play important roles in the respiration (**Table 2**) and ethylene production rates of fruits (**Table 3**) [19]. When fruit is coated with exogenous waxes, natural gases, and humidity, exchanges between the fruit and its respective microclimates are disrupted. The respiration, ethylene production, and gas permeability of the coated fruit will alter the gas concentrations within the cellular fruit, decreasing O_2 and increasing CO_2 and C_2H_4. By measuring the concentration of gases in the fruit's intercellular spaces and/or cavity, the fruit gets into an equilibrium of internal gases (**Figure 5**). Ethylene and CO_2 accumulation at different levels in "Solo" papaya during on-tree ripening [20] are shown in **Table 4**.

Many tropical fruits, such as mangosteen, longan, longkong, rambutan, durian, and salah, contain two parts: the pericarp (peel) and aril (flesh), which individually develop during fruit maturation. The aril of some tropical fruits is derived either from the funiculus (durian and lychee) or integument (mangosteen and rambutan) of the seed [14]. During maturation and ripening, the pericarp and the aril mature

Commodity	Respiration rates (mL CO_2·kg^{-1}·h^{-1})			
	13°C		25°C	
	Max	Min	Max	Min
"Rongrien" rambutan	20.40	—	48.00	—
"Monthong" durian	58.39	6.72	106.05	28.70
Mangosteen	—	—	20.50	—
"Nam Dok Mai" mango	26.00	11.00	72.50	17.05
Longkong	36.37	—	81.66	—
"Khao Pan" pummelo	4.00	—	9.25	—

Table 2.
Respiration rates of some tropical fruits at 13°C and 25°C (adopted from Kosiyachinda and Tansiriyakul [19]).

Commodity	Ethylene production rates (μL·kg^{-1}·h^{-1})	
	13°C	25°C
"Rongrien" rambutan	0.78	0.66
"Monthong" durian	2.40	9.52
Mangosteen	0.22	3.60
"Nam Dok Mai" mango	0.67	—
Longkong	2.19	1.39
"Khao Pan" pummelo	0.026	0.026

Table 3.
Ethylene production rates of some tropical fruits at 13°C and 25°C (adopted from Kosiyachinda and Tansiriyakul [19]).

independently at different levels. Fruits like the mango, whose flesh develops from the ovary wall, soften, change color, and release their natural aromas as they ripen regularly and consistently. However, the pericarp (dusk) of the durian fruit releases most of the climacteric ethylene during whole fruit ripening, which in turn causes the aril to ripen. Double climacteric blooming occurs in several cultivars (fruit ripening and dehiscence). The respiration and ethylene production rates of durian pulp during ripening are much lower than those from the pericarp [21]. The endogenous ethylene produced in the husk is required for whole fruit and pulp ripening [22]. Thus, a coating of fresh-cut durian must undergo proper ripening before husk removal because unripe pulp often fails to ripen regularly [23]. Interestingly, stages of maturity are crucial for fruit coating. In "Nam Dok Mai" mango, for example, the mature green fruit was induced to get anaerobic conditions and produce off flavor at a high concentration of composite coating, but the treated fruit was typical when the ripe fruit was coated at the same concentration [24].

3. Basic materials for fruit coatings

After harvesting, coating fruits with shellac or fat-related substances was a frequent procedure in the past. Later, fat-based plant extracts from bananas, pineapple leaves,

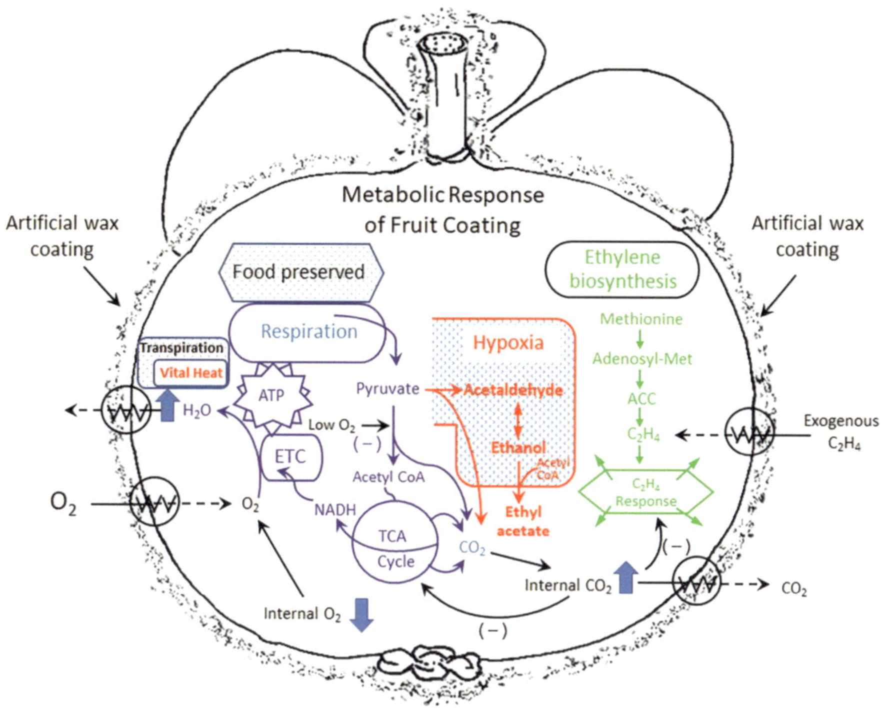

Figure 5.
Dynamic and metabolism changes in coated fruits and the surroundings.

Ripening stage	Concentrations of gases in the fruit cavity	
	Carbon dioxide (%)	Ethylene ($\mu L \cdot L^{-1}$)
Mature green	1.8	0
Fully ripe	5.5	2.8
Overripe	5.0	2.3

Table 4.
Internal CO_2 and C_2H_4 concentrations in the fruit cavity of "solo" papaya during fruit ripening (adopted from Akamine and Goo [20]).

and carnauba trees were utilized. In contrast, coating materials often consist of a single complex molecule that is effective at preventing water loss but has poor gas exchange. Therefore, it is vital to understand the composition before searching for an acceptable covering. There are three major chemical properties of edible coatings.

3.1 Polysaccharides

The polysaccharide-based coating comes from modified or extracted polysaccharides from natural products, typically including many starches from plants (corn, wheat, rice, cassava, and potato), glucomannan, galactomannan, inulin, plant cellulose and pectin, plant gum, alginate from brown seaweeds, pullulan from the fungus *Aureobasidium pullulans*, and chitosan from shrimp. The edible film made of polysaccharides is transparent, strong mechanically, and impervious to lipids. From those materials, chitosan, a plant disease elicitor, has been studied for coating many fruits. With stored longans (*Dimocarpus longan* Lour. cv. "Daw"), chitosan coating at 1.0% and 1.5% revealed a delay in pericarp browning and an effective retardant of disease growth with less than 4% disease incidence at 4°C [25]. Gac fruit (*Momordica cochinchinensis* Spreng) at the yellow stage coated with 0.5%, and 1.0% chitosan delayed fungal infection and enhanced fruit appearance [26]. Chitosan coatings have been successfully used to maintain the quality of many tropical fruits such as pineapple [27], banana [28], and papaya [29].

The polysaccharide-based coating is useful for coating freshly cut fruits because it can prevent fat remnants from leaking out. However, the constraint of the polysaccharide coatings is a lack of control over water loss and O_2 and CO_2 exchanges [30]. Moreover, if the logistics of handling fresh produce involve a high-temperature fluctuation, or if the fresh produce is immediately removed from the cold storage and placed at room temperature, the coating may be in a reversible phase, resulting in the peeling of the coating from condensed water droplets on the surface. Researchers have attempted to create an edible polysaccharide coating or film using antioxidants from plants. Typically, this is done to enhance the physical and chemical properties of the film and coating for safe consumption.

3.2 Proteins/oligo peptides

Proteins from plants, animals, and microorganisms are biopolymers that can be utilized to make films with tunable physical and functional properties when mixed with plasticizers or other components. The proteins such as gelatin, casein, collagen, whey proteins, egg white, etc. have been studied. Covalent unions (side chain cross-linking) and electrostatic or ionic interactions between protein chains contribute to coating formation [31]. Protein films' mechanical and hydrophobic barrier properties, and hence their suitability for food packaging applications, are strongly influenced by the final chain contacts and bonds. The major bonding processes are controlled by production conditions such as pH, salt addition, heating, enzyme action, drying, and reactions to food-grade chemicals [32]. Protein films have been enhanced with antimicrobials and antioxidants, among other active compounds. Some protein extracts have been studied for fruit coatings. Park et al. [33] used a corn-zein coating to delay fruit ripening in tomato fruit, while Avena-Bustillos et al. [34] used a casein (milk) protein coating to reduce weight loss in zucchini. Furthermore, zein and gelatin coatings could delay ripening in mangoes stored at 32°C [35].

Although the protein-based coating belongs to the hydrophilic group, the covalent cross-links, and electrostatic networks boost the coating's structural stability, limiting the coating's reversible phase when droplets condense on the surface during logistics.

3.3 Lipids

Fatty acid derivatives, as a significant component of the natural cuticle, play an important role in preventing water loss and gas exchanges in fruits. The properties of flexibility, hydrophobicity, and cohesion that edible films require are provided by lipids [36]. The quality of fruits can be maintained by edible coatings made of lipids, which are effective barriers against moisture, O_2, and CO_2, but not C_2H_4. Edible coatings such as carnauba wax, shellac, bee wax, and some plant oils based on lipids can cover fruits and vegetables. Films and edible coatings made from fat have gained appeal due to their functional and nutritional benefits. Shellac and carnauba have long been applied to postharvest fruits. Ten percent shellac coating prevented fresh weight loss and disease infection, and reduced the respiration and ethylene production rates of gac fruit at 25°C storage, while 15% shellac coating led to a high accumulation of acetaldehyde since day nine [37]. Shellac and carnauba emulsions were coated on "Nova" mandarins (*Citrus reticulata*) at 20°C storage. The carnauba waxes resulted in minor weight loss compared to the uncoated control and shellac coating, but shellac-coated fruit showed the highest fruit shine. The highest levels of CO_2 and the lowest level of O_2 were found in shellac-coated fruit, resulting in the highest ethanol content in the juice due to induced anaerobic respiration [38].

Many lipid-based coatings can provide the best water transpiration prevention due mainly to their strong hydrophobic properties, but for fresh produce, the switch from aerobic to anaerobic respiration caused by too low O_2 and/or high CO_2 in the coated fruit must be considered. Although long fatty acid derivatives are the most abundant in natural cuticles, attachment to sterols, terpenoids, polysaccharides, and phenolics may modify the complex structure, increasing gas exchange permeability.

4. Recent approaches to enhance the efficiency of coatings for fruit preservation

Fruit coatings have been improved and developed, mostly relying on complex cuticle structures, fruit types, proposed uses, and raw material availability. Modified materials and techniques are elucidated in this chapter.

4.1 Modified materials

4.1.1 Plant residues

Plants are a massive source of fruit-coating materials, particularly from agricultural waste. New materials from plant cell structures, including cell wall components and stored chemical substances such as gum, polysaccharides, proteins, and lipids, have been extracted and researched for food packaging [39].

Recently, agricultural wastes have been used to produce some coating components in carboxymethyl cellulose (CMC). Ketrodsakul [40] developed a process to extract crude cellulose from corn stems, an agricultural waste left after corn harvesting in central Thailand, and turn them into CMC. To improve the quality of CMC from corn stems (**Figure 6A**), Clorox was used to remove green chlorophylls from the samples (**Figure 6B**). The chemical spectrum analysis using Fourier transform infrared spectroscopy (FT-IR) shows that the CMC from corn stem has characteristics of sub-residues similar to a commercial CMC (**Figure 6C**). The modified CMC can be used as a material coated on mango fruit surfaces, demonstrating complete covering over the stomata of mango peel (**Figure 7B**), compared to the uncoated control (**Figure 7A**).

Supapvanich et al. [41] used *Aloe vera* gel coatings on fresh-cut "Taaptimjaan"' rose apples stored at 4°C. *A. vera* coating preserved the white index and slowed the browning and chilling injury symptoms, particularly at 75% (v/v). Modified atmosphere coatings by *A. vera* dips delayed the increases in phenolic concentration and polyphenol oxidase (PPO) activity.

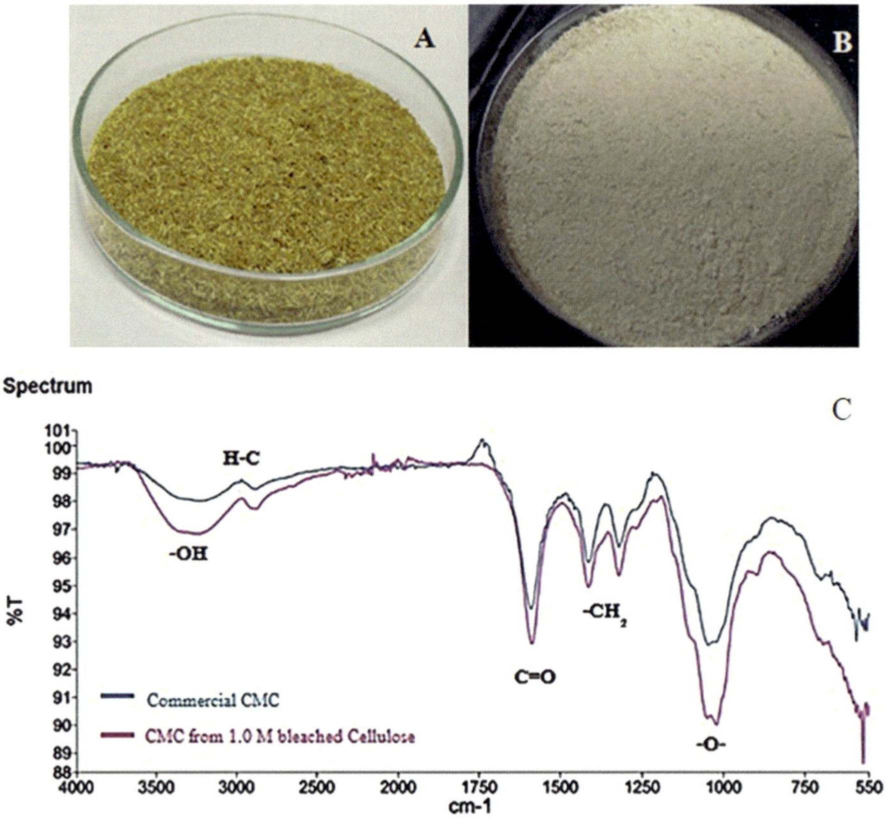

Figure 6.
Visual appearances of CMC from corn stem cellulose, extracted with 1.0 M NaOH (A) and then with Clorox treatment for 12 h (B), and the FT-IR spectrums of a commercial CMC compared with the CMC from corn stem cellulose extracted with 1.0 M NaOH and bleached with Clorox (C).

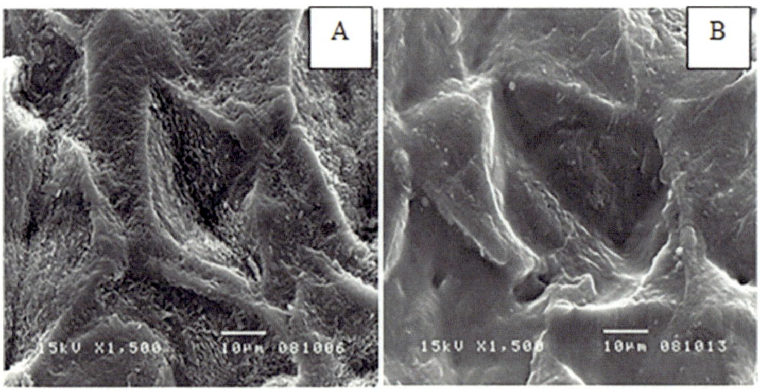

Figure 7.
Scanning electron microscopes (1500 x) of stomata on uncoated mango peel (A) and 2% corn-CMC coated mango peel (B).

4.1.2 Animal residues

New coating materials derived from animal production waste have been gradually discovered. Sericin, for example, a natural protein from silk industry wastewater, is hydrolyzed and used for food coatings. The FDA approves sericin and its derivatives as "GRAS" substances [42]. Sericin coatings on fresh-cut mango [43] and apples [44] had a lower water loss and browning index than the control at low-temperature storage by decreasing the activities of browning-related enzymes, mainly PPO. Because sericin hydrolysates contain serine (30–33%), glycine (19%), and aspartic acid (17.8%) [45], holding high hydroxyl (–OH) groups that could absorb water, leading to a reduction in water loss and inhibiting the browning-related enzymes in the fresh-cut produce.

4.1.3 Microorganism products

Some bacteria cultivated under specific conditions can produce edible polymers as by-products. For instance, Acetobacter can oxidize sugars, sugar alcohols, and ethanol and produce acetic acid as the primary end product that generally contains bacterial exopolysaccharides. Acetobacter species such as *Acetobacter xylinus* [46] are capable of synthesizing cellulose and have many uses in some fermented food products, which produce soluble polystyrene and contain rhamnose, glucose, mannose, and glucuronic acid as their acetane-related structures. Bacterial cellulose (BC) is one of the promising biomaterials that can be developed as a food packaging plastic material and is produced through the fermentation of high carbohydrate-containing substrates such as agricultural and industrial waste [46, 47]. Yanti et al. [47] produced CMC from a modified BC film and used glycerol as a plasticizer.

There is currently very little information about studying fruit coated with BC-based materials. However, an antimicrobial composite edible film from fermented cheese whey with *Candida tropicalis* was found to inhibit *Pseudomonas aeruginosa*, *in vitro* [48]. Thus, edible BC films can potentially be used for fresh-cut and intact fruit coating.

4.1.4 Composite coating materials

Composite coatings consist of two or more biopolymers, which can minimize the disadvantages of each component. As a result, composite coatings can contain active components such as antibiotics, metal nanoparticles, essential oils, and antioxidants to improve the function of coatings in maintaining the quality of fruits. Because of differences in the polarities of materials, suitable emulsifiers or plasticizers may require a better mixture of the solutions. When compared to chitosan coating alone, the layer of chitosan integrated with tiny Tween-20 provides smooth and complete coverage on mangoes (**Figure 8**).

4.1.4.1 Polysaccharide and polysaccharide composite coating

Because of their hydrophilic properties, polysaccharide-based materials for fruit coatings, including homopolysaccharides and heteropolysaccharides, have low barrier

Figure 8.
Mango fruit coated with chitosan and chitosan+Tween-20.

properties to water vapor, CO_2, and O_2. It has been demonstrated that combining different polysaccharides improves coating functions.

Chitosan coatings, cross-linked to hydroxypropyl methylcellulose (HPMC) by incorporating neem oil, reduced the number of hydroxyl moieties. Thus, decreasing hydrophilic characteristics could improve the moisture barrier of coatings. The emulsion of neem oil in chitosan cross-linked to HPMC presented a higher contact angle than the chitosan solution on the surface of pitaya, which showed a slight hydrophobic characteristic. Composite chitosan coating with HMC reduced weight loss and delayed the senescence of pitaya fruit compared to chitosan coating [49]. Chitosan combined with *A. vera* retarded the weight loss of blueberry fruit compared to chitosan coating alone due to the hydrophobic character of the *A. vera* liquid fraction. Furthermore, the composite coating based on chitosan and *A. vera* delayed microbial spoilage via the antifungal activity of both chitosan and *A. vera.* [50]. Furthermore, κ-carrageenan has been introduced to improve chitosan-based composite coating based on the interaction between oppositely charged polysaccharides. Because of the formation of a hydrogen-bond network between chitosan and κ-carrageenan, the composite coating has better water vapor barrier properties [51]. The composite coating reduced the physiological weight loss, improved the accumulation of phenolics, and suppressed the activities of the major chlorophyll-degrading enzymes, resulting in the retention of chlorophyll content in the bracts of dragon fruit and their green color (**Figure 9**) [52].

On the other hand, sodium alginate combined with hydroxyethyl cellulose (HEC) generated a continuous and smooth coating layer on the strawberry fruit surface. Meanwhile, sodium alginate alone could not form a continuous film on the fruit, and hydroxyethyl cellulose alone displayed a coating layer with wrinkles and was multi-porous [53]. The single coatings could not reduce the weight loss caused, but the sodium alginate-HEC composite coating significantly decreased the weight loss of strawberry fruit. Furthermore, this composite coating was a good gas barrier that retarded the loss of phenolics and flavonoids due to the degradation of strawberries.

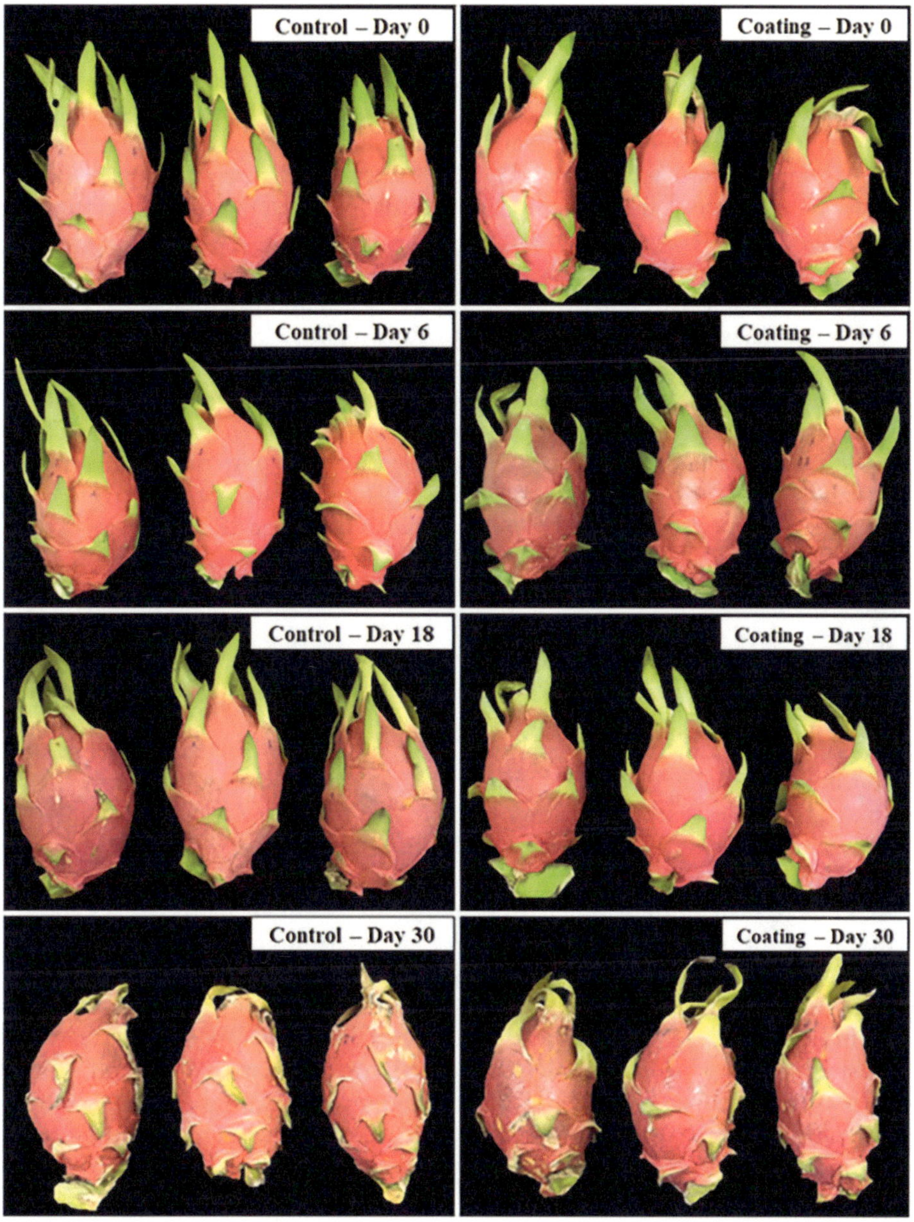

Figure 9.
Visual appearance of dragon fruit coated with chitosan- and κ carrageenan-based composite coating compared to the uncoated control dragon fruit during storage at 10°C.

4.1.4.2 Polysaccharide and lipid composite coatings

Composite coatings based on polysaccharides and lipids are probably used to enhance the moisture and gas barrier properties due to the hydrophobic characteristic of lipids. Fagundes et al. [54] discovered that a composite coating of HPMC and beeswax containing an antifungal compound had 2.5 times lower water vapor permeability than chitosan coating, resulting in effective weight loss and respiration rate reductions in cherry tomatoes. The composite coating on strawberry fruit reduced weight loss by 15–20%, while the chitosan coating reduced weight loss by 11% compared to uncoated fruit [55]. At 22°C, a composite coating based on wheat straw arabinoxylan and oat bran-glucan stearic acid ester applied to apples reduced weight loss by 1.2 times, compared to uncoated fruit. Aside from that, arabinoxylan and β-glucan in the stearic acid ester composite coating inhibited microbial contamination while preserving fruit sensory quality [56].

Plant oils have recently been used to combine with polysaccharides to form complex composite coatings. Olive oil, containing high levels of monounsaturated fatty acids and antioxidants, was emulsified with chitosan and alginate [57]. Both composite coatings of chitosan and alginate emulsions with olive oil reduced the fig fruit's respiration rate and fungal decay. Furthermore, chitosan-olive oil coating and alginate-olive oil coating reduced the weight loss of figs by 15.69% and 22.66%. In addition, 1.0% sucrose fatty acid ester (SFE) slightly reduced the fruit softening and respiration rates of gac fruit during 16 days of storage at 25°C [58]. The sucrose moieties esterified to the fatty acid materials could provide more gas permeability to the SFE coating.

4.1.4.3 Protein and lipid composite coatings

Shellac has been used as a fruit coating for protecting fruits from water transpiration, gas, and microbial spoilage. However, shellac is unstable due to the polymerization of the structure's hydroxyl and carboxyl groups. The electrostatic interaction between the negative charge of shellac and the positive charge of gelatin could protect the shellac's active site, thus reducing the esterification process and enhancing the shellac's stability. The composite film based on 60% shellac and 40% gelatin acted as an effective barrier to prevent moisture and gas movement, resulting in slight decreases in weight loss, firmness, and maintaining the quality of bananas at low temperatures for more than 30 days, compared with uncoated fruits [59]. Furthermore, the hydrophilic nature of whey protein from flaxseed was used to improve the hydrophobic property of bee wax by blending both to form a composite coating. The composite coating significantly reduced the water vapor permeability but increased the oxygen permeability of the whey protein isolate coating. The composite coating reduced the shriveling of the plum due to weight loss and delayed the softening of the fruit [60]. The stability of composite coatings depends on the amount of lipid added. A whey protein isolate-based composite coating containing 10% lipid showed fewer defects compared to the coating containing 5% lipid during the 15 days of storage at 5°C.

4.1.4.4 Composite coating containing growth regulators and plant extracts

Recently, the incorporation of plant residues and extracts as growth regulators, natural antioxidants, and antimicrobials into biopolymers to develop composite coatings has been increased to enhance the quality and shelf life of many fruits [61]. A pomegranate pericarp extract (PPE), containing enriched phenolics, was incorporated with chitosan-pullulan to formulate a composite coating for mangoes in cold storage. The chitosan-pullulan composite coating enriched with PPE effectively retained the fruit's phenolic content, flavonoid content, and antioxidant activity due to the barrier property of the coating against moisture and gas transference [62]. Nguyen et al. [63] developed the composite film of passion fruit peel pectin combined with chitosan and then incorporated *Piper betle* L. leaf extract (PLE) for the preservation of purple eggplants. Despite the fact that the addition of PLE increased the water vapor permeability of the pectin/chitosan composite film due to the increased concentration of polar groups, the composite coating outperformed the control film against bacteria. Moreover, alginate combined with 0.45 $mg.L^{-1}$ longkong peel extract (LPE)-silver particles coating prevented severe browning, weight loss, and decay incidence during storage of longkong fruit during storage at 13°C and 90–95% RH by limiting the growth of fruit browning mostly via decreasing peroxidase (POD) and PPO activity [64].

The polyphenols in *Cleistocalyx operculatus* (Roxb.) fruit (CFE) were successfully extracted and added to chitosan and gum Arabic edible coatings for banana fruit. The chitosan-gum-CFE-based composite coating showed high effectiveness in improving the freshness of bananas stored at ambient conditions. The surface structure of the banana showed a wrinkle and crack structure (observed by scanning electron microscopy) for uncoated bananas and a smooth surface for bananas coated with the composite coating [65]. "Nam Dok Mai" mango (*Mangifera indica* L.) is usually encountered with postharvest decay due to anthracnose's invasion during 25°C storage. Double layers of chitosan, which contains positive charges, and sodium alginate, which contains negative charges coated on "Nam Dok Mai" mango fruit delayed the peel color changes, and retarded the decay. The untreated control developed disease black spots on day six, and the symptoms worsened throughout storage. Interestingly, mangoes coated with both materials revealed 12.5% of the disease symptoms on day eight and then were steady until day 12 [24]. The chitosan- and κ-carrageenan-based composite coating was more effective in retaining the chlorophyll content and nutritional quality of dragon fruit when combined with GA_3 or MeJA pretreatment. This composite coating, combined with hot water treatment, controlled the diseases by regulating H_2O_2 accumulation and antioxidant enzyme activities and maintained the overall quality of dragon fruit [66]. Moreover, fruit dehiscence during ripening is a crucial postharvest problem in "Chanthaburi II" durian, a new hybrid cultivar of Thailand. Fruit coating with 1% chitosan+100 $mg{\cdot}L^{-1}$ GA_3 effectively reduced the dusk dehiscence [67] (**Figure 10**). Gibberellic acid induces vegetative development in the plant parts that perform against ethylene responses.

Figure 10.
Fruit dehiscence in "Chanthaburi II" coated with 1% chitosan (B), 100 mg·L−1 GA3 (C), or 1% chitosan+100 mg·L−1 GA3, compared to the uncoated control (A) on day eight at 25°C.

4.2 Modified techniques

4.2.1 Nanotechnology

Nowadays, nanotechnology is considered the most promising innovative technique in food packaging due to its high safety and quality impact. Nanomaterials can be prepared using modified techniques. The materials show a higher effect than ordinary materials because of their smaller size and adhesive forces [68]. In food packaging, nanomaterials can be mixed in the polymer matrix to the increase gas barrier properties of films and coatings, or designed to be an active component in coatings.

4.2.1.1 Biopolymer nanocomposite coatings

Candeuba wax solid lipid nanoparticles (267–344 nm) were used as coatings on guava fruit, compared with xanthan gum coating at a low temperature. The coating based on 65 $g \cdot L^{-1}$ solid lipid nanoparticles had lower permeabilities to O_2 and CO_2, responsible for reducing the respiration rate of guava fruit and maintaining the nutritional quality of the fruit for five weeks. Furthermore, 65 $g \cdot L^{-1}$ of solid lipid nanoparticle-based coating retained total phenolic and ascorbic acid content by delaying the oxidative reaction in guava fruit. However, the coating based on 75 $g \cdot L^{-1}$ of solid lipid nanoparticles resulted in anaerobic respiration, which caused physiological damage to the fruit [69].

Chitosan nanoparticles with a low molecular weight were successfully created and used as a coating for banana fruit. Cavendish bananas coated with chitosan nanoparticles showed uniform and smooth skin. A chitosan nanoparticle-based coating delayed the ripening of banana fruit by two to three days, compared to the uncoated control [70]. Chitosan nanoparticles added to a *Moringa oleifera* plant extract or aloe vera gel also had a significant impact on the firmness, ethylene rate, respiration, and total phenolic content of Cavendish bananas during storage. Banana fruit coated with this composite coating showed a lower weight loss and a higher score for consumer evaluation compared to the fruit coated with aloe vera or *M. oleifera* plant extract alone [71].

4.2.1.2 Nanoparticles incorporated in coatings

Antibacterial ZnO nanoparticles were combined in a κ-carrageenan solution for coating bananas (*Musa* sp., AAA group) during storage at ambient temperature. The nano-ZnO treatment significantly reduced the weight loss of banana fruit, while the κ-carrageenan-based coating reduced the fruit's respiration. Furthermore, κ-carrageenan-based coating combined with nano-ZnO delayed peel color changes by maintaining chlorophyll content, reduced weight loss, retained firmness, and reduced fruit disease incidence [72] (**Figure 11**). In addition, 500 $mg \cdot L^{-1}$ nano-ZnO was mixed in a 10% shellac solution to improve the postharvest quality of gac fruit (*M. cochinchinensis* Spreng) at 25°C. Throughout the 12 days, the nano-ZnO coating effectively inhibited disease infection and severity on fruit [37].

Hmmam et al. [73] developed carboxymethyl cellulose (CMC) and guar gum-based silver nanoparticles (AgNPs) coatings for "Seddik" mango fruit. Nanoparticles were formed at an average size of 84.8 to 213 nm for CMC-AgNPs and 61.7 to 132 nm for guar gum-AgNPs. The guar gum-AgNP coating significantly reduced the weight loss and respiration rate of mango fruit during storage, compared to the CMC-AgNP

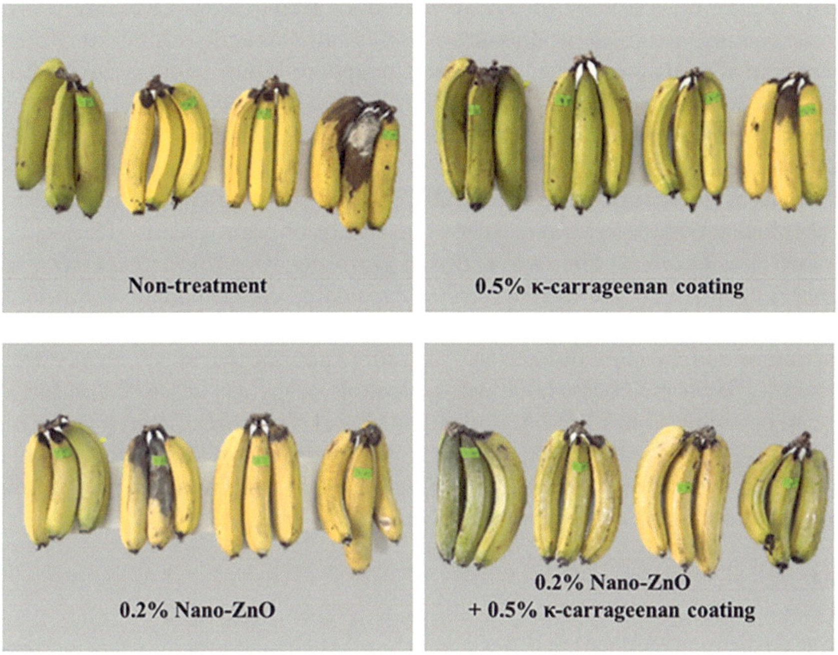

Figure 11.
Appearance of banana fruit treated with various coatings on day 10 at ambient temperature.

coating and uncoated fruit. The application of CMC- or guar gum-based AgNP coatings retarded the ripening and prolonged the postharvest life of mango fruit. Vieira et al. [74] fabricated an active coating based on hydroxypropyl methylcellulose (HPMC) and silver nanoparticles to extend the papaya's shelf life. HPMC, glycerol, and silver nanoparticles were well dispersed into the nanocomposite film due to the chemical bonds between HPMC chains and AgNPs. AgNPs did not affect the water vapor, oxygen, or carbon dioxide permeabilities. The coating based on HPMC and 0.25% AgNPs retained color and firmness, reduced weight loss, and delayed the change to soluble solids of papaya fruit during storage.

4.2.2 Encapsulation

Incorporating bioactive chemicals into food items confers numerous advantages for food preservation and the development of functional foods. However, bioactive compounds may cause a quick loss of function or be evaporated through the air. Encapsulation with edible coatings is a possibly advanced technology that can mitigate the disadvantages of employing bioactive chemicals by storing the compounds and managing the release control [75]. Depending on the qualities and objectives of the bioactive chemical, various encapsulation methods may be utilized. Most of the target compounds are volatiles or essential oils. These methods are more successful than a direct application on the food surface because edible coatings prevent the agents from migrating away from the surface, retaining a high concentration of bioactive compounds where needed. The encapsulators should contain electrostatic charges as well as emulsifiers such as dextran, oligosaccharides, oligopeptides, glycerol, phospholipids, etc.

Many reports have studied the encapsulation of active compounds integrated into some coatings for fresh produce storage. Cinnamaldehyde exhibits antifungal functions but is easily evaporated into the atmosphere. Thus, using an inclusion complex method, cinnamaldehyde was encapsulated in β-cyclodextrin to produce a complex that can be used to preserve fresh-cut produce. The 25:70 cinnamaldehyde/β-cyclodextrin ratio demonstrated the highest encapsulation efficiency and capacity, whereas, in the first three hours, the 25:75 ratio had faster control release. Antimicrobial activity was tested against two strains of gram-positive (*Staphylococcus aureus* and *Bacillus cereus*) and two strains of gram-negative (*Escherichia coli* and *P. aeruginosa*) bacteria. β-cyclodextrin with encapsulated cinnamaldehyde inhibited all tested bacterial strains [76]. Much work has been done on encapsulating fruit aroma/flavor compounds through interactions with some polysaccharides such as starch. For example, flavor molecules (aldehydes, alcohols, terpenes, ketones, and fatty acids) can be wrapped in a left-handed single helical structure of starch (with high linear amylose). Alternatively, the interaction between starch and flavor compounds included polar interactions. The hydrogen bonds are formed between the hydroxyl groups of starch and flavor compounds [77].

4.2.3 Multilayer coating

Typical coating techniques may not cover the whole fruit correctly, causing improper permeability of gases and water vapor between the coated fruit and the atmosphere. Thus, efforts are being made to find multi-component edible coatings that are rationally developed to boost the overall performance of edible coatings. The technique of multilayer coating or layer-by-layer (LBL) electrostatic deposition is one method that uses thin multilayers to improve the performance of edible coatings. With the LBL method, coating characteristics and functionality may be efficiently controlled by alternating the deposition of polyelectrolytes with opposite charges onto fruit surfaces [78]. LBL coating forms integrated thin films by alternating layers of various materials carrying different charges or functional groups. The first layer often holds polyelectrolytes with positive charges; thus, the second polyelectrolyte layer should have a negative charge opposite the first layer. Each additional layer flips the polarity of the charge on the surface. Repeating these steps multiple times creates a multilayered LBL coating (**Figure 12**).

The method of constructing LBL edible coatings allows for the combination of the best characteristics of various coating materials. For example, antibacterial polysaccharides can be combined with well-adhesive proteins, or active polysaccharides can be combined with polysaccharides that improve adhesion and texture. The feasibility of using the LBL edible coating method for complete surface coverage is elucidated by the contact angles of a coating droplet on the surface (**Figure 13**). Chitosan holding positive charges is in orange, whereas polystyrene sulfonate (PSS) having negative charges is in a clear drop. Each coating layer began with chitosan and ended with PSS. The contact angle of a chitosan droplet is less than 90° (flat shape) on all surface coatings, indicating high adhesive force between the layers. On the other hand, a droplet's contact angle is over 90° (round shape), showing high cohesion between PSS molecules.

Prior to storage at 25°C, a multilayer coating of oppositely charged chitosan (CTS: +) and polystyrene sulfonate (PSS: -) was treated on mature green "Nam Dok Mai" mangoes. Fruit coated at 3½ layers delayed ripening and reduced disease infection without off flavors, whereas fruit coated at 5½ layers had fermentation disorders at the end of storage [79]. Subsequently, allyl isothiocyanate (AIT), a natural antifungal compound, was integrated into the first layer of a multilayer coating of 0.5% CTS and 0.5%

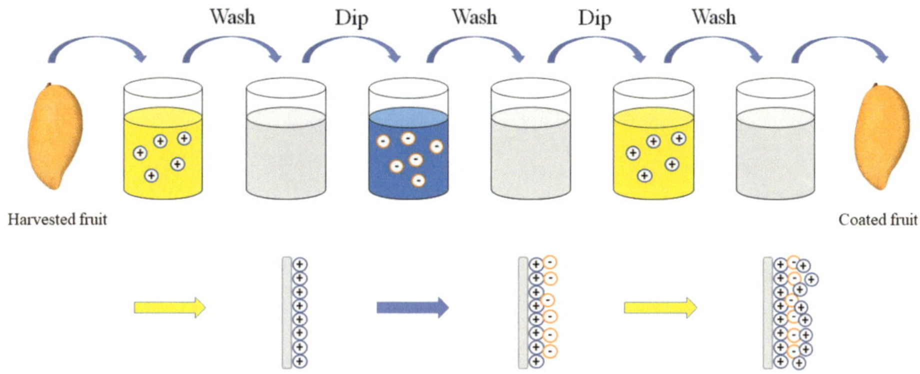

Figure 12.
The coating process with layer-by-layer electrostatic of opposite coating materials using a dipping and washing procedure (adopted from Costa et al. [78]).

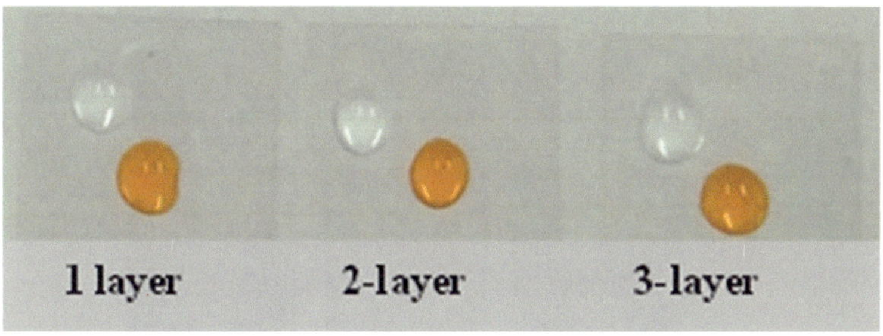

Figure 13.
Contact angle of a drop of chitosan (orange) and PSS (clear) on different chitosan/PSS coating layers.

PSS. The concentrations above 0.15% AIT effectively inhibited, *in vitro*, the mycelial growth of *Colletotrichum gloeosporioides*. The multicoating delayed changes in weight loss, firmness, and antioxidant capacities of mango. Furthermore, mangoes coated with 0.5% CTS and 0.5% PSS + 0.15% AIT (**Figure 14C**) significantly reduced anthracnose disease severity in *C. gloeosporioides*-inoculated fruit (**Figure 14A** and **B**) [80].

Figure 15A shows an *in vitro* culture of *Colletotrichum gloeosporioides* on different PDA media with 1% chitosan, 500 mg·L^{-1} prochloraz (a commercial fungicide), and some plant extracts. The fungal growth is wholly inhibited by 5000 μL·L^{-1} galangal extract or 1000 μL·L^{-1} sweet-flag extract. To reduce anthracnose during 25°C incubation, "Nam Dok Mai" mango fruit was treated with a double coating plus a sweet flag extract [24]. Mango fruit coated with 1% chitosan+3500 μL·L^{-1} sweet flag extract for the first layer and 0.1% sodium alginate for the second layer effectively inhibited disease infection for nine days at 25°C (**Figure 15B**) [24].

5. Modulation of fruit coatings: Fields for future research

5.1 Advantages and limitations of fruit coatings

Fruits have a natural cuticle that varies depending on the type and stage of development. The lack of protective fruit cuticles is caused by improper maturity [81]

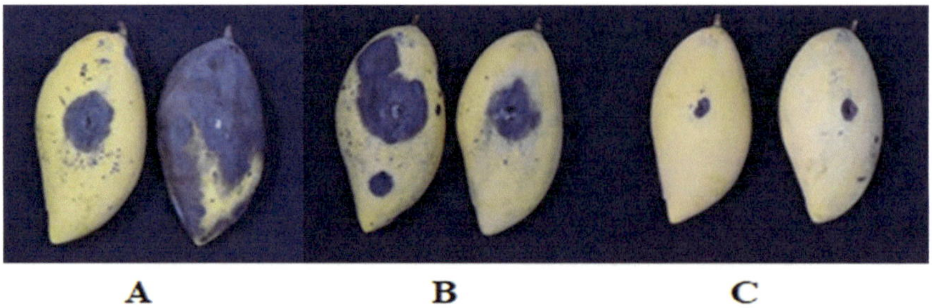

Figure 14.
Disease growth of Colletotrichum gloeosporioides inoculated "Nam Dok Mai" mangoes multilayer-coated with 0.5% chitosan/ 0.5% PSS (B), and 0.5% chitosan/0.5% PSS + 0.15% AIT (C), compared to the uncoated control (A) on day 10 at 25°C storage.

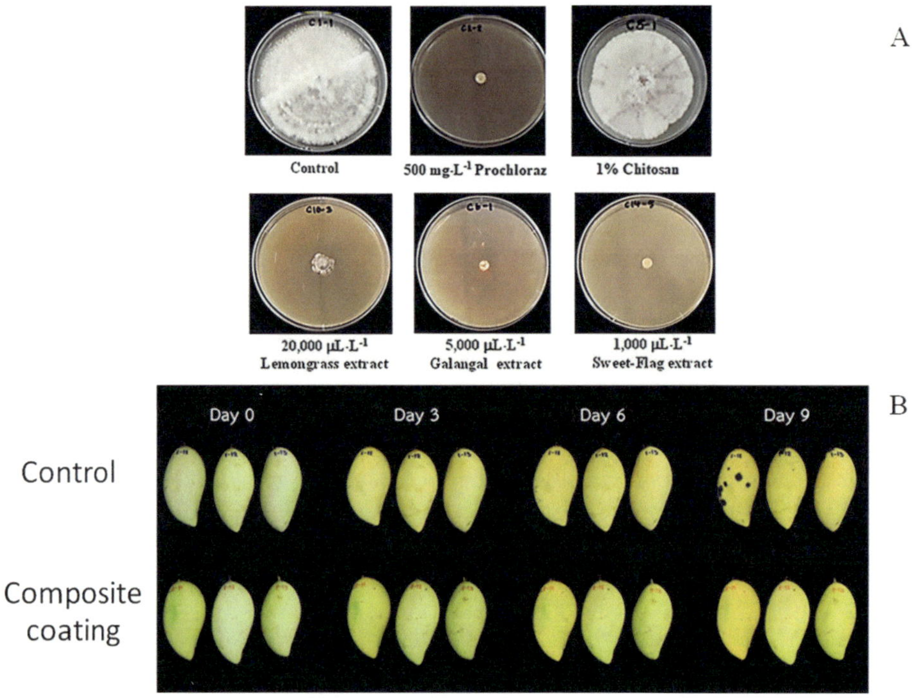

Figure 15.
In vitro *cultures of Colletotrichum gloeosporioides on different chemicals and plant extract PDA media on day 13 at 25°C (A), and "Nam Dok Mai" mangoes coated with a double coating of 0.5% chitosan+3500 µL·L^{-1} sweet-flag extract and 0.1% sodium alginate (lower row), compared to the uncoated control (upper row) during 25°C storage (B).*

or postharvest handling [82]. Coating as a part of packaging can add a polymer layer to the fruit surface. Aside from adding visual luster (**Figure 16**), fruit coating can extend shelf life and days to decay, reduce chilling injury and browning, and delay ripening by preventing water loss and creating modified environments inside the coated fruits.

Some concerns have been raised, however, about fruit coating. There is a possibility that the concentrations and components of coatings cause the fruit to undergo anaerobic respiration due to the inefficiency of its respiratory and transpiratory systems [24]. It is possible that the fruit's failure to ripen was due to unsuitable coatings. Some coating materials may cause a "plastic-like" film to form on the fruit's surface (**Figure 17A**), or they may cause a toxic response in the surface tissues of the fruit peel (**Figure 17B**). In some cases, a combination of postharvest treatments is required for effective fruit quality preservation. Coating alone may not be sufficient for preventing water loss in the long-term storage of some fresh commodities, such as rambutans [83] that contain a high number of spinterns. The sucrose fatty acid ester coatings, for example, cannot maintain the buying quality of rambutans (**Figure 18**), so those additional postharvest handlings, such as MAP at low-temperature storage, would be intensively required. To effectively reduce postharvest disease during low-temperature storage, Nguyen et al. [84] used hot water treatment prior to the chitosan- and κ-carrageenan-based composite coating of dragon fruit.

Some coating materials, particularly polysaccharide-based coatings, can become reversible during temperature fluctuations. Furthermore, in some cases in which the internal atmosphere is modified, fruit coating can reduce or inhibit the production of natural fragrances such as ester volatile compounds [85].

5.2 Developments in fruit coating research

Fruit coting is a practical method of handling fresh produce after harvest. The coating material chosen is determined by the intended use, coating types and techniques, and fruit types (high or low respiration rates). Nowadays, most fruit coating materials are edible and can be directly consumed safely. A trend in the industry is to find new coating materials, especially from agricultural waste [39, 86], that are flexible at various temperatures (up and down). As a result, advanced polymers should be used to improve coating efficiency, particularly in disease prevention. Another interesting issue is adding additional volatiles to coating materials, as fruit coating may reduce the release of the fruit's aroma or volatiles. MA or hypoxic conditions could inhibit some volatile

Figure 16.
Visual appearances of uncoated gac fruit (left) and fruit coated with 8% shellac+nano-silver particles (right).

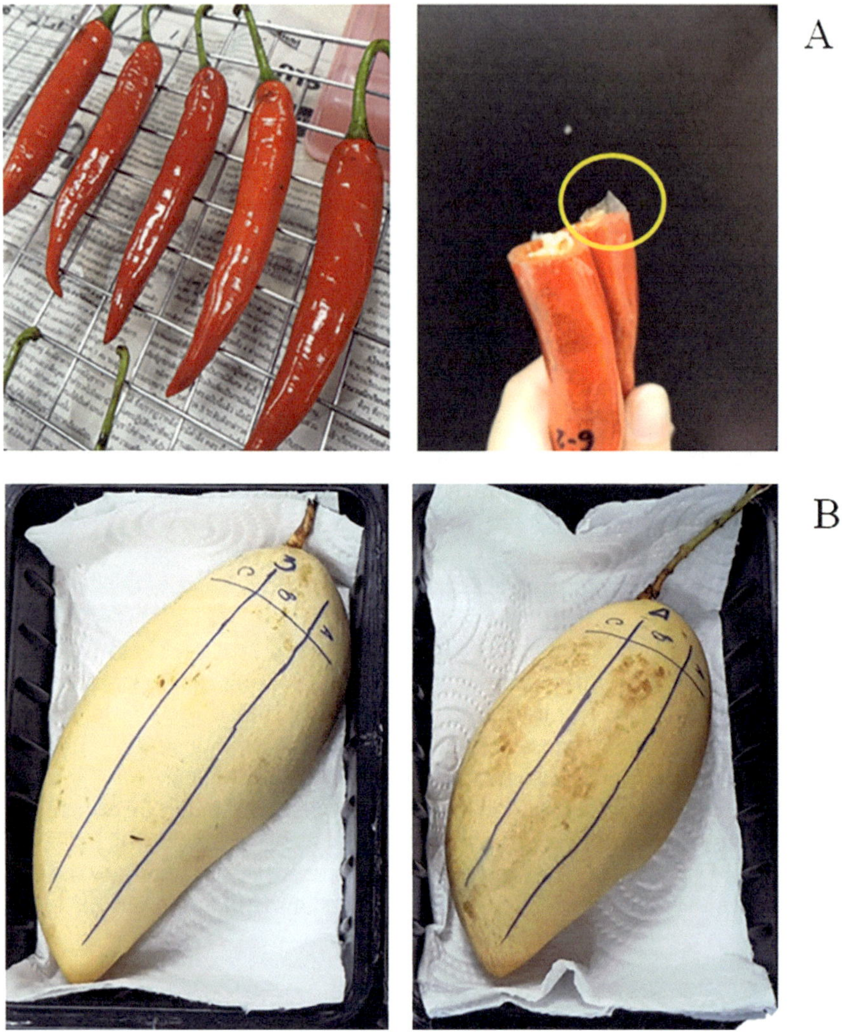

Figure 17.
A "plastic-like" appearance of chili fruit coating (A) and peel toxicity tests of coating materials at different concentrations (B).

production, particularly ethylene, depending on volatiles such as ester compounds [85]. Encapsulating some natural volatiles into the coating could prevent microorganisms and entice customers' aroma preferences as indicated in aroma sensory research [87].

Furthermore, in the final step, the product's feasibility should be tested prior to retailing. **Figure 19** from our collaborative research with the Food and Agribusiness, Trade, and Investment Queensland shows the cold chain logistics of "Murcott" mandarins treated with different coatings and MA storage shipped from Queensland, New Zealand, to Bangkok, Thailand. The alterations of temperature and relative humidity of the atmosphere during cold chain shipping and transport were recorded. The fruit quality was monitored in Thailand after arrival and subsequent storage in Bangkok in 2015, and the merchandise has since been sold on shelves in many modern trade stores in Thailand.

Figure 18.
Rambutan fruit coated with different coating materials of sucrose fatty acid esters (sucrose myristate (M-1695), sucrose palmitate (P-1670), and sucrose stearate (S-1670)) for 12 days at 13°C.

Figure 19.
The study of cold chain logistics of coated "Murcott" mandarins shipped from Queensland, New Zealand, to Bangkok, Thailand, in 2015.

6. Conclusions

The effectiveness of fruit coating is correlated with the intended purposes of use, which may include improving glossy, disease resistance, gas and moisture permeability, or some combination of these. Furthermore, the fruit's nature, such as structure, types, maturity stages, and physiological metabolism, has different behaviors. The information is crucial for formulating the coating for each product. Some fruits are not conducive to standard MAP but lend themselves well to fruit coating. For instance, durian fruit has many tiny, sharp spines that can pierce plastic wrapping. Even with coating, the shelf life of some fruits, such as rambutan, is limited by the presence of many stomata on the peel. Coating materials and processes are now being researched to enhance them, including composite combinations, encapsulation of nanoparticles, and multilayer coating. Eventually, fruit coating is both environmentally beneficial and economically effective.

Acknowledgements

We appreciate the Postharvest Technology Innovation Center, Ministry of Higher Education, Science, Research, and Innovation, Bangkok, Thailand for some research data.

Author details

Chalermchai Wongs-Aree[1,2]*, Hanh Thi Nguyen[3] and Sompoch Noichinda[4]

1 Division of Postharvest Technology, School of Bioresources and Technology, King Mongkut's University of Technology Thonburi, Bangkok, Thailand

2 Postharvest Technology Innovation Center, Ministry of Higher Education, Science, Research and Innovation, Bangkok, Thailand

3 Department of Postharvest Technology, Faculty of Food Science and Technology, Vietnam National University of Agriculture, Hanoi, Vietnam

4 Division of Agro-Industrial Technology, Faculty of Applied Science, King Mongkut's University of Technology North Bangkok, Bangkok, Thailand

*Address all correspondence to: chalermchai.won@kmutt.ac.th

References

[1] Kirci M, Isaksson O, Seifert R. Managing perishability in the fruit and vegetable supply chains. Sustainability. 2022;**14**:5378

[2] Lufu R, Ambaw A, Opara UL. Water loss of fresh fruit: Influencing pre-harvest, harvest and postharvest factors. Scientia Horticulturae. 2020;**272**:109519

[3] Prasad K, Siddiqui MW, Sharma RR, Gaurav AK, Neha P, Meena NK. Edible coatings and their effect on postharvest fruit quality. In: Siddiqui MW, Rahman MS, Wani AA, editors. Innovative Packaging of Fruits and Vegetables: Strategies for Safety and Quality Maintenance. New York: CRC Press; 2018. pp. 201-264

[4] Domínguez E, Heredia-Guerrero JA, Heredia A. The biophysical design of plant cuticles: An overview. New Phytologist. 2011;**189**:938-949

[5] Segado P, Domínguez E, Heredia A. Ultrastructure of the epidermal cell wall and cuticle of tomato fruit (*Solanum lycopersicum* L.) during development. Plant Physiology. 2016;**170**:935-946

[6] Malinovsky FG, Fangel JU, Willats WG. The role of the cell wall in plant immunity. Frontiers in Plant Science. 2014;**5**:178

[7] Heredia A. Biophysical and biochemical characteristics of cutin, a plant barrier biopolymer. Biochimica et Biophysica Acta. 2003;**1620**:1-7

[8] Yeats TH, Rose JK. The formation and function of plant cuticles. Plant Physiology. 2013;**163**:5-20

[9] Huang H, Lian Q, Wang L, Shan Y, Li F, Chang SK, et al. Chemical composition of the cuticular membrane in guava fruit (*Psidium guajava* L.) affects barrier property to transpiration. Plant Physiology and Biochemistry. 2020;**155**:589-595

[10] Yingsanga P, Srilaong V, McGlasson B, Kabanoff E, Kanlayanarat S, Noichinda S. Morphological differences associated with water loss in rambutan fruit cv Rongrein and see-Chompoo. Acta Horticulturae. 2006;**712**:453-460

[11] Pereira E, Silva RGB, Spagnol WA, Silveira JV. Water loss in table grapes: Model development and validation under dynamic storage conditions. Food Science and Technology (Campinas). 2018;**38**:473-479

[12] Lentzou D, Xanthopoulos G, Templalexis C, Kaltsa A. The transpiration and respiration as mechanisms of water loss in cold storage of figs. Food Research. 2021;**5**:109-118

[13] Khalid S, Ullah MI, Khalid MS, Shahid M, Sultana SR. Relationship between moisture loss and physicochemical quality attributes of tomato under different storage conditions. Postharvest Technology. 2022;**10**:71-79

[14] Wongs-Aree C, Noichinda S. Postharvest quality properties of potential tropical fruits related to their unique structural characters. In: Florkowski WJ, Banks NH, Shewfelt RL, Prussia SE, editors. Postharvest Handling: A Systems Approach. 4th ed. Oxford: Academic Press; 2022. pp. 277-316

[15] Landrigan M, Safaris V, Morris SC, McGlasson BW. Structural aspects of rambutan (*Nephilium lappaceum*)

fruits and their relation to postharvest browning. Horticultural Science. 1994;**69**:571-579

[16] Wongs-Aree C, Noichinda S. Rambutan and longkong. In: de Freitas ST, Pareek S, editors. Postharvest Physiological Disorders in Fruits and Vegetables, Boca Raton: CRC Press; 2019. p. 551-588

[17] Nishimwe G, Kosgei JC, Okoth EM, Asudi GO, Rimberia F. Evaluation of the characteristics of new papaya hybrid lines in Kenya. African Journal of Biotechnology. 2019;**18**(2):58-67

[18] Woolley JT. Maintenance of air in intercellular spaces of plants. Plant Physiology. 1983;**72**:989-991

[19] Kosiyachinda S, Tansiriyakul S. Respiration rate and ethylene production of fresh fruit, vegetables and cut flowers. Food. 1998;**18**(1):1-10 (with English abstract)

[20] Akamine EK, Goo T. Concentrations of carbon dioxide and ethylene in the cavity of attached papaya fruit. HortScience. 1979;**14**(2):138-139

[21] Brooncherm P, Siriphanich J. Postharvest physiology of durian pulp and husk. Kasetsart Journal (Natural Science). 1991;**25**:119-125

[22] Amornputti S, Ketsa S, van Doorn WG. 1-Methylcyclopropene (1-MCP) inhibits ethylene production of durian fruit which is correlated with a decrease in ACC oxidase activity in the peel. Postharvest Biology and Technology. 2016;**114**:69-75

[23] Boonthanakorn J, Daud W, Aontee A, Wongs-Aree C. Quality preservation of fresh-cut durian cv. 'Monthong' using microperforated PET/PE films. Food packaging and shelf. Life. 2020;**23**:100452

[24] Promboon J. Development of Double-Layer Coating Containing Natural Extracts on 'Nam Dok Mai' Mango for Reducing Anthracnose Disease during Retail [Thesis]. Bangkok: King Mongkut's University of Technology Thonburi; 2019

[25] Vangnai T, Wongs-Aree C, Nimitkeatkai H, Kanlayanarat S. Quality maintaining of ´Daw´ longan using chitosan coating. Acta Horticulturae. 2006;**712**:599-604

[26] Win S, Mejunpet N, Buanong M, Kanlayanarat S, Wongs-Aree C. Postharvest quality alteration of gac fruit harvested at different maturities and coated with chitosan. International Food Research Journal. 2015;**22**(6):2219-2224

[27] Basumatary IB, Mukherjee A, Katiyar V, Dutta J, Kumar S. Chitosan-based active coating for pineapple preservation: Evaluation of antimicrobial efficacy and shelf-life extension. LWT. 2022;**168**:113940

[28] Hossain MS, Iqbal A. Effect of shrimp chitosan coating on postharvest quality of banana (*Musa sapientum* L.) fruits. International Food Research Journal. 2016;**23**:277-283

[29] González-Aguilar GA, Valenzuela-Soto E, Lizardi-Mendoza J, Goycoolea F, Martínez-Téllez MA, Villegas-Ochoa MA, et al. Effect of chitosan coating in preventing deterioration and preserving the quality of fresh-cut papaya 'Maradol'. Science of Food and Agriculture. 2009;**89**:15-23

[30] Felicia WXL, Rovina K, Nur'Aqilah MN, Vonnie JM, Erna KH, Misson M, et al. Recent advancements of polysaccharides to enhance quality and delay ripening of fresh produce: A review. Polymers (Basel). 2022;**14**:1341

[31] Wittaya T. Protein-based edible films: Characteristics and improvement of properties, (chapter 3). In: Eissa AA, editor. Structure and Function of Food Engineering. Oxford: London, UKInTech Open; 2012. pp. 43-70

[32] Chen H, Wang J, Cheng Y, Wang C, Liu H, Bian H, et al. Application of protein-based films and coatings for food packaging: A review. Polymers. 2019;**11**:10.3390

[33] Park HJ, Chinnan MS, Shewfelt RL. Effect of corn-zein film coating on the storage life of tomatoes. Food Processing and Preservation. 1994;**18**:317-331

[34] Avena-Bustillos RJ, Krochta JM, Saltveit ME, Rojas-Villegas R, Sauceda-Perez JA. Optimisation of edible coating formulations on zucchini to reduce water loss. Food Engineering. 1994;**21**:197-214

[35] Gol NB, Ramana Rao TV. Influence of zein and gelatin coatings on the postharvest quality and shelf life extension of mango (*Mangifera indica* L.). Fruits. 2014;**69**:101-115

[36] Morillon V, Debeaufort F, Blond G, Capelle M. Factors affecting the moisture permeability of lipid-based edible films: A review. Critical Reviews in Food Science and Nutrition. 2002;**42**:67-89

[37] Art S. Influence of Shellac and Chitosan Coating Combined with Nano Zinc Oxide on the Postharvest Quality of Gac Fruit (*Momordica cochinchinensis* Spreng) [Thesis]. Bangkok: King Mongkut's University of Technology Thonburi; 2019

[38] Miranda M, Sun X, Ference C, Plotto A, Bai J, Wood D, et al. American Society of Horticultural Science. 2021;**146**(1):40-49

[39] Merino D, Quilez-Molina AI, Perotto G, Bassani A, Spigno G, Athanassiou A. A second life for fruit and vegetable waste: A review on bioplastic films and coatings for potential food protection applications. Green Chemistry. 2022;**24**:4703-4727

[40] Ketrodsakul A. Investigation on the Properties of Cellulose from Baby Corn Plants for Developing of a Coating Material for Fruits: A Case Study of Mango cv. 'Nam Dok Mai No. 4' [Thesis]. Bangkok: King Mongkut's University of Technology Thonburi; 2015

[41] Supapvanich S, Mitrsang P, Srinorkham P, Boonyaritthongchai P, Wongs-Aree C. Effects of fresh Aloe vera gel coating on browning alleviation of fresh cut wax apple (*Syzygium samarangenese*) fruit cv. Taaptimjaan. Food Science and Technology. 2016;**53**(6):2844-2850

[42] Food and Drug Administration (FDA). Notice of inventory. http://www.fda.gov/downloads/Food/IngredientsPackagingLabeling/GRAS/NoticeInventory/UCM267039;2001.

[43] Chimvaree C, Wongs-Aree C, Supapvanich S, Charoenrat T, Tepsorn R, Boonyaritthongchai P. Effect of sericin coating on reducing browning of fresh-cut mango cv. 'Nam Dok Mai No. 4'. Agricultural and Natural Resources. 2019;**53**(5):521-526

[44] Thongsook T, Tiyaboonchai W. Inhibitory effect of sericin on polyphenol oxidase and its application as edible coating. International Journal of Food Science and Technology. 2011;**46**(10):2052-2061

[45] Kato N, Sato S, Yamanaka A, Yamada H, Fuwa N, Nomura M. Silk protein, sericin, inhibits lipid peroxidation and tyrosinase activity.

Bioscience, Biotechnology, and Biochemistry. 1998;**62**:145-147

[46] Tsouko E, Kourmentza C, Ladakis D, Kopsahelis N, Mandala I, Papanikolaou S, et al. Bacterial cellulose production from industrial waste and by-product streams. International Journal of Molecular Sciences. 2015;**5**(16):14832-14849

[47] Yanti NA, Ahmad SW, Ramadhan LOAN, Jamili M, Walhidayah T, Mamangkey J. Properties and application of edible modified bacterial cellulose film based sago liquid waste as food packaging. Polymers (Basel). 2021;**13**(20):3570

[48] Utama GL, Dinika I, Nurmilah S, Masruchin N, Nurhadi B, Balia RL. Characterisation of antimicrobial composite edible film formulated from fermented cheese whey and cassava peel starch. Membranes. 2022;**12**(6):636

[49] Hernández-Valencia CG, Román-Guerrero A, Aguilar-Santamaría Á, Cira, Shirai K. Cross-linking chitosan into hydroxypropyl methylcellulose for the preparation of neem oil coating for postharvest storage of pitaya (*Stenocereus pruinosus*). Molecules. 2019;**24**:219

[50] Vieira JM, Flores-López ML, de Rodríguez DJ, Sousa MC, Vicente AA, Martins JT. Effect of chitosan–Aloe vera coating on postharvest quality of blueberry (*Vaccinium corymbosum*) fruit. Postharvest Biology and Technology. 2016;**116**:88-97

[51] Shahbazi M, Rajabzadeh G, Ettelaie R, Rafe A. Kinetic study of κ-carrageenan degradation and its impact on mechanical and structural properties of chitosan/κ-carrageenan film. Carbohydrate Polymers. 2016;**142**:167-176

[52] Nguyen HT, Boonyaritthongchai P, Buanong M, Supapvanich S, Wongs-Aree C. Edible coating of chitosan ionically combined with κ-carrageenan maintains the bract and postharvest attributes of dragon fruit (*Hylocereus undatus*). International Food Research Journal. 2021;**28**(4):682-694

[53] Liu C, Jin T, Liu W, Hao W, Yan L, Zheng L. Effects of hydroxyethyl cellulose and sodium alginate edible coating containing asparagus waste extract on postharvest quality of strawberry fruit. LWT - Food Science and Technology. 2021;**148**:111770

[54] Fagundes C, Palou L, Monteiro AR, Pérez-Gago MB. Effect of antifungal hydroxypropyl methylcellulose-beeswax edible coatings on gray mold development and quality attributes of cold-stored cherry tomato fruit. Postharvest Biology and Technology. 2014;**92**:1-8

[55] Velickova E, Winkelhausen E, Kuzmanova S, Alves VD, Moldão-Martins M. Impact of chitosan-beeswax edible coatings on the quality of fresh strawberries (*Fragaria ananassa* cv Camarosa) under commercial storage conditions. LWT - Food Science and Technology. 2013;**52**:80-92

[56] Ali U, Basu S, Mazumder K. Improved postharvest quality of apple (rich red) by composite coating based on arabinoxylan and beta-glucan stearic acid ester. International Journal of Biological Macromolecules. 2020;**151**:618-627

[57] Vieira TM, Moldao-Martins M, Alves VD. Composite coatings of chitosan and alginate emulsions with olive oil to enhance postharvest quality and shelf life of fresh figs (*Ficus carica* L. cv. 'Pingo De Mel'). Food. 2021;**10**(4):718

[58] Win S, Mejunpet N, Buanong M, Kanlayanarat S, Wongs-Aree C. Changes

in postharvest quality of Gac fruit (*Momordica cochinchinensis* Spreng) coated with sucrose fatty acid esters. Acta Horticulturae. 2015;**1088**:237-241

[59] Soradech S, Nunthanid J, Limmatvapirat S, Luangtana-anan M. Utilisation of shellac and gelatin composite film for coating to extend the shelf life of banana. Food Control. 2017;**73**:1310-1317

[60] Reinoso E, Mittal GS, Lim L-T. Influence of whey protein composite coatings on plum (*Prunus domestica* L.) fruit quality. Food and Bioprocess Technology. 2008;**1**:314-325

[61] Shiekh KA, Ngiwngam K, Tongdeesoontorn W. Polysaccharide-based active coatings incorporated with bioactive compounds for reducing postharvest losses of fresh fruits. Coatings. 2022;**12**:10.3390

[62] Kumar N, Pratibha N, Petkoska AT, Al-Hilifi SA, Fawole OA. Effect of chitosan–pullulan composite edible coating functionalised with pomegranate peel extract on the shelf life of mango (*Mangifera indica*). Coatings. 2021;**11**:764

[63] Nguyen TTT, Le TQ, Nguyen TTA, Nguyen LTM, Nguyen DTC, Van Tran T. Characterisations and antibacterial activities of passion fruit peel pectin/chitosan composite films incorporated *Piper betle* L. leaf extract for preservation of purple eggplants. Heliyon. 2022;**8**:e10096

[64] Lichanporn I, Techavuthiporn C, Wongs-Aree C. Effect of silver particle-longkong peel extract coating on postharvest decay and browning in longkong fruit. Horticulture Journal. 2020;**89**(3):328-336

[65] Le KH, La DD, Nguyen PTM, Nguyen MD-B, Vo ATK, Nguyen MTH, et al. Fabrication of *Cleistocalyx operculatus* extracts/chitosan/gum arabic composite as an edible coating for preservation of banana. Progress in Organic Coatings. 2021;**161**:106550

[66] Nguyen HT, Boonyaritthongchai P, Buanong M, Supapvanich S, Wongs-Aree C. Chitosan- and κ-carrageenan-based composite coating on dragon fruit (*Hylocereus undatus*) pretreated with plant growth regulators maintains bract chlorophyll and fruit edibility. Scientia Horticulturae. 2021;**281**:109916

[67] Wongs-Aree C, Tongchitpakdee S, Mongkolchaiyaphruek A, Penchaiya P, Wattanawan C. Postharvest Handling and Packaging to Manage Quality of Fresh-Cut and Intact Produce of Prominent Perishable Crops, PL.P.1/2560 [Full research report]. Bangkok: Postharvest Technology Innovation Center; 2020

[68] Carrillo JY, Raphael E, Dobrynin AV. Adhesion of nanoparticles. Langmuir. 2010;**26**:12973-12979

[69] García-Betanzos CI, Hernández-Sánchez H, Bernal-Couoh TF, Quintanar-Guerrero D, de la Luz Zambrano-Zaragoza M. Physicochemical, total phenols and pectin methylesterase changes on quality maintenance on guava fruit (*Psidium guajava* L.) coated with candeuba wax solid lipid nanoparticles-xanthan gum. Food Research International. 2017;**101**:218-227

[70] Esyanti RR, Zaskia H, Amalia A, Nugrahapraja H. Chitosan nanoparticle-based coating as postharvest technology in banana. Journal of Physics: Conference Series. 2019;**1204**:012109

[71] Odetayo T, Sithole L, Shezi S, Nomngongo P, Tesfay S, Ngobese NZ. Effect of nanoparticle-enriched coatings

on the shelf life of Cavendish bananas. Scientia Horticulturae. 2022;**304**:111312

[72] Nguyen HT, Nguyen NT, Wongs-Aree C, Nguyen TBT. Effect of kappa-carrageenan-based coating combined with nano-ZnO on quality of banana (*Musa acuminata* AAA group) during storage at ambient condition. In: International Conference of Transforming System and Strengthening Development for Agricultural Sustainability. Bangkok: The International Society for Southeast Asian Agricultural Science; 2021

[73] Hmmam I, Zaid NM, Mamdouh B, Abdallatif A, Abd-Elfattah M, Ali M. Storage behavior of "Seddik" mango fruit coated with CMC and guar gum-based silver nanoparticles. Horticulturae. 2021;**7**(3):44

[74] Vieira ACF, de Matos FJ, Menezes NMC, Monteiro AR, Valencia GA. Active coatings based on hydroxypropyl methylcellulose and silver nanoparticles to extend the papaya (*Carica papaya* L.) shelf life. International Journal of Biological Macromolecules. 2020;**164**:489-498

[75] Quirós-Sauceda AE, Ayala-Zavala JF, Olivas GI, González-Aguilar GA. Edible coatings as encapsulating matrices for bioactive compounds: A review. Food Science and Technology. 2014;**51**:1674-1685

[76] Chimvaree C, Tepsorn R, Supapvanich S, Wongs-Aree C, Srilaong V, Boonyaritthongchai P. Encapsulation of cinnamaldehyde from cinnamon essential oils in cyclodextrin. IOP Conference Series: Earth and Environmental Science. 2020;**515**:012034

[77] Buljeta I, Pichler A, Ivic I, Šimunovic J, Kopjar M. Encapsulation of fruit flavor compounds through interaction with polysaccharides. Molecules. 2021;**26**:4207

[78] Costa RR, Mano JF. Polyelectrolyte multilayered assemblies in biomedical technologies. Chemical Society Reviews. 2014;**43**(10):3453-3479

[79] Hadthamard N, Chaumpluk P, Buanong M, Boonyaritthongchai P, Wongs-Aree C. Effects of multilayer coating of chitosan and polystyrene sulfonate on quality of 'Nam Dok Mai No.4′ mango. International Journal of Agricultural and Biosystems Engineering. 2019;**13**(3):42-48

[80] Hadthamard N, Chaumpluk P, Buanong M, Boonyaritthongchai P, Wongs-Aree C. Integrated allyl isothiocyanate effectively reduces anthracnose and maintains the marketing quality of chitosan-polystyrene sulfonate multicoated mango. Plant Cell Biotechnology and Molecular Biology. 2021;**22**:124-138

[81] Martin LBB, Rose JKC. There's more than one way to skin a fruit: Formation and functions of fruit cuticles. Experimental Botany. 2014;**65**:4639-4651

[82] Lara I, Heredia A, Domínguez E. Shelf life potential and the fruit cuticle: The unexpected player. Frontiers in Plant Science. 2019;**10**:10.3389

[83] Brown BI, Wilson PR. Exploratory study of postharvest treatments on rambutan (*Nephelium lappaceum*) 1986/1987 season. Rare Fruit Council of Australian Newsletter. 1988;**48**:16-18

[84] Nguyen HT, Boonyaritthongchai P, Buanong M, Supapvanich S, Wongs-Aree C. Postharvest hot water treatment followed by chitosan- and κ-carrageenan-based composite coating induces the disease resistance and preserves the quality in

dragon fruit (*Hylocereus undatus*). International Journal of Fruit Science. 2020;**20**:S2030-S2044

[85] Wongs-Aree C, Noichinda S. Glycolysis fermentative by-products and secondary metabolites involved in plant adaptation under hypoxia during pre- and postharvest (chapter 4). In: Das K, Biradar MS, editors. Hypoxia and Annoxia. London: IntechOpen; 2018. pp. 59-72

[86] Choudhary U, Bhinchhar BK, Paswan VK, Kharkwal S, Yadav SP, Singh P. Utilization of agro-industrial wastes as edible coating and films for food packaging materials. In: Var I, Uzunlu S, editors. A Glance at Food Processing Applications. London: IntechOpen; 2021. pp. 1-18

[87] Chambers E, Koppel K. Associations of volatile compounds with sensory aroma and flavor: The complex nature of flavor. Molecules. 2013;**18**:18054887

Chapter 3

Alternative Green and Novel Postharvest Treatments for Minimally Processed Fruits and Vegetables

Mildred Osei-Kwarteng, Gustav K. Mahunu, Moomin Abu and Maurice Apaliya

Abstract

Minimally processed fresh produce is ready to eat and subjected to minimal technology before consumption. Fresh fruits and vegetables (FFVs) are minimally processed commodities that are metabolically active and undergo physiological processes such as ripening and senescence, reducing their quality and shelf life. Postharvest technologies maintain the quality and prolong the shelf life of harvested produce, without which the quality deteriorates such that significant economic loss ensues due to water and nutrients loss, physiological deterioration, biochemical changes, and microbial degeneration. Conventional postharvest treatments such as temperature management, and chemical and gaseous treatments are widely known for controlling postharvest issues in FFVs. However, there are novel and green alternative safe methods that are employed to maintain the postharvest quality and prolong the shelf life of FFVs. This chapter focuses on seven common alternative novel and green postharvest treatments: nitric oxide, ozone, methyl jasmonate, salicylic acid, oxalic acid, calcium, and heat treatments. These treatments are explained and some of their current application on FFVs are discussed and tabularized indicating the optimum treatment conditions reported in the latest scientific publications.

Keywords: calcium, heat, methyl jasmonate, nitric oxide, oxalic acid, salicylic acid, ozone

1. Introduction

Fresh fruits and vegetables (FFVs) are considered minimally processed because they are ready to eat and subjected to minimal technology before consumption. FFVs are a source of micro- and macronutrients, and secondary metabolites that possess antioxidant and reactive oxygen species (ROS) scavenging abilities for good health [1–4]. FFVs are living tissues and are metabolically active experiencing senescence, and ripening processes that necessitate quality preservation and the reduction of storage losses [4–8]. Postharvest technologies control and prolong the shelf life of harvested produce, without which the quality (i.e., nutritional, appearance, and pathogenic safety) deteriorates such that significant economic losses ensue [5, 6] due to water and nutrient loss, physiological deterioration, biochemical changes, and microbial degeneration [6, 9]. Hence, optimum postharvest treatments must be employed to impede physiological processes (e.g., senescence and maturation) and minimize the occurrence of microbial contamination to preserve quality loss [5, 6, 10]. Temperature management, physical (e.g., heat, irradiation, and edible coatings) chemical (antimicrobials, antioxidants, and antibrowning), and gaseous treatments (e.g., chlorine dioxide) are some conventional postharvest treatments for FFVs [5, 9]. Nevertheless, there are currently novel and green postharvest treatments that are used to maintain and enhance the quality, and prolong the postharvest shelf life of fresh produce.

Novel postharvest treatments are new postharvest treatments that, among other functions enhance bioactive molecule contents or compounds, retard senescence and ripening, prevent postharvest diseases, prolong shelf life, and ensure the overall safety of fresh produce for consumption [6]. On the other hand, green postharvest treatments are those that are environmentally friendly, such that their application is important and socially acceptable to the industry and consumers at large in maintaining and enhancing the postharvest quality, and prolonging the storage life of fresh produce. Green postharvest treatments include non-chemical applications like biological control agents, natural compounds, biomaterials, irradiation (ultraviolet), and heat treatments [6, 11–13]. Therefore, some novel postharvest treatments are green, but not all novel treatments are green. The quality attributes (i.e., nutritional value, appearance, texture, flavour, and chemical, toxicological, and microbial safety) and shelf life of FFVs are maintained via various novel and green postharvest treatments [5]. This chapter presents seven postharvest treatments (i.e., nitric oxide, ozone, methyl jasmonate, salicylic acid, oxalic acid, calcium, and heat treatment) for FFVs. All seven are novel, but four are green (i.e., salicylic acid, oxalic acid, ozone, and heat treatments) [6, 8, 13].

2. Green and novel postharvest treatments and traditional postharvest techniques

Novel postharvest treatments are postharvest treatments other than conventional or already known postharvest treatments whereas green postharvest treatments are those that are environmentally friendly and socially acceptable [6, 12].

Some of the latest novel postharvest treatments include the application of nitric oxide, ozone, salicylic acid, oxalic acid, methyl jasmonates, calcium, and heat [6]. Notwithstanding, various compounds, including AVG (Aminoethoxyvinylglycine), 1MCP (1-Methylcyclopropene), and PAs (Polyamines) and other postharvest treatments have been previously used to regulate postharvest quality in some FFVs at different degrees of effectiveness. The aforementioned chemicals had limitations, such as uneven ripening, post-ripening disorders, and non-viability for commercial use. Conventional postharvest harvest techniques that are used to prolong the shelf life of fruits and vegetables include cold storage, modified and controlled atmosphere storage, suppression of ethylene biosynthesis or ethylene inaction by PAs, AVG, and 1-MCP application [6].

Green and novel postharvest treatments extend the shelf life of freshly harvested produce by maintaining their quality [6]. We focused on the latest applications of nitric oxide, ozone, methyl jasmonate, salicylic acid, oxalic acid, calcium, and heat treatments on FFVs, with emphasis on quality and shelf life under storage. The seven were chosen because of their extensive use on FFVs [6, 8].

2.1 Nitric oxide (NO)

Nitric oxide (NO) is an important signalling molecule mediating several pre- and postharvest developmental and physiological activities in horticultural crops [6, 14].

NO is a bioactive, mobile gaseous molecule that also mediates various abiotic and biotic stress responses [15]. NO can protect against various stressful impacts, scavenge free oxygen radicals, counteract oxidative damage, inhibit, or suppress ethylene biosynthesis, delay ripening and senescence, and enhance the resistance to diseases [6, 14–17]. NO has a more profound effect on non-climacteric horticultural crops (especially fruits) than climacteric crops. Consequently, the climacteric phase (i.e., the surge in ethylene production and an increased respiration rate) of many horticultural crops can be impeded by endogenous NO production. Also, endogenous NO reduces yellowing, chlorophyll degradation and extends the shelf life of fruits and vegetables [6]. Notwithstanding, NO is volatile and exhibits reactive oxygen species toxicity. Hence, super-optimal concentrations can be harmful. Thus, it is necessary that the correct threshold levels are employed to obtain the desirable ripening modulation and extension of shelf life in FFVs [6]. It is worth noting that NO is more effective in its gaseous or donor form in suppressing ethylene production and respiration rate, reducing softness, retarding colour development and metabolism, reducing chlorophyll degradation, and delaying senescence [6].

2.1.1 NO treatments of fresh fruits and vegetables

NO is a good alternative for extending the shelf life of fresh horticultural produce. The application of optimum levels of NO has been shown to impede senescence and ripening processes in many horticultural crops, including FFVs. Donor compounds (e.g., sodium nitroprusside (SNP)) can also be incorporated into biological systems to release NO gas under controlled environmental conditions [6]. Due to the highly diffusible nature of NO, NO was earlier considered an environmental pollutant as it caused, among other things the reduction of ozone in the stratosphere [6, 14]. Also, the emission of NO is related to the greenhouse effect [6, 14, 18]. Moreover, NO gas has a short lifetime, such that in the presence of oxygen (O_2), it can be converted to the noxious gas nitrogen dioxide (NO_2), which can degrade the quality of FFVs. Consequently, NO gas should be placed in airtight containers to minimize contact with oxygen. Further, NO must be diluted by flushing with nitrogen (N_2) after fumigation to avoid damage to FFVs [17].

Postharvest immersion of FFVs such as plum, longan, apple, and broccoli in NO or SNP impeded internal browning. Also, exogenous application of NO via fumigation (In an O_2-free environment or NO-releasing agents such as “N-tert-butyl-α-phenylnitrone” and “3-morpholino sydnonimine”) remarkedly delayed maturation and ripening, controlled postharvest pest and prolonged the shelf life of FFVs such as apples, bananas, peaches, strawberries, and some other vegetables [6, 16]. Additionally, NO maintained the levels of polyphenols, soluble solids ascorbic acid in sliced apples, longan, fresh-cut apples, litchi, peach, lettuce, and broccoli and generally improved the quality and postharvest shelf life of some stored fruits and vegetables [6, 17, 19].

A significant reduction of chlorophyll degradation, delayed postharvest yellowing, accumulation of malondialdehyde, and reduced lipid peroxidation were observed in broccoli florets [6]. Treatment of cucumber with 25 microlitres per litre NO also reduced deterioration and exhibited radical scavenging activities compared to the control during storage. Furthermore, NO significantly increased the antioxidative process in cucumber fruits. Browning on the cut surfaces of leafy vegetables was effectively inhibited by NO treatments [6]. **Tables 1** and **2** present treatments of some fruits and vegetables with NO or NO donors, respectively.

2.2 Ozone treatments

Ozone is a highly reactive form of oxygen that decomposes easily into diatomic oxygen. Ozone can function as a potential oxidant and disinfectant when it reacts with targeted organic matter and microorganisms. Historically, it has been used as a water disinfectant. Ozone attained "GRAS" (generally regarded as safe) status and was approved as an antimicrobial additive by the United States Food and Drug Administration (FDA) in 2001 [1, 6, 21].

The influence of ozone on postharvest disease control and storage has been investigated in some fruits and vegetables for shelf life extension and preservation [1, 6, 13, 21–26]. Ozonated water/aqueous ozone has been used for disinfecting vegetables, while gaseous ozone is used for the sanitization and preservation of vegetables during storage. Gaseous ozone is a less effective antimicrobial agent than aqueous ozone [1, 21]. Moreover, ozone can be potentially used for the decontamination of surfaces of freshly harvested produce, degradation of ethylene, odour elimination in mixed storage, spore elimination in storage rooms, and reduction of pesticide levels over the fresh produce [1, 6, 13, 21, 25, 26]. Nevertheless, ozone should be properly used to avoid negative effects such as loss of sensory quality although ozone application is an eco-friendly technology [1, 21]. Consequently, ozone treatments are mostly specific for various fresh produce due to the intrinsic characteristics of fresh produce and the extrinsic factors that affect ozone efficiency [1, 21]. **Table 3** provides information on the ozone treatment of some fresh fruits and vegetables.

2.3 Salicylic acid treatments

Salicylic acid (SA) is a plant hormone that acts as a signalling molecule against environmental and pathogenic stress. SA influences various physiological events in plants [6, 34–36]. SA plays a key role in the retardation of fruit ripening, and the inhibition of ethylene biosynthesis, promoting pathogen resistance, activating antioxidant systems, consequently, maintaining postharvest quality and prolonging the shelf life of fruits and vegetables. Moreover, SA is an effective enhancer of biocontrol agents like antagonist yeast in controlling rot and decay [6, 34]. SA and its derivatives (particularly Methyl salicylate; MeSA) have been generally recognized as safe (GRAS) for fruits and vegetables and are environmentally friendly [34, 35, 37].

Several studies have demonstrated the effect of postharvest SA application (exogenous) in fruits and vegetables [3, 6, 34, 38]. The dose of SA for exogenous postharvest application is varied for various fruits and vegetables, however a general non-toxic range for fruits and vegetables is 0.5–2.0 mM. Higher concentrations can damage fruit skin and cause fungal attacks [37]. **Table 4** presents information on some current applications of SA on some fruits and vegetables.

FVs	Cultivar	NO/NO donor concentrations	Storage condition	Number of days(d)/ hours (h)/weeks of storage	Inferences	References
Apple (*Malus sylvestris* var. domestica Borkh.)	"Fuji"	10 uL L^{-1}	20 ± 0.5°C	50 d	Reduced fruit firmness	[6]
Banana (*Musa spp*.)	"Brazil"	5 mM SNP (slices)	24°C	5 d	Delayed or retarded pulp softening, decreased the activities of PG, PE, and EGase and β-Gal	[6]
Kiwifruit (*Actinidia chinensis* Planch)	"Xuxiang"	1.2 μmol L^{-1} NO dip	20°C	13 d	Delayed fruit softening	[6]
Mango (*Mangifera indica* L.)	"Kensington Pride"	20 μL L^{-1} NO	13°C	21 d	Retarded fruit softening with decreased activities of exo-PG, endo-PG, and Egase	[6]
Plum (*Prunus salicina* Lindell)	"Amber Jewel"	10 and 20 μL L^{-1}	21 ± 1°C or 0°C followed by ripening at 21 ± 1°C	10 or 5, 6, & 7 weeks	Retarded fruit softening	[6]
	"Damili"	1 mM SNP	2°C, 85%–90% RH	90 d &120 d	Delayed fruit softening	[6]
Peach (*Prunus persica* L.)	"Feicheng"	5 and 10 μL L^{-1} NO	5°C & 25°C	35 d & 7 d	Suppressed ethylene production and ACO activity, higher MACC and ACC content, but did not affect ACS activity	[6]
Longan (*Dimocarpus longan* Lour.)	"Shixia"	1 mM SNP	28°C	6 d	Increased SSC and AA; delayed pericarp browning during storage	[6, 15]

Ascorbic acid (AA), 1-aminocyclopropane-1-carboxylicacid synthase (ACS), fresh fruits (FFs), sodium nitroprusside (SNP), polygalacturonase (PG), pectin esterase (PE), endo-1,4-β-D-glucanase (Egase), β-galactosidase (β-Gal), soluble solids content (SSC).

Table 1.
Nitric oxide (NO) treatments on some fresh fruits.

FVs	Cultivar	NO/NO donor concentrations	Storage condition	Number of days(d)/hours (h) of storage	Inferences	References
Tomato (*Solanum lycopersicum* L.)	"Myrock"	200 μL L^{-1} NO	20°C	18 d (mature green) or 10 d (breaker red)	Decreased and delayed ethylene production as well as the expression of LeACO1, LeACOH2, and LeACO4 gene during the ripening stage	[6]
Lettuce Butterhead (*Lactuca sativa* L.)	"Cosmopolia"	100 & 200 ppm NO gas (fumigation)	4°C & 12°C respectively	1 to 2 h respectively	Delayed senescence and significantly prolonged the shelf life of fresh-cut lettuce	[15]
Pointed gourd (*Trichosanthes dioica* Roxb.)	"Rajendra Parwal-1"	2 mM	12°C & simulated ambient storage	14 d plus simulated ambient storage for 3 d	Significant improvement in postharvest shelf life via the maintenance of chlorophyll, phenolics, antioxidant activity. and membrane integrity	[20]

Table 2.
Nitric oxide (NO) treatments on some fresh vegetables (FVs).

FFVs	Cultivar/ variety	Ozone concentration (optimum treatment condition)	Storage condition	Number of days (d) of storage	Inferences	References
Guava (*Psidium guajava* L.), pineapple (*Ananas comosus* L.), and banana (Musa spp.)	—	8 ± 0.2 ml/s for 10 minutes (guava) and 20 minutes (pineapple and banana) (ozonated water)	Room temperature	—	Enhanced the antioxidant capacity but reduced the vitamin C content	[1, 2]
Apple (*Malus domestica*)	"Fuji"	1.4 mg L–1 for 5 minutes (ozonated water)	4 ± 1°C and 90% RH	12 d	Reduced microbial load and quality deterioration, enhanced the antioxidant capacity and shelf life	[1, 27]
Kiwi (Actinidia deliciosa)	"Hayward"	1 mg/L) for 10 minutes (gaseous ozone)	4°C, 80%–85% RH	12 d	Enhanced the antioxidant capacity and improved the shelf life	[28]
Grapes (*Vitis vinifera*) and apples (*Malus domestica*)	—	450 ppb (gaseous ozone)	97% RH and 20°C	12 d	Significant reduction in lesion size, height of aerial mycelium growth, and decay incidence	[21, 25]
Lettuce (*Lactuca sativa*)	"Green leaf"	2 ppm (ozonated water) and 2 minutes exposure time	4°C	12 d	Maintained sensory quality and reduced microbial load	[29]
Cabbage (*Brassica oleracea* L.)	—	1.4 mg/L (aqueous) for 5 minutes and 10 minutes	4°C	12 d	Removed pesticide residues, enhanced the storability of fresh-cut cabbage (10 minutes application), and inhibited microbial growth (10 minutes application)	[1, 30]
Spinach (*Spinacia oleracea* L.)	—	0.8 mg/L for 30 seconds	12°C and > 95% RH.	13 d	Reduced yellowing and microbial population, maintained compositional characteristics, and extended shelf life	[1, 31]
Green bell pepper (*Capsicum annuum* L.)	—	>2.4 mg/L to 3 mg/L for 5 minutes	(5 ± 0.5°C, 85% ± 5% RH	14 d	Reduced microbial load, retained quality characteristics, and prolonged shelf life	[1, 32]
Tomato (*Lycopersicon esculentum* L.)	"Thomas"	0.4 mg L^{-1} for 3 minutes	5°C	10 d	Retained firmness, reduced microbial load, and reduced the consumption of fructose and glucose	[1, 33]

FFVs	Cultivar/ variety	Ozone concentration (optimum treatment condition)	Storage condition	Number of days (d) of storage	Inferences	References
Carrots (*Daucus carota* L.)	"Vitabrite"	15 $\mu l{\cdot}L^{-1}$ in a total flow of 0.5 litres per minute (gaseous ozone)	2°C	28 d	Reduction in daily microbial (*Botrytis cinerea* and *Sclerotinia sclerotiorum*) growth rate and physiological damage	[21, 24]
Carrots (*Daucus carota* L.)	—	450 ppb (gaseous ozone)	97% RH and 20°C	2 d	Reduced lesion size, aerial mycelium height, and microbial (*B. cinerea* and *S. sclerotiorum*) growth rate	[21, 25]

Relative humidity (RH).

Table 3.
Ozone treatment on some fresh fruits and vegetables.

FFVs	Cultivar/variety	Salicylic acid concentration (optimum treatment condition)	Storage condition	Number of days (d) of storage	Inferences	References
Pummelo (*Citrus maxima* Merr.)	"Jinshayou"	0.3%	20 ± 2°C	90 d	Maintained higher postharvest storability, enhanced antioxidant capacity, and gave best overall quality	[38]
Pear (*Pyrus pyrifolia* × *Pyrus communis*)	"Punjab Beauty"	2.0 mM (SA)+enriched beeswax (2.0%)	Cold storage	67 d cold storage and 20 d supermarket	Delayed respiration, reduced weight loss, and maintained fruit firmness	[39]
Apricot (*Prunus armeniaca* L.)	"Saimaiti"		4°C and 90%–95% relative humidity.	35 d	Inhibited ethylene biosynthesis and decreased cell degrading enzyme activities	[40]
Satsuma mandarin (*Citrus unshiu*)	—	2 mM (dipped for 2 minutes)	2–16°C; relative humidity: 90%–95%	50 & 120 d	Reduced rot rate, maintained fruit firmness, and increased defence-related metabolites	[41]
Goji (*Lycium barbarum* L.)	—	2 mM/L	0°C	—	Decreased respiration and weight loss, ethylene production, promoted the accumulation of bioactive compounds, and enhanced the antioxidant capacity	[3]
Tomato (*Solanum lycopersicum* L.)	"BSS-488" and "Hisar Arun"	(0.75 mM)	25 ± 1°C C and 75% ± 5%)	15 d	Delayed ripening and cell wall degradation	[37]

Table 4.
Salicylic acid (SA) treatment on some fresh fruits and vegetables (FFVs).

2.4 Oxalic acid treatments

Oxalic acid (OA) is a common organic acid found in plants. OA can enhance the postharvest life and quality of fruits and vegetables. OA retards ripening and senescence, controls post-harvest diseases, inhibits enzymatic browning, reduces decay, and alleviates chilling injury in fruits and vegetables [6, 42]. Exogenous OA induces systemic resistance against fungal, bacterial, and viral diseases in plants. Further, endogenous OA induces intrinsic heat tolerance and increases antioxidant capacity in plants [6]. The recent application of OA for the improvement of postharvest life and quality control during storage has been successful for some fruits and vegetables [6]. OA is abundant in some plant species, such as beets and beet greens (*Beta vulgaris* L.), bell peppers (*Capsicum annuum* L.), spinach (*Spinacia oleracea* L.), swiss chard (*Beta vulgaris* L. cv. Cicla), poppy seeds (*Papaver somniferum*), purslane (*Portulaca oleracea* L.), sorrel (*Oxalis corniculata* L.), and rhubarb (*Rheum officinale* Baill) [6].

Aqueous OA solutions have been applied to fruits and vegetables, including apples, banana, kiwi, mango, peach, tomatoes, lettuce, endives, and other vegetables at millimolar concentrations to delay ripening, quality deterioration, control of postharvest diseases, and alleviate chilling injuries [6, 43, 44]. In such cases the storage life of the various OA treatments were extended [6, 45]. **Table 5** shows the effect of OA treatments on the storage life of some fruits and vegetables.

2.5 Calcium Ca^{2+} treatments

Calcium (Ca^{2+}) delays ripening and senescence-related processes by regulating signalling responses and inhibiting ethylene biosynthesis and respiration in fruits and vegetables [6, 36]. Ca^2+ is also believed to be the most important mineral element determining fruit quality [52]. Postharvest Ca^{2+} treatments prevent loss of flavour and nutritional value, enhance antioxidant capacity, reduce physiological disorders and decay incidence, and increase cell wall strength, thus prolonging the shelf life of FFVs [6, 52, 53]. Fruits and vegetables may be dipped, washed, vacuumed or pressure infiltrated, mixed with wax coatings, or electrostatic powder coatings [6, 52].

Optimum Ca^{2+} application rates either alone or in combination with other techniques may be different for specific fruits and vegetables [6]. Other properties of Ca^{2+}, such as the form and source of the application are also based on the interaction with the type of fruit or vegetable [6]. For example, pressure infiltration with calcium chloride solution is more effective in increasing the calcium concentration of apple fruits than vacuum infiltration and dipping. However, excessive Ca^{2+}uptake may cause injury [52]. **Table 6** provides information on some of the latest applications of calcium treatments on fresh fruits and vegetables.

2.6 Heat treatments

Heat or thermal treatment of fresh fruits and vegetables is an efficient, easy, safe, and cost-effective method for controlling postharvest decay and maintaining quality [7]. Postharvest heat treatment has been used to maintain firmness, preserve colour, prevent overripening, alleviate chilling injury, control insect infestation, and improve the shelf life of fruits and vegetables [7, 59]. Heat treatments include hot water treatment or dips, short hot water rinsing and brushing, hot air or steam treatments [6, 7, 11]. These methods retain the quality of fresh produce during prolonged cold storage, reduce rot development, and provide security [6, 7, 11, 60, 61]. Heat treatments can

FFVs	Cultivar	OA concentrations	Storage condition	Number of days(d)/hours (h) of	Inferences	References
Banana (*Musa* spp.)	"Brazil"	20 mM for 10 minutes	23 ± 2°C and 75–90% RH	24	Reduced fruit deterioration, respiration rate and ethylene production, oxidative injury and delayed the decreases in firmness, and hue angle during storage; thus, generally inhibited postharvest ripening of banana	[6, 46]
Mango (*Mangifera indica* L.)	"Samar Bahisht Chaunsa"	5 mM/L	32 ± 3°C for 7 days ;12 ± 1°C	For 28 days	Reduced ethylene production, respiration rate, and activity of exo-polygalacturonase (exoPG) enzyme; maintained higher fruit firmness and pectin esterase (PE) activity in mango fruit during ripening and cold storage period	[6, 42]
Peach (*Prunus persica* L.)	"Bayuecui"	1 and 5 mM	25°C	4	Higher flesh firmness, lower respiration, maintained membrane integrity, and delayed fruit ripening process	[6, 47]
Pear (*Pyrus* spp.)	"Le Conte"	5 mM	0 ± 1°C with 90%–95% R.H	90	Decreased ethylene production, hue angle and delayed ripening and fruit decay; decreased decay and total loss percentage after cold storage and 5 days during marketing	[6, 48]
Plum (*Prunus salicina* Lindl.)	"Damili"	5 mM/L for 3 minutes and packed into polyethylene bags	25°C for 12 days, or at 2°C for 20 days and 25°C for 12 days.	12 (25°C), 20 (2°C), and 12 (25°C)	Decreased ethylene production, ripening, senescence, and stress injury	[6, 49]

FFVs	Cultivar	OA concentrations	Storage condition	Number of days(d)/hours (h) of	Inferences	References
Sweet cherry (*Prunus avium* L.)	"Cristalina" and "Prime Giant"	1 mM/10 L	Cold storage at 2°C and RH of 85% in darkness	20	Extended the storability, increased the content of bioactive compounds and antioxidant activity	[6, 50]
Tomato (*Solanum lycopersicum* L.)	"Pusa Gaurav", "Pusa Rohini", and 18 others	3 mM	20°C	15 minutes	Prolonged the shelf life and reduced weight loss	[6, 51]
Asparagus (*Asparagus officinalis* L.)	"Grande Vegalim" and "Purple Passion"	1 mM and 3 mM	5°C	12	Reduced respiration, preserved the visual quality, and improved the appearance of spears	[6, 43]

Table 5.
Oxalic acid (OA) treatment on some fresh fruits and vegetables (FFVs).

FFVs	Cultivar	Calcium concentration (optimum treatment condition)	Storage condition	Number of days (d) of storage	Inferences	References
Jujube (*Ziziphus jujuba* Mill.)	—	1% calcium nitrate and 1% calcium chloride (immersion for 5 minutes)	4°C	50 d	Preserved fruit quality (biochemical and organoleptic) after storage	[54]
Banana (*Musa* spp.)	"Grand Nain"	2% $CaCl_2$	20 ± 2°C and 60%–70% RH	8 d	Retarded ripening and retained quality	[55]
Papaya (*Carica papaya* L.)	"Huanong 1"	Calcium chloride (5% $CaCl_2$ solution immersion for 15 minutes)	25 ± 1°C	12 d	Delayed ripening and softening of fruit	[53]
Peach (*Prunus persica*)	"Texas A 69"	4% $CaCl_2$ solution for 10 minutes	±8–10°C and 80–85% RH	30 d	Retained quality attributes, reduced weight loss, disease incidence, and increased fruit firmness	[56]
Apple (*Malus domestica*)	Local	3% $CaCl_2$ for 60 minutes	25°C and 85%–90%	15 d	Enhanced the quality and shelf life	[57]
Broccoli (*Brassica oleracea* var. *italica*)	"Imperial"	2% $CaCl_2$ + 2 mM SA	5°C and 95% RH	21 d	Reduced weight loss and maintained quality (bioactive components and antioxidants).	[36]
Tomato (*Solanum lycopersicum* L.)	"Izmir"	6% $CaCl_2$ for 10 minutes. postharvest coating treatments with either 10% Arabic gum or 50% cactus mucilage for 3 minutes	(10 ± 1°C) and RH 85% ± 3	35 d	Increased fruit firmness, titratable acidity, reduced the percentage of weight loss and percent of decayed fruits	[58]

Table 6.
Calcium (Ca^{2+}) treatments on some fresh fruits and vegetables (FFVs).

also be combined with other methods such as safe GRAS chemicals (e.g., bicarbonate salts and ethanol), Modified Atmosphere Packaging (MAP), Controlled Atmosphere (CA), microbial biocontrol agents, plant extracts (e.g., leaf extracts of essential oils, and edible coatings to provide an effective control system against postharvest decay development, chilling injury, and quality loss [6, 7, 61–63].

Hot Water Treatment (HWT) is the simplest among heat treatments, and it has three categories: batch, continuous, and drainage systems: Hot water treatment may be applied either solely or in combination with other treatments in controlling postharvest disease infestation and rot development of FFVs [6, 60]. Several studies on hot water treatment have been reported on fruits such as banana, mango, papaya, nectarine, peach, papaya, lime, and melon. In these studies, hot water treatments were between 42°C and 60°C for a maximum and minimum duration of 30 and six (6) minutes, respectively, depending on the type of fruit. These treatments mostly controlled disease and rot causing pathogens. Nonetheless, longer exposure times caused injuries to some fruits [6]. Generally, fruits and vegetables tolerate water temperatures of 50–60°C for a duration of 10 minutes. The duration of dipping and immersion can last longer (>1 hour), and temperatures can be less than 50°C for insects disinfestation of FFVs. However, for antifungal treatments, temperatures are greater than 50°C, and the dipping time is shorter (minutes) [7].

Hot water rinsing and brushing is another heat method that cleans and disinfects freshly harvested produce at a relatively high temperature of 45–62°C. The produce is passed over revolving brushes for a short time (15–25 seconds). This is a commercial method that reduces decay development, maintains fruit quality, and prolongs the shelf life of treated fruits and vegetables. Cold storage after hot water rinsing and brushing was found to intensify the effect of the treatment [6].

Vapour or moist hot air, is a heat treatment whereby a fine mist and air are forced under circulation, which forms heated, vapour-saturated air that raises the temperature of the commodity to a required level for a specified duration. By means of condensation of vapour on the produce, latent heat is released, resulting in a quick but even increase in temperature, thus preventing damage [6]. Temperatures normally range from 40 to 50°C [7]. The fresh produce is cooled immediately after the treatment to prevent heat injury to the product [6, 7]. This treatment normally reduces decay susceptibility by killing insects' eggs and larvae [6, 7], enabling some fresh produce (e.g., basil) to be stored at temperatures that usually result in considerable injuries [6].

Steam heating involves a moist hot air treatment that uses water at about 100°C for 2–3 seconds. Steam-based systems such as steam jets coupled with electrical steam-drying elements and reflectors produce high-temperature heating above the rotating produce for 3 seconds. After the treatment, the produce is hydro-cooled. Cooled produce (e.g., carrots) showed a significant reduction in sensitivity to post-storage soft rots caused by *Sclerotinia sclerotiorum* [6]. Information on the effect of some heat treatments for FFVs are provided in **Table 7**.

2.7 Methyl Jasmonates treatments

Methyl Jasmonate (MeJa) is an endogenous phytohormone, a signalling molecule, and a volatile compound vital for regulating and mediating various processes and defence responses against biotic and abiotic stresses. It is a derivative of jasmonic acid. MeJa has been used to prevent postharvest diseases, increase bioactive compounds, maintain fruit quality, and prolong the shelf life of fresh produce [6, 59, 69–72].

FFVs	Cultivar/Variety	Heat treatment (optimum treatment condition)	Storage condition	Number of days(d) of storage	Inferences	References
Peach (*Prunus persica* L.)	-	Hot water (HW) 45°C in distilled water for 10 minutes	0 ± 1°C and 90–95% RH	35 d	Reduced chilling injury index, decreased reactive oxygen species (ROS) accumulation, and increased the activity of ROS-scavenging enzyme	[64]
Mango (*Mangifera indica*)	"Chenab Gold"	HW 48°C for 60 minutes Vapour (47°C for 25 minutes)	25°C ± 1; 60%–65% RH	21 d	Slowed weight loss, better skin colour, and maintained biochemical attributes	[65]
Murcott Mandarins (*Citrus reticulata blanco*)	"Blanco"	50°C heat for 2.5 minutes	25°C	13 d	Delayed mold growth, retained quality, overall acceptability	[66]
Newhall navel Orange (*Citrus sinensis* L. Osbeck)	"Newhall"	Hot air flow (37°C for 48 h) at 85%–95% RH	6 ± 0.5°C and 85–90% RH	120 d	Improved total soluble solid, titrable acids, total sugar, and vitamin C contents; reduced respiration rate, delayed fruit deterioration, reduced oxidative damage, and lipid peroxidation	[11]
Sweet pepper (*Capsicum annuum* L.)	"Winner"	45°C for 15 minutes + 0.05 mmol L^{-1} + methyl salicylate (MS)	Cold storage	—	Reduced chilling injury, mass loss, and maintained a better quality	[61]

FFVs	Cultivar/Variety	Heat treatment (optimum treatment condition)	Storage condition	Number of days(d) of storage	Inferences	References
Cucumber (*Cucumis sativus*)	—	Hot water temperature of 44°C, and immersion time of 10 minutes + LAE (N-alpha-Lauroyl-L-arginine ethyl ester hydrochloride) concentration of 1.00 g L^{-1},	4 +/− 1°C	—	Reduced chilling injury in cucumber fruit via enhancing antioxidant enzymes activities, PA, proline, and GABA contents to maintain ROS homeostasis during cold storage	[67]
Eggplant (*Solanum melongena* L.)	"Senryo"	45°C for 10 minutes	1°C	15 d	Alleviated chilling injury and enhanced antioxidant activity	[68]

Polyamines (PA), gama amino butyric acid (GABA), reactive oxygen species (ROS).

Table 7.
Heat treatments of some fresh fruits and vegetables (FFVs).

FFVs	Cultivar/variety	Methyl Jasmonates concentration (optimum treatment condition)	Storage condition	Number of days(d) of storage	Inferences	References
Persimmon (*Diospyros kaki*)	"Karaj"	16 and 24 μL L^{-1}	0–1°C	120 d	Decreased chilling injuries, preserved physio-chemical properties, increased antioxidant capacity, and maintained phenolic compounds	[73]
Kiwi (*Actinidia deliciosa*)	"Hayward"	1.0 mM	0 ± 0.5°C and 90 ± 5% RH	180 d	Maintained flesh firmness, reduced weight loss, reduced respiration, and enhanced antioxidant capacity	[74]
Mandarin (*Citrus nobilis* L. X *C. deliciosa* L)	"Kinnow"	0.001 μmol L^{-1}	—	75 d	Decreased weight loss, spoilage, firmness, maintained quality, and prolonged shelf life	[72]
Mandarin (*Citrus nobilis* L. X *C. deliciosa* L)	"Kinnow"	1 mM	6 ± 1°C and 90 ± 5% RH	60 d	Maintained the highest level of bioactive compounds	[75]

Table 8.
Methyl jasmonates treatments on some fresh fruits (FFs).

Several studies have demonstrated the application of exogenous MeJa on fruits and vegetables with regard to postharvest quality, senescence and ripening, chilling stress and pathogen infection. These investigations showed that MeJa treatments altered the characteristics of harvested fresh produce, largely by enhancing the antioxidant capacity, volatiles production, phenolics and flavonoids content, and reduced chilling injury, particularly in fruits [59, 69, 71]. **Table 8** provides information on the application of MeJa on some fresh fruits and vegetables.

3. Conclusion

This chapter provided information on four alternative green (ozone, salicylic acid oxalic acid, and heat) and seven novel (nitric oxide, ozone, salicylic acid, oxalic acid, methyl jasmonates, calcium, and heat) postharvest treatments. The optimum concentrations applied, storage conditions, storage duration, and inferences established by some researchers were also provided accordingly. The overall effects of these postharvest treatments were basically to maintain quality, enhance the antioxidant capacity by maintaining high levels of bioactive compounds, control postharvest infestations and, consequently, prolong the postharvest shelf life of fresh fruits and vegetables. Based on the insights obtained from these seven latest novel and green postharvest methods, we recommend that the appropriate method of application, concentrations, and other critical considerations (e.g., storage duration and conditions) must be adhered to when these treatments are applied to specific fresh fruits and vegetables. Further, the compiled optimum conditions of these postharvest treatments are a good starting point for future studies as well as for commercial applications.

Conflict of interest

The authors declare no conflict of interest.

Notes/thanks/other declarations

The authors are grateful to our corresponding person or author service manager, Ms. Maja Bozicevic, for the continuous support and patience during the preparation of this chapter.

Author details

Mildred Osei-Kwarteng[1*], Gustav K. Mahunu[1], Moomin Abu[1] and Maurice Apaliya[2]

1 University for Development Studies, Tamale, Ghana

2 Cape Coast Technical University, Cape Coast, Ghana

*Address all correspondence to: osei.kwarteng@uds.edu.gh; misokwart@yahoo.com

References

[1] Botondi R, Barone M, Grasso C. A review into the effectiveness of ozone technology for improving the safety and preserving the quality of fresh-cut fruits and vegetables. Food. 2021;**10**:748. DOI: 10.3390/foods10040748

[2] Alothman M, Kaur B, Fazilah A, Bhat R, Karim AA. Ozone-induced changes of antioxidant capacity of fresh-cut tropical fruits. Innovative Food Science & Emerging Technologies. 2010;**11**:666-671. DOI: 10.1016/j.ifset.2010.08.008

[3] Zhang H, Ma Z, Wang J, Wang P, Lu D, Deng S, et al. Treatment with exogenous salicylic acid maintains quality, increases bioactive compounds, and enhances the antioxidant capacity of fresh goji (*Lycium barbarum* L.) fruit during storage. LWT—Food Science and Technology. 2021;**140**:110837. DOI: 10.1016/j.lwt.2020.110837

[4] Palumbo M, Attolico G, Capozzi V, Cozzolino R, Corvino A, Chiara MLV de, et al. Emerging postharvest technologies to enhance the shelf-life of fruit and vegetables: An overview. Foods. 2022;**11**:3925. DOI: 10.3390/foods11233925

[5] Mahajan PV, Caleb OJ, Singh Z, Watkins CB, Geyer M. Postharvest treatments of fresh produce. Philosophical Transactions. Series A, Mathematical, Physical, and Engineering Sciences. 2014;**372**:20130309. DOI: 10.1098/rsta.2013.0309

[6] Pareek S, editor. Novel Postharvest Treatments of Fresh Produce. Boca Raton: CRC Press; 2018

[7] Siddiqui MW, editor. Postharvest Disinfection of Fruits and Vegetables. Cambridge, MA: Academic Press; 2018

[8] Zhang W, Jiang Y, Zhang Z. The role of different natural organic acids in postharvest fruit quality management and its mechanism. Food Frontiers. 2023:1-7. DOI: 10.1002/fft2.245

[9] Sun X, Baldwin E, Bai J. Applications of gaseous chlorine dioxide on postharvest handling and storage of fruits and vegetables—A review. Food Control. 2019;**95**:18-26. DOI: 10.1016/j.foodcont.2018.07.044

[10] González-Aguilar GA, Ayala-Zavala JF, Olivas GI, La Rosa LA de, Álvarez-Parrilla E. Preserving quality of fresh-cut products using safe technologies. Journal für Verbraucherschutz und Lebensmittelsicherheit 2010;**5**:65-72. DOI: 10.1007/s00003-009-0315-6

[11] Wan C, Kahramanoğlu İ, Chen J, Gan Z, Chen C. Effects of hot air treatments on postharvest storage of newhall navel orange. Plants (Basel). 2020;**9**:170. DOI: 10.3390/plants9020170

[12] Papoutsis K, Mathioudakis MM, Hasperué JH, Ziogas V. Non-chemical treatments for preventing the postharvest fungal rotting of citrus caused by *Penicillium digitatum* (green mold) and *Penicillium italicum* (blue mold). Trends in Food Science and Technology. 2019;**86**:479-491. DOI: 10.1016/j.tifs.2019.02.053

[13] Islam F, Saeed F, Afzaal M, Ahmad A, Hussain M, Khalid MA, et al. Applications of green technologies-based approaches for food safety enhancement: A comprehensive review. Food Science & Nutrition. 2022;**10**:2855-2867. DOI: 10.1002/fsn3.2915

[14] Soegiarto L, Wills R, Seberry JA, Leshem YY. Nitric oxide degradation

in oxygen atmospheres and rate of uptake by horticultural produce. Postharvest Biology and Technology. 2003;**28**:327-331. DOI: 10.1016/S0925-5214(02)00199-0

[15] Iakimova ET, Woltering EJ. Nitric oxide prevents wound-induced browning and delays senescence through inhibition of hydrogen peroxide accumulation in fresh-cut lettuce. Innovative Food Science & Emerging Technologies. 2015;**30**:157-169. DOI: 10.1016/j.ifset.2015.06.001

[16] Yang X, Liu Y-B. Nitric oxide fumigation for postharvest pest control on lettuce. Pest Management Science. 2019;**75**:390-395. DOI: 10.1002/ps.5123

[17] Liu Y, Chen T, Tao N, Yan T, Wang Q, Li Q. Nitric oxide is essential to keep the postharvest quality of fruits and vegetables. Horticulturae. 2023;**9**:135. DOI: 10.3390/horticulturae9020135

[18] Fan C, Leng Y, Zhang Q, Zhao X, Gao W, Duan P, et al. Synergistically mitigating nitric oxide emission by co-applications of biochar and nitrification inhibitor in a tropical agricultural soil. Environmental Research. 2022;**214**:113989. DOI: 10.1016/j.envres.2022.113989

[19] Shahkoomahally S, Ramezanian A, Farahnaky A. Postharvest nitric oxide treatment of persimmon (*Diospyros kaki* l.) improves fruit quality during storage. Fruits. 2015;**70**:63-68. DOI: 10.1051/fruits/2014045

[20] Siddiqui MW, Homa F, Lata D, Ahmad MS. Exogenous nitric oxide delays ripening and maintains postharvest quality of pointed gourd during storage. Journal of Plant Growth Regulation. 2021;**40**:2371-2378. DOI: 10.1007/s00344-020-10270-6

[21] Sarron E, Gadonna-Widehem P, Aussenac T. Ozone treatments for preserving fresh vegetables quality: A critical review. Food. 2021;**10**:605. DOI: 10.3390/foods10030605

[22] Miller FA, Silva CLM, Brandão TRS. A review on ozone-based treatments for fruit and vegetables preservation. Food Engineering Reviews. 2013;**5**:77-106. DOI: 10.1007/s12393-013-9064-5

[23] Garcia A, Mount JR, Davidson PM. Ozone and chlorine treatment of minimally processed lettuce. Journal of Food Science. 2003;**68**:2747-2751. DOI: 10.1111/j.1365-2621.2003.tb05799.x

[24] Liew CL, Prange RK. Effect of ozone and storage temperature on postharvest diseases and physiology of carrots (*Daucus carota* L.). Jashs. 1994;**1198**:563-567. DOI: 10.21273/JASHS.119.3.563

[25] Sharpe D, Fan L, McRae K, Walker B, MacKay R, Doucette C. Effects of ozone treatment on *Botrytis cinerea* and *Sclerotinia sclerotiorum* in relation to horticultural product quality. Journal of Food Science. 2009;**74**:M250-M257. DOI: 10.1111/j.1750-3841.2009.01234.x

[26] Vijay Rakesh Reddy S, Sudhakar Rao DV, Sharma RR, Preethi P, Pandiselvam R. Role of ozone in post-harvest disinfection and processing of horticultural crops: A Review. Ozone Science and Engineering. 2022;**44**:127-146. DOI: 10.1080/01919512.2021.1994367

[27] Liu C, Ma T, Hu W, Tian M, Sun L. Effects of aqueous ozone treatments on microbial load reduction and shelf life extension of fresh-cut apple. International Journal of Food Science and Technology. 2016;**51**:1099-1109. DOI: 10.1111/ijfs.13078

[28] Wang Y, Li Y, Yang S, Wu Z, Shen Y. Gaseous ozone treatment prolongs the

shelf-life of fresh-cut kiwifruit by maintaining its ascorbic acid content. LWT - Food Science and Technology. 2022;**172**:114196. DOI: 10.1016/j.lwt.2022.114196

[29] Ölmez H, Akbas MY. Optimization of ozone treatment of fresh-cut green leaf lettuce. Journal of Food Engineering. 2009;**90**:487-494. DOI: 10.1016/j.jfoodeng.2008.07.026

[30] Liu C, Chen C, Jiang A, Zhang Y, Zhao Q, Hu W. Effects of aqueous ozone treatment on microbial growth, quality, and pesticide residue of fresh-cut cabbage. Food Science & Nutrition. 2021;**9**:52-61. DOI: 10.1002/fsn3.1870

[31] Papachristodoulou M, Koukounaras A, Siomos AS, Liakou A, Gerasopoulos D. The effects of ozonated water on the microbial counts and the shelf life attributes of fresh-cut spinach. Journal of Food Processing & Preservation. 2018;**42**:e13404. DOI: 10.1111/jfpp.13404

[32] Ummat V, Singh AK, Sidhu GK. Effect of aqueous ozone on quality and shelf life of shredded green bell pepper (*Capsicum annuum*). Journal of Food Processing & Preservation. 2018;**42**. DOI: 10.1111/jfpp.13718

[33] Aguayo E, Escalona V, Silveira AC, Artés F. Quality of tomato slices disinfected with ozonated water. Food Science and Technology International. 2014;**20**:227-235. DOI: 10.1177/1082013213482846

[34] Chen C, Sun C, Wang Y, Gong H, Zhang A, Yang Y, et al. The preharvest and postharvest application of salicylic acid and its derivatives on storage of fruit and vegetables: A review. Scientia Horticulturae. 2023;**312**:111858. DOI: 10.1016/j.scienta.2023.111858

[35] Chen Y, Sun J, Lin H, Lin M, Lin Y, Wang H, et al. Salicylic acid treatment suppresses *Phomopsis longanae* Chi-induced disease development of postharvest longan fruit by modulating membrane lipid metabolism. Postharvest Biology and Technology. 2020;**164**:111168. DOI: 10.1016/j.postharvbio.2020.111168

[36] El-Beltagi HS, Ali MR, Ramadan KMA, Anwar R, Shalaby TA, Rezk AA, et al. Exogenous postharvest application of calcium chloride and salicylic acid to maintain the quality of broccoli florets. Plants (Basel). 2022;**11**:1513. DOI: 10.3390/plants11111513

[37] Kumar N, Tokas J, Raghavendra M, Singal HR. Impact of exogenous salicylic acid treatment on the cell wall metabolism and ripening process in postharvest tomato fruit stored at ambient temperature. International Journal of Food Science and Technology. 2021;**56**:2961-2972. DOI: 10.1111/ijfs.14936

[38] Huang Q, Huang L, Chen J, Zhang Y, Kai W, Chen C. Maintenance of postharvest storability and overall quality of 'Jinshayou' pummelo fruit by salicylic acid treatment. Frontiers in Plant Science. 2022;**13**:1086375. DOI: 10.3389/fpls.2022.1086375

[39] Sinha A, Gill P, Jawandha SK, Kaur P, Grewal SK. Salicylic acid enriched beeswax coatings suppress fruit softening in pears by modulation of cell wall degrading enzymes under different storage conditions. Food Packaging and Shelf Life. 2022;**32**:100821. DOI: 10.1016/j.fpsl.2022.100821

[40] Li Y, He H, Hou Y, Kelimu A, Wu F, Zhao Y, et al. Salicylic acid treatment delays apricot (*Prunus armeniaca* L.) fruit softening by inhibiting ethylene biosynthesis and cell wall degradation. Scientia Horticulturae. 2022;**300**:111061. DOI: 10.1016/j.scienta.2022.111061

[41] Zhu F, Chen J, Xiao X, Zhang M, Yun Z, Zeng Y, et al. Salicylic acid treatment reduces the rot of postharvest citrus fruit by inducing the accumulation of H_2O_2, primary metabolites and lipophilic polymethoxylated flavones. Food Chemistry. 2016;**207**:68-74. DOI: 10.1016/j.foodchem.2016.03.077

[42] Razzaq K, Khan AS, Malik AU, Shahid M, Ullah S. Effect of oxalic acid application on Samar Bahisht Chaunsa mango during ripening and postharvest. LWT - Food Science and Technology. 2015;**63**:152-160. DOI: 10.1016/j.lwt.2015.03.069

[43] Barberis A, Cefola M, Pace B, Azara E, Spissu Y, Serra PA, et al. Postharvest application of oxalic acid to preserve overall appearance and nutritional quality of fresh-cut green and purple asparagus during cold storage: A combined electrochemical and mass-spectrometry analysis approach. Postharvest Biology and Technology. 2019;**148**:158-167. DOI: 10.1016/j.postharvbio.2018.10.016

[44] Li P, Zheng X, Chowdhury MGF, Cordasco K, Brecht JK. Prestorage application of oxalic acid to alleviate chilling injury in mango fruit. Horts. 2015;**50**:1795-1800. DOI: 10.21273/HORTSCI.50.12.1795

[45] Wang F, Yang Q, Zhao Q, Li X. Cold shock treatment with oxalic acid could alleviate chilling injury in green bell pepper by enhancing antioxidant enzyme activity and regulating proline metabolism. Scientia Horticulturae. 2022;**295**:110783. DOI: 10.1016/j.scienta.2021.110783

[46] Huang H, Jing G, Guo L, Zhang D, Yang B, Duan X, et al. Effect of oxalic acid on ripening attributes of banana fruit during storage. Postharvest Biology and Technology. 2013;**84**:22-27. DOI: 10.1016/j.postharvbio.2013.04.002

[47] Zheng X, Tian S, Meng X, Li B. Physiological and biochemical responses in peach fruit to oxalic acid treatment during storage at room temperature. Food Chemistry. 2007;**104**:156-162. DOI: 10.1016/j.foodchem.2006.11.015

[48] Tarabih ME. Improving storability of Le Conte pear fruit using Aminoethoxyvinylglycine (AVG) and oxalic acid (OA) under cold storage conditions. Asian Journal of Crop Science. 2014;**6**:320-333. DOI: 10.3923/ajcs.2014.320.333

[49] Wu F, Zhang D, Zhang H, Jiang G, Su X, Qu H, et al. Physiological and biochemical response of harvested plum fruit to oxalic acid during ripening or shelf-life. Foodservice Research International. 2011;**44**:1299-1305. DOI: 10.1016/j.foodres.2010.12.027

[50] Valero D, Díaz-Mula HM, Zapata PJ, Castillo S, Guillén F, Martínez-Romero D, et al. Postharvest treatments with salicylic acid, acetylsalicylic acid or oxalic acid delayed ripening and enhanced bioactive compounds and antioxidant capacity in sweet cherry. Journal of Agricultural and Food Chemistry. 2011;**59**:5483-5489. DOI: 10.1021/jf200873j

[51] Kant K, Arora A, Singh VP, Kumar R. Effect of exogenous application of salicylic acid and oxalic acid on post harvest shelf-life of tomato (*Solanum lycopersicon* L.). Indian Journal of Plant Physiology. 2013;**18**:15-21. DOI: 10.1007/s40502-013-0004-4

[52] Conway WS, Sams CE, Hickey KD. Pre- and postharvest calcium treatment of apple fruit and its effect on quality. Acta Horticulture. 2002;**2002**:413-419. DOI: 10.17660/ActaHortic.2002.594.53

[53] Gao Q, Tan Q, Song Z, Chen W, Li X, Zhu X. Calcium chloride postharvest treatment delays the ripening and softening of papaya fruit. Journal of Food Processing & Preservation. 2020;**44**. DOI: 10.1111/jfpp.14604

[54] Moradinezhad F, Ghesmati M, Khayyat M. Postharvest calcium salt treatment of fresh jujube fruit and its effects on biochemical characteristics and quality after cold storage. Journal of Horticultural Research. 2019;**27**:39-46. DOI: 10.2478/johr-2019-0009

[55] Al-Qurashi AD, Awad MA. Postharvest gibberellic acid, 6-benzylaminopurine and calcium chloride dipping affect quality, antioxidant compounds, radical scavenging capacity and enzymes activities of 'Grand Nain' bananas during shelf life. Scientia Horticulturae. 2019;**253**:187-194. DOI: 10.1016/j.scienta.2019.04.044

[56] Rahman MU, Sajid M, Rab A, Ali S, Shahid MO, Alam A, et al. Impact of calcium chloride concentrations and storage duration on quality attributes of peach (*Prunus persica*). Russian Agricultural Sciences. 2016;**42**:130-136. DOI: 10.3103/S1068367416020099

[57] Gemeda F. Effect of post-harvest calcium chloride treatment on quality and shelf life of apple (*Malus domestica*). IJFET. 2021;**5**:79. DOI: 10.11648/j.ijfet.20210502.16

[58] Sati F, Qubbaj T. Effect of calcium chloride post-harvest treatment in combination with plant natural substance coating on fruit quality and storability of tomato (*Solanum lycopersicum*) fruits during cold storage. 2021. doi:10.5073/JABFQ.2021.094.012

[59] Barman K, Sharma S, Siddiqui MW. Emerging Postharvest Treatment of Fruits and Vegetables. Oakville, ON, Waretown, NJ: Apple Academic Press; 2018

[60] Kahramanoğlu İ, Chen C, Chen Y, Chen J, Gan Z, Wan C. Improving storability of "Nanfeng" mandarins by treating with postharvest hot water dipping. Journal of Food Quality. 2020;**2020**:1-12. DOI: 10.1155/2020/8524952

[61] Rehman RNU, Malik AU, Khan AS, Hasan MU, Anwar R, Ali S, et al. Combined application of hot water treatment and methyl salicylate mitigates chilling injury in sweet pepper (*Capsicum annuum* L.) fruits. Scientia Horticulturae. 2021;**283**:110113. DOI: 10.1016/j.scienta.2021.110113

[62] Bahadırlı NP, Kahramanoğlu İ, Wan C. Exposure to volatile essential oils of Myrtle (*Myrtus communis* L.) leaves for improving the postharvest storability of fresh loquat fruits. Journal of Food Quality. 2020;**2020**:1-10. DOI: 10.1155/2020/8857669

[63] Kahramanoğlu İ, Usanmaz S, Alas T, Okatan V, Wan C. Combined effect of hot water dipping and *Cistus creticus* L. leaf extracts on the storage quality of fresh Valencia oranges. Folia Horticulturae. 2020;**32**:337-350. DOI: 10.2478/fhort-2020-0029

[64] Zhao L, Zhao Y, Wang L, Hou Y, Bao Y, Jia Z, et al. Hot water treatment improves peach fruit cold resistance through PpHSFA4c-mediated HSF-HSP and ROS pathways. Postharvest Biology and Technology. 2023;**199**:112272. DOI: 10.1016/j.postharvbio.2023.112272

[65] Javed S, Fu H, Ali A, Nadeem A, Amin M, Razzaq K, et al. Comparative response of mango fruit towards pre- and post-storage quarantine heat Treatments. Agronomy. 2022;**12**:1476. DOI: 10.3390/agronomy12061476

[66] Queb-González DB, Lopez-Malo A, Sosa-Morales ME, Villa-Rojas R. Postharvest heat treatments to inhibit *Penicillium digitatum* growth and maintain quality of Mandarin (Citrus reticulata blanco). Heliyon. 2020;**6**:e03166. DOI: 10.1016/j.heliyon.2020.e03166

[67] Hou Y, Liu L, Liu Y, Zhao L, Bao Y, Zheng Y, et al. N-α-Lauroyl-L-arginine ethyl ester hydrochloride combined with hot water treatment alleviates chilling injury of postharvest cucumber fruit. Scientia Horticulturae. 2023;**315**:111986. DOI: 10.1016/j.scienta.2023.111986

[68] Kantakhoo J, Ose K, Imahori Y. Effects of hot water treatment to alleviate chilling injury and enhance phenolic metabolism in eggplant fruit during low temperature storage. Scientia Horticulturae. 2022;**304**:111325. DOI: 10.1016/j.scienta.2022.111325

[69] Lata D, Anand A, Ozturk B, Ilahy R, Ahmad MS, Siddiqui MW. Methyl jasmonate and its application for improving postharvest quality of fruits. In: Aftab T, Yusuf M, editors. Jasmonates and Salicylates Signaling in Plants. Cham: Springer International Publishing; 2021. pp. 239-254. DOI: 10.1007/978-3-030-75805-9_11

[70] Zuñiga PE, Castañeda Y, Arrey-Salas O, Fuentes L, Aburto F, Figueroa CR. Methyl jasmonate applications from flowering to ripe fruit stages of strawberry (Fragaria × ananassa 'Camarosa') reinforce the fruit antioxidant response at post-harvest. Frontiers in Plant Science. 2020;**11**:538. DOI: 10.3389/fpls.2020.00538

[71] Wang S-Y, Shi X-C, Liu F-Q, Laborda P. Effects of exogenous methyl jasmonate on quality and preservation of postharvest fruits: A review. Food Chemistry. 2021;**353**:129482. DOI: 10.1016/j.foodchem.2021.129482

[72] Baswal A, Dhaliwal H, Singh Z, Mahajan B, Gill K. Postharvest application of methyl jasmonate, 1-methylcyclopropene and salicylic acid extends the cold storage life and maintain the quality of 'Kinnow' mandarin (*Citrus nobilis* L. X C. deliciosa L.) fruit. Postharvest Biology and Technology. 2020;**161**:111064. DOI: 10.1016/j.postharvbio.2019.111064

[73] Bagheri M, Esna-Ashari M. Effects of postharvest methyl jasmonate treatment on persimmon quality during cold storage. Scientia Horticulturae. 2022;**294**:110756. DOI: 10.1016/j.scienta.2021.110756

[74] Öztürk B, Yücedağ F. Effects of methyl jasmonate on quality properties and phytochemical compounds of kiwifruit (*Actinidiadeliciosa* cv. 'Hayward') during cold storage and shelf life. Turkish Journal of Agriculture and Forestry. 2021;**45**:154-164. DOI: 10.3906/tar-2004-69

[75] Baswal AK, Dhaliwal HS, Singh Z, Mahajan BV. Post-harvest application of methyl jasmonate, 1-methylcyclopropene and salicylic acid elevates health-promoting compounds in cold-stored 'Kinnow' mandarin (*Citrus nobilis* Lour x C. *deliciosa Tenora*) Fruit. International Journal of Fruit Science. 2021;**21**:147-157. DOI: 10.1080/15538362.2020.1860865

Chapter 4

Volatile Organic Compounds Produced by Microbes in the Management of Postharvest Diseases of Fruits

Subbaraman Sriram, Pooja Shekar Patel,
Darisi Venkata Sudhakar Rao
and Kodthalu Seetharamaiah Shivashankara

Abstract

Nutritional security can be achieved only with the proper intake of fruits and vegetables. However, on an average 30% of the fruit produce are lost between harvest and consumption due to post-harvest spoilage. About 30–40% of total fruits production is lost after harvest. Main causes of postharvest loss include lack of temperature management, rough handling, poor packaging material, and lack of education about the need to maintain quality. There are many ways in which the post-harvest spoilage is managed. Use of chemicals in post-harvest management has direct effect on the consumers and there is a need for alternative strategies. Use of microbial biological control agents have been successfully adopted for soil borne diseases. Registration and biosafety issues make it difficult to use them against post-harvest diseases. Use of volatile organic compounds (VOCs) from bioagents for the post-harvest management provides an opportunity to explore the use of bioagents without having contact with fruits. Many classes of chemicals are produced as volatiles by microbial agents. This chapter describes the potential of VOCs in managing post-harvest diseases, their characterization and identification, biosynthesis, volatiles reported from bacterial, fungal and yeast bioagents, success stories of their use as potential bioagents.

Keywords: volatile organic compounds, bacteria, yeasts, fungi, fruits, fruit rot, post-harvest spoilage

1. Introduction

Fruit farming is one of the most important and long-standing traditions throughout the world. The cultivation of fruit crops has a significant impact on the overall well-being of humans and the state of the nation. Fruit crop production can be viewed as an open and complex system in which factors including the sowing or planting method, environmental conditions, soil type, crop management, and their interactions affect the growth, development, and future yield [1].

Fruits and vegetable consumption provides nutritional security and ensures that malnutrition is addressed efficiently. However, the entire product produced do not reach the table of consumers. Due to delicate nature, the post-harvest loss interrupts the reach of fruits to the consumers. Post-harvest loss (PHL) is defined as a measurable quantitative and qualitative loss of an edible food product from harvest to consumption. Increasing post-harvest loss has been highlighted as a global problem in food industries across nations. Food production would need to expand by 70% from present levels by 2050, according to projections, and the situation is significantly more dire in developing nations with poor productivity [2, 3]. As per FAO reports, approximately 14% of produced food was lost after harvest during storage in 2019 with a global net worth of approximately one trillion US dollars but at present up to 30% postharvest loss has been observed in vegetables and fruits.

"An apple a day keeps the doctor away" is a popular saying that emphasizes the importance of fruit crops for the human diet. A diet rich in fruits and vegetables and low in saturated fats is healthy and protective against cardiovascular diseases and certain cancers [4–6]. The World Health Organization (WHO) recommends a daily intake of at least 400 g of fruits and vegetables per person [7].

Huge pre– and postharvest losses in fruits are caused by various diseases and unfavorable environment leading to the total failure of the crops. It has been estimated that phytopathogenic fungi cause more than 50 per cent of total post-harvest losses. Fruits are prone to number of fungal, bacterial and viral diseases which significantly affect its quality and production. However, fungal diseases inflict huge losses to the crop. Fruits are highly susceptible to postharvest spoilage because of high perishable nature. During the storage conditions numbers of fungi are known to cause spoilage of fruits. Under storage conditions considerable losses occur due to the rots caused by different species belonging to the genera viz., *Alternaria*, *Aspergillus*, *Bipolaris*, *Botryodiplodia*, *Botrytis*, *Colletotrichum*, *Curvularia*, *Fusarium*, *Penicillium*, *Rhizopus*, *etc*. [8]. Among these pathogens, green and blue molds, caused by *Penicillium digitatum* (Pers.: Fr.) Sacc. and *P. italicum* Wehmer, are the most economically important postharvest diseases of citrus in all production areas [9]. The losses due to penicillium decay are variable and depend upon climatic conditions, orchard factors, citrus cultivar and the extent of physical injury to the fruit during harvest and subsequent handling [10]. It has been estimated that fruit rots caused by *Penicillium* spp. accounted for 55–80% of total postharvest decay observed in oranges and mandarins during the entire commercialization season, and for 30–55% of decay observed in storage rooms in citrus packing houses [11]. Synthetic fungicides are primarily used for the control of postharvest diseases of fruits and vegetables [12]. However, the trend around the world is shifting towards reduced use of fungicides on produce and thus, there is a strong public and scientific interest in safer and eco-friendly alternatives to reduce the high loss due to decay of harvested commodities. Furthermore, the growing awareness about health hazards and environmental deterioration due to chemical use has necessitated the switching to new non-chemical strategies for the control of postharvest diseases of fruits and vegetables [13].

Among the different biological approaches, use of the microbial antagonists are quite promising and gaining popularity. Microbial bioagents can be used in management of postharvest diseases of fruits and vegetables in two ways. They are use of microorganisms which already exist on the produce itself, which can be promoted and managed or those that can be artificially introduced against postharvest pathogens [12]. Several modes of action have been suggested to explain the biocontrol activity of microbial antagonists, that include competition [14–18] antibiosis, mycoparasitism, cell wall degrading enzymes, and induced resistance [19–21].

Antibiotics are microbial toxins which at low concentrations can kill other microbes. Bacteria produced volatile antibiotics viz. hydrogen cyanide, aldehydes, alcohols, ketones, and sulphides and non-volatile antibiotics: polyketides (diacetyl phloroglucinol and mupirocin), heterocyclic nitrogenous compounds (phenazine derivatives: pyocyanin, phenazine-1-carboxylic acid, and hydroxy phenazines) and phenylpyrrole antibiotic (pyrrolnitrin) [22].

There is growing importance for the non-chemical control methods to reduce postharvest decay throughout the world. Volatile organic compounds released by microbial antagonists have shown greater potential and it substitutes to synthetic fungicides for the control of postharvest decay of fruit [23]. Therefore, bioagents have gained the considerable attention and emerge to be a promising as well as a viable alternative to chemical management practices.

2. Volatile organic compounds (VOCs)

Microorganisms have potential of synthesizing numerous volatile substances called as microbial volatile organic compounds (VOCs) with low boiling points, small molecular masses (on average 300 Da) that quickly evaporate at normal temperature and pressure [24]. Both plants and microorganisms produce VOCs that enable them to communicate intra- and inter-specifically. By emitting VOCs, plants defend themselves against herbivores and pathogens, compete with other plants, and/or feed microbial populations. Microorganisms emit VOCs to communicate or attack each other [25]. Production of VOCs with antimicrobial activity has been described in filamentous fungi [26, 27], bacteria [28, 29], yeasts [30, 31], *Streptomyces* spp. [32, 33], and higher plants [34].

The majority of microbial VOCs have distinct smells [35]. Some of them, which are found in wine, beer, and other fermented foods, have pleasant flavors preferred by people. On the other hand, VOCs are connected to wastelands, deterioration, sewerage facilities, dirty socks, and water-damaged structures also. Hundreds of distinct volatiles, including mixtures of alcohols, ketones, esters, tiny alkenes, thiols, monoterpenes, and sesquiterpenes, can be released simultaneously by any type of microbe [36].

The most interesting developments encompassing about volatile organic compounds come from the study of endophytes, i.e., the microbe that colonize inside the plant tissues without causing any negative effects. Bacterial, fungal and yeast endophytes constitute the plant microbiome. Among them, *Muscodor albus* which is a nonsporulating, filamentous and VOC-emitting, endophytic, "stinky white" fungus isolated from the spice tree *Cinnamomum zeylanicum* and was antagonistic to pathogenic bacteria and fungi. More than 28 VOCs were identified from the laboratory cultures of *M. albus*, comprising acids, alcohols, esters, ketones, and lipids [37]. Therefore, application of these antimicrobial properties of cocktail of VOCs to control microbial contamination has been termed Mycofumigation [38].

3. Identification of VOCs

The identification and quantification of VOCs have to overcome many technical challenges. As, VOCs are highly evaporative, during sampling, handling and assay procedures the chances of occurrence of compound loss is more [39]. In addition to that the VOCs synthesis at low concentrations in complex mixtures. VOCs identification, collection and quantification depends on either the compounds of interest, the required sensitivity, the intended application, the cost and the ease of use. In classical work, the extraction, separation, and identification steps were separate. Earlier days, the fungal VOCs were identified using the steam distillation method followed by liquid extraction and concentration [40, 41]. Present-day use of Solid phase microextraction (SPME) based gas chromatography (GC) combined with a mass spectrometer (GC–MS) [42], Solvent extraction/liquid–liquid extraction (LLE), stir bar sorptive extraction (SBSE), dynamic headspace (DHS) approaches are used for separation and identification of compounds [43].

Besides these methods, electronic nose or artificial nose combined with multisensory array, an information-processing unit, pattern recognition software, and reference library databases are also used in the identification of VOCs. These resultant electronic fingerprints express unique aroma that helps to detect odor profiles without separation of the mixture into its components [44].

4. Biosynthesis of VOCs

Biosynthesis of VOCs mainly depends on the availability of carbon, nitrogen, sulfur and energy received from the primary metabolism. Therefore, the primary metabolite are considered as a building block of secondary metabolite and exhibits major impact on the concentration of any secondary metabolite, including VOCs. This demonstrates the high degree of connectivity between primary and secondary metabolism. VOCS are divided into different classes based on the biosynthetic origin. The important classes include terpenoids phenylpropanoids/ benzenoids, fatty acid derivatives and amino acid derivatives in addition to a few species–/genus-specific compounds which may not be represented in those major classes. Hence, precursors for VOCs basically originate from the primary metabolism (glycolysis, tricarboxylic acid and pentose phosphate pathway) which helps in the synthesis of VOCs. The four major VOC biosynthetic pathways are the mevalonic acid (MVA), shikimate/phenylalanine, lipoxygenase (LOX) and the methylerythritol phosphate (MEP) pathways lead to the emission of benzenoids/ phenylpropanoids, sesquiterpenes, monoterpenes, hemiterpenes, diterpenes, volatile carotenoid derivatives and methyl jasmonate/green leaf volatiles [45].

Though VOCs are considered as secondary metabolites, some researchers argue that they are degradation products of fatty acids, biotransformation products of amino acids, or incidental breakdown products of fungal extracellular enzymes acting on exogenous substrates. The biosynthetic pathways for geosmin which is a vital odor present in soils produced by many streptomycetes, cyanobacteria and fungi [46] and 1-octen-3-ol which is a breakdown product of linoleic acid, also known as mushroom alcohol [34] have been elucidated. Less is known about the biosynthetic pathways that fungi use to produce VOCs [38].

How enzymes are putatively involved in VOCs synthesis has been reported utilizing bioinformatic methods using either terpene or alcohol or ester synthases as an example. The combination of volatilome data collected at various developmental stages with transcriptional data of selected genes makes an effective method for locating enzymes which are likely to be involved in fungal VOC production . Representative pathways for different VOCs are given in **Figure 1**.

The selected references on the identification of VOCs for post-harvest pathogens of fruits are listed in **Table 1**.

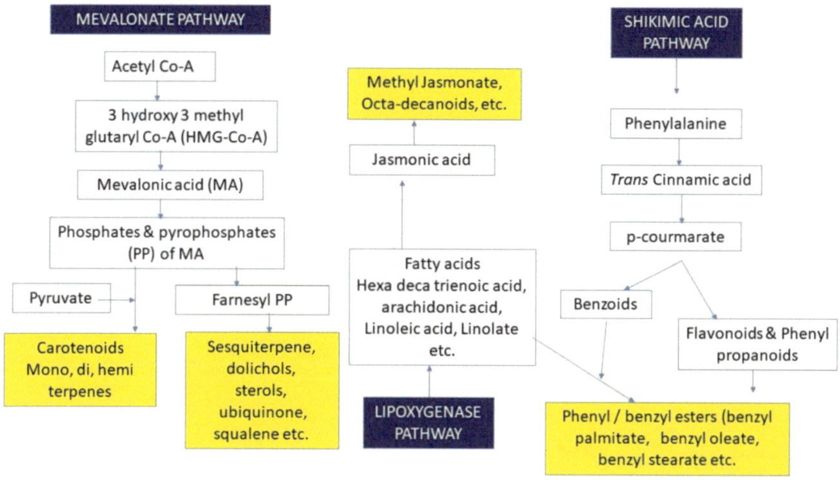

Figure 1.
Three main pathways required for the production of volatile organic compounds (VOCs). (yellow boxes are the volatile organic compounds and blue colored boxes indicate the name of the pathway (Modified figure based on Dudareva et al. [47] and Kaddes et al. [45]).

S. No	Micro organism	VOCs	Effect	References
1	*Bacillus amyloliquefaciens*	2,5-Dimethyl pyrazine 2-Dodecanone	Antifungal activity against *Fusarium* sp. and *Colletotrichum gloeosporioides*	[48]
		Acetoin	Antifungal activity against *Botrytis cinerea*, *Alternaria brassicicola*, *A. brassicae* and *Sclerotinia sclerotiorum*	[49]
		1,3 Pentadiene	Antifungal activity against *Monilinia laxa*, *Monilinia fructicola* and *B. cinerea*	[50]
		1-(2-Aminophenyl) ethanone	Antifungal activity against *Peronophythora litchi*	[51]
		Benzothiazole	Induction of fruit defenses	[51]
		Acetic acid 2-phenylether ester oleic acid styrene	*Alternaria alternata, Colletotrichum gloesporioides, Penicillium expansum*	Patel et al. (Unpublished data)
2	*Bacillus atrophaeus*	Chloroacetic acid	Antifungal activity against *C. gloeosporioides*	[52]
3	*Bacillus megaterium*	Heneicosane and Heptacosane	Antifungal activity against *B. cinerea*, *F. solani*, *Rhizoctonia solani*, *S. sclerotiorum* and *Verticillium dahliae*	[53]
4	*Bacillus mycoides*	Dimethyl disulfide	Antifungal activity against *C. gloeosporioides*	[54]
5	Bacillus subtilis	2,4-Di-tert-butylphenol, 1-Octanol, Benzothiazole, Benzoic acid Benzaldehyde 3-Methyl butanal	Antifungal activity against *B. cinerea*, *C. gloeosporioides*, *P. expansum*, *M. fructicola* and *A. alternata*	[55]

S. No	Micro organism	VOCs	Effect	References
6	*Bacillus velezensis*	Pyrazine(2,5-dimethyl) Benzothiazole 4-Chloro-3-methyl Phenol-2,4-bis (1,1dimethyl ethyl)	Antifungal activity against *Alternaria solani*, *B. cinerea*, *Valsa mali*, *M. fructicola*, *F. oxysporum f.* sp. *capsicum* and *Colletotrichum lindemuthianum*	[56]
7	*Hanseniaspora uvarum*	Ethyl caproate Ethyl acetate	Increase strawberry fruit flavor and defense	[57]
8	*Hanseniaspora opuntiae*	acetic acid 2-phenyl ether ester styrene β- Phyllandrene thujopsene	*A. alternata*, *C. gloesporioides*, *P. expansum*	Patel et al. (Unpublished data)
9	*Muscodor albus*	Naphthalene	Antifungal activity against *Aspergillus fumigatus* and *A. ochraceus*	[58]
10	*Muscodor brasiliensis*	Pogostol 2-Phenyl ethyl acetate Phenyl ethyl alcohol	Antifungal activity against *P. digitatum*	[59]
11	*Trichoderma virens*	Thujopsene δ-Cadinene	Induction of resistance against *B. cinerea*	[60]

Table 1.
Effect of MVOCs against postharvest pathogens.

5. Application of microbial volatile organic compounds

Poveda [25] reviewed the application of microbial volatile organic compounds in plants viz. plant growth promotion, tolerance to abiotic stress, induction of defense mechanism, anti-microbial activity against phytopathogens, attractants or repellents to insect pests and post-harvest disease management.

i. Plant growth promotion: Acetoin which is a MVOCs produced by *B. amyloliquefaciens* [49] capable of promoting growth in lettuce by an increase in the number of lateral roots, dry weight, root growth, shoot length and chlorophyll content. MVOC 2,3-butanediol synthesized by *B. amyloliquefaciens* [47, 61], *B. mojavensis* [62] or *B. subtilis* [49], have the ability to promote the growth of *A. thaliana.* Other MVOCs produced from different *Bacillus* sp. such as 2-heptanone 2-ethyl-1-hexanol and tetrahydrofuran-3-ol promotes *A. thaliana* by increasing the endogenous levels of action of auxins and strigolactones, in tomato [60]. The fungal origin MVOCs (*Trichoderma* genus) have also been described with the ability to promote plant growth in *A. thaliana*. By exposure to δ-cadinene produced by *T. virens* [63], leads to an increase in root branching and an increase in total biomass, chlorophyll content and acceleration of flowering by isobutyl alcohol, isopentyl alcohol and 3-methylbutanal produced from *T. viride* [64] in *A. thaliana.*

ii. Enhancing tolerance to abiotic stress: The different species of *Trichoderma* are capable of inducing salt tolerance in *A. thaliana* plants through protection against oxidative damage, by reducing the accumulation of H_2O_2 accumulation under salt stress but the exact MVOCs involved have not been determined [65]. Under salt stress, myristic acid, phenol-2-methoxy, stearic acid and tetracontane emitted by *Pseudomonas simiae* are capable to reduce Na+, and increase K+ and P, content in roots of soybean seedlings, because of an increase in the expression of peroxidase, catalase, vegetative storage protein and nitrite reductase genes [66]. Increased in tolerance to salinity in *A. thaliana* by downregulating the expression of *high-affinity K+ transporter 1* in roots and upregulation in shoots by the 2,3-Butanediol compound released by *B. subtilis* (Zhang *et al*. 2008). Acetoin released by *B. amyloliquefaciens* is able to increase tolerance against salinity in *M. piperita*, by increasing SA content [67].

iii. Induction of defense mechanism: MVOCs induces the plant defense by activating signaling pathways within the plants. Methyl benzoate which is MVOC emitted by *Cladosporium* sp. *via* JA-signaling pathway and m-cresol VOC from *Ampelomyces* sp. activates SA- and JA-signaling pathways [68]. The MVOCs released from *Fusarium oxysporum* induce systemic resistance in *A. thaliana* by activating SA-signaling pathway against *Pseudomonas syringae* pv. *tomato* [69]; The MVOC emitted by *B. subtilis* and *B. amyloliquefaciens* are capable of activating a systemic resistance in *A. thaliana* mediated by ethylene- (ET) signaling pathway against *Erwinia carotovora* subsp. *carotovora* [70]. Same MVOC from *Enterobacter aerogenes* involved in the induction of plant resistance against the Northern corn leaf blight fungus *Setosphaeria turcica* [71].

iv. Inhibition of phytopathogens: The direct antibiosis of different MVOCs against plant pathogens is one of the most studied benefits for plants. The MVOCs produced by *B. amyloliquefaciens*, such as 2-undecanone, 2-tridecanone and heptadecane inhibits the mobility, biofilm formation and root colonization of pathogenic bacteria. In case of fungi, it inhibits the number of membrane lipids present in the mycelium of the pathogen. It causes abnormal morphology of appressoria, suppresses the mycelial growth and sporulation [72].
MVOs released from *Pseudomonas putida* such as 1-undecene and dimethyldisulfide showed oomicitidal activity against *Phytophthora cactorum*, *P. nicotianae* and *Pythium ultimum in vitro* [53], Dimethyl hexadecylamine from *Arthrobacter agilis* is able to inhibit the growth of *Phytophthora cinnamomi* by decreasing the number of membrane lipids present in the mycelium of the pathogen [73], Bacteriostatic effect also reported by MVOCs toluene, ethyl benzene, m-xylene and benzothiazole from *P. fluorescens*, restricting the growth and virulence [74].

v. Attractant or repellent effect on insect pests: In plants MVOCs against insects acts as a direct attractant or repellent of natural enemies. The emission of 2-methyl-1-butanol and 3-methyl- 1-butanol by the yeast *Aureobasidium pullulans* causes the active attraction of predatory insect-pest wasps such as the western yellowjacket (*Vespula pensylvanica*) and the German yellowjacket (*V. germanica*) [75] 2,3-butanediol emitted from *Enterbacter aerogenes* increases the attraction of parasitoid insect-pest wasps such as *Cotesia marginiventris* during the interaction with maize roots [70]. MVOCs, such as 1-octen-3-ol and 3-octanol released by *Fusarium verticillioides* acts as a repellant against insect-pest *Sitophilus zeamais* in stored maize kernels [76].

6. MVOCs in the management of postharvest diseases of fruits

6.1 VOCs produced by fungi

The use of VOCs for the control of postharvest diseases of fruits is well represented by the report [77]. on an endophytic fungus, *Muscodor albus*, discovered in *C. zeylanicum* in a Honduras botanical garden. In the late 1990s, an isolate of *M. albus* was identified as a good fungal bioagent for the biofumigation of fruits and vegetables after harvest to control apple and peach decay [78], green mold and sour rot of lemon [79] and gray mold of table grapes [80]. The ability of *M. albus* to manage postharvest diseases depends on the medium that supports the growth of an endophytic fungus which in turn greatly influences the quality, emission and the effectiveness of VOCs. It has also been reported that more than 28 VOCs of five groups of organic compounds such as acids, alcohols, esters, and lipids was identified through GC/MS analysis from an endophytic fungus [37]. About 48.5 per cent of 2-methyl-1-butanol serving a major component along with second and third compound as isobutyric acid (14.9%) and ethyl propionate (9.63%) were identified from *M. albus* inoculated in an autoclaved rye [78].

Even though *Oxyporus latemarginatus* is a plant pathogen affecting trees, an isolate of this species EF06 that produced VOCs was isolated from the healthy tissues of pepper plants which acts as a potential biological agent [81]. *O. latemarginatus* EF06 tested in half-plate divided Petri plates produced VOCs and inhibited the mycelial growth of *Alternaria alternata*, *B. cinerea*, *C. gloeosporioides*, *Fusarium oxysporum* f. sp. *lycopersici*, and *R. solani*. The gaseous VOCs produced by *O. latemarginatus* EF069 multiplied in a wheat bran–rice hull cultures of 50 g upon exposure to apple fruits in closed container suppressed 98.4 per cent development of *Botrytis* lesions at 20°C. The hexane extract of wheat bran–rice hull cultures of *O. latemarginatus* EF069 was used for the identification of an antifungal VOCs by repeated silica gel column chromatography and identified as 5-pentyl-2-furaldehyde (PTF). The purified PTF were effective against various plant pathogens. The mycofumigation with EF069 was also effective in reducing postharvest decay of apple fruits caused by *B. cinerea* [82].

The most common fungal pathogen during storage conditions is *P. expansum* [83]. But there are few reports of antifungal VOCs extracted from this species. VOCs from *P. expansum* R82 have been used to control postharvest diseases of fruits. *P. expansum* R82 tested in double Petri dish assays was found to be able to inhibit mycelial growth of postharvest pathogens *viz.*, *B. cinerea*, *Monilinia* spp., *C. acutatum* and other strains of *P. expansum* by producing VOCs [84]. The SPME-GC analysis revealed the presence of geosmin and phenethyl alcohol (PEA) as the major terpenoid and alcohol VOCs produced respectively by the *P. expansum* R82. Synthetic PEA does not show any inhibitory activity when tested *in vitro* against the pathogens. Without any direct contact of *P. expansum* to the fruits, the fungus could be grown in a separate chamber and the produced VOCs can be transferred to the storage room through pump and further can be exposed to fruits to control postharvest diseases [85].

The different VOCs produced by *Ceratocystis fimbriata* showed strong bioactivity against a wide range of pathogens including postharvest diseases such as peach brown rot and citrus green mold. The volatiles exposure of *C. fimbriata in vitro* to the peach and citrus fruits against *M. fructicola* and *P. digitatum* pathogens showed strong inhibition towards mycelial growth, conidial production and spore germination and the *in vivo* exposure lead to the reduction of disease over control of 92 and 97 per cent respectively. Exposure to VOCs of *C. fimbriata* lead to misshapen hyphae and conidia when observed under scanning electron microscopy also their pathogenicity was greatly reduced. The VOCs were identified as butyl acetate, ethyl acetate and ethanol by head space GC–MS [86].

In spite of the fact that the storage temperature may affect either VOC emission or control activity, storage at 20°C was effective compared to 5°C then period of exposure of VOCs can increase from 4 to 24 h, obtaining the same level of efficiency [79].

6.2 VOCs produced by bacteria

Different species of *Bacillus* and *Pseudomonas* have displayed inhibitory effect against the growth of postharvest pathogens of fruits with multiple modes of action, including the production of VOCs.

Bacillus subtilis strain CL2 showed antagonistic effect upon producing the VOCs *in vitro* `against wolfberry postharvest pathogens by inhibiting the hyphal growth of *Mucor circinelloides* LB1, *Fusarium arcuatisporum* LB5, *Alternaria iridiaustralis* LB7, and *Colletotrichum fioriniae* LB8 using two-sealed-base-plates method. The mycelial morphology of the inhibited pathogens were deformed, twisted, folded, and shrunken when observed under scanning electron microscope. The VOCs could also reduce the weight loss and decay incidence rate of wolfberry fruits infected by the postharvest pathogens. The headspace-gas chromatography-ion mobility spectrometry analysis, revealed the production of seven VOCs by strain CL2. Among them, 2,3-butanedione and 3-methylbutyric acid are the main antifungal active substances [87].

The volatilome produced by three strains of *Bacillus velezensis* (BUZ-14, I3 and I5) displayed inhibitory effect *in vitro* against *B. cinerea*, *M. fructicola*, *M. laxa*, *Penicillium italicum*, *P. digitatum* and *P. expansum.* Among three strains I3 and I5 showed 100 per cent inhibition of *B. cinerea*. The volatile metabolites of I3 also reduced 50 per cent inhibition of grapes gray mold and BUZ-14 decreased apricots brown rot severity by reducing the *M. fructicola* infection from 60 to 4 mm. The main volatiles responsible for showing its antifungal activity identified from SPME coupled with GC–MS ranged from 12 to 15 compounds including 2-nonanone, 2-undecanone, 2-heptanone, 1-butanol, acetoin, benzaldehyde, butyl formate, diacetyl, nonane, or pyrazine, benzaldehyde and diacetyl. Among those VOCs diacetyl was able to control 60 per cent infection of gray mold in table grapes and blue rot in mandarins with only 0.02 mL L^{-1} concentration. The diacetyl and benzaldehyde VOCs have been identified as promising compounds and can be applied in active packaging during the postharvest storage, transit and trade of fruit crops. However, prior to the application of any VOCs, it is crucial to determine the active dose as well as the phytotoxicity of the volatiles, since some of the fruits such as apricots and apples have proven to be highly sensitive [88].

In planta prophylactic fumigation of mango fruits cv. *Amrapali* with 24 h exposure of either endophytic bacteria *Pseudomonas putida* BP25 or the identified volatile from the bacteria i.e., synthetic VOC 2- ethyl-5-methylpyrazine at 25°C showed a reduction of more than 76 per cent of anthracnose severity on fruit. Additionally, the physico-chemical properties of fumigated fruits were also increased representing a new compound for the postharvest management of mango during its commercialization [51].

The downy blight of litchi caused by the oomycete pathogen *P. litchii* severely suppress the production and quality of litchi fruit. Fumigation of litchi fruits with *B. amyloliquefaciens* PP19, *Exiguobacterium acetylicum* SI17 and *B. pumilus* PI26 volatiles reduced the disease severity of downy blight. The volatile profiles identified from the above-mentioned bacterial strains *viz.*, 1-(2-Aminophenyl) ethenone, benzothiazole and α-farnesene displayed inhibitory effect against the downy blight and serves as promising VOC for the postharvest diseases control of litchi fruits [89].

The VOCs produced from *B. thuringiensis* and *B. pumilus* decreased the mango anthracnose infections by 88.5 per cent [90]. When VOCs emitted from *B. subtilis* and *B. amyloliquefaciens* were assayed alone or in combination for their antifungal property against *Penicillium* infection on citrus, the volatile profiles of *B. amyloliquefaciens* (8 VOCs) and *B. subtilis* (21 VOCs) reduced the disease incidence of citrus by *P. crustosum*, but no synergetic effect was exhibited in citrus fruit treated with the antagonist combination. The volatile profile included N-containing compounds, alcohols and ketones were common in both of them. There were morphological abnormalities in *P. crustosum* when exposed to VOCs. There was a swelling in the hyphae, sporangium and conidia [91].

Upon exposure of VOCs produced by the *B. subtilis*, there was a retraction of protoplasm with in the hyphal tips and separation of an empty hyphal segments in germinating conidia of *B. cinerea*. In addition, due to the protoplasm retraction pathogen conidia could not germinate, even after the withdrawal of VOCs, indicating that the protoplasm damage may be irreversible and lethal for pathogens [28].

Streptomyces spp. is group of actinomycete capable of reducing the growth of certain fungal pathogens such as citrus *P. italicum* [33], gray mold of tomato fruit [92] and strawberries [32] caused by *B. cinerea* due to the ability of emitting VOCs. Volatiles of *S. platensis* F-1 displayed a strong reduction in strawberry gray mold incidence by 73 per cent. Among 16 volatiles identified from *S. platenesis* F-1, VOCs phenylethyl alcohol and (+)-epi-bicycle sesquiphellandrene were previously been detected in *M. albus* [37] and in the *Kleina odora* essential oil [93] respectively. Patel *et al.* (Unpublished data) reported that acetic acid 2-phenylether ester, styrene, β-Phyllandrene and thujopsene were most abundant VOCs released from *B. amyloliquefaciens* against postharvest pathogens of grapes.

6.3 VOCs produced by yeast

Yeasts as bioagents have been extensively studied since they own many features that satisfy the requirements for being biocontrol agents in fruits [94, 95]. Yeast species usually require a simple nutritional diet, colonizes on dry surfaces for longer periods and can grow rapidly on less expensive substrates in bioreactors [96]. Importantly, they do not produce any kind of allergenic spores or mycotoxins as many fungi or antibiotics which might be produced by bacterial antagonists [97, 98]. In addition, yeasts are a major constituent of the epiphytic microbial community of fruits and vegetables and also phenotypically adapted to this niche.

Candida intermedia 410 inhibited incidence of strawberry gray mold by the production of volatiles. It has been confirmed that VOCs production were the only probable mechanism against *B. cinerera* because there was no direct contact between the pathogen and yeast cells. The most abundant compounds identified from *C. intermedia* were 1,3,5,7-cyclooctatetraene, 3- methyl-1-butanol, 2-nonanone, and phenethyl alcohol and confirmed that yeast can be potentially developed as a biofumigant for the control of gray mold of strawberries [99].

Strains belonging to different species of yeasts such as *Saccharomyces cerevisiae*, *Wickerhamomyces anomalus* and *Metschnikowia pulcherrima* showed both *in vitro* and *in vivo* inhibitory effect due to the emission of VOCs on *B. cinerea* causing postharvest bunch rot of table grapes [100].

The VOCs produced by *S. cerevisiae* inhibited *Phyllosticta citricarpa*, causing black spot of citrus. Individual exposure of 3-methyl-1-butanol and 2-methyl-1-butanol controlled the development of new lesions close to 90%, even after removing the fruits from the VOC influence and displayed effective inhibition of mycelial growth, appressorium formation and germination of conidia [101].

The psychrotrophic, non-pectinolytic yeast *Candida sake* grown in apple juice act as a potential biocontrol agent. The antifungal volatile organic compounds produced by *C. sake* inhibited the growth of five postharvest pathogens of apple (*P. expansum*, *B. cinerea*, *A. alternata*, *A. tenuissima* and *A. arborescens*). The VOCs were also effective on *in vivo* assays to control *P. expansum* in Red Delicious apples [102]. Patel *et al.* (Unpublished data) reported that acetic acid 2-phenylether ester, styrene, β-Phyllandrene and thujopsene were most abundant VOCs released *Hanseniaspora opuntiae* against postharvest pathogens of grapes.

7. Safety of microbial volatile organic compounds

The disease control using VOCs from microbes would be safer to human health and environment as the yeast and their products are Generally Recognized As Safe (GRAS) by the U.S. Food and Drug Administration (FDA), and the yeast is classified as Biosafety Level 1 by U.S. Centres of Disease Control and Prevention (CDC/OHS, 2009), as it is not a human pathogen, it generally does not produce mycotoxins, antibiotics, or other molecules that are unacceptable in foods [100]. Many antimicrobial VOCs, such as decyl alcohol, nonanal, acetoin and phenylethyl alcohol, are already in use as additives in foods and cosmetics and their use can be extended to control postharvest diseases in fruits and vegetables [103]. However, they are reports on the issues related to safety of these mVOCs to human being. For example some mVOCs reported as allergenic and asthmatic agents such as 1-octen-3-ol [104]. Many of the reports arise from analysis of environmental samples from moist and damp rooms or closed places. As mentioned by Piechulla and Degenhardt [105] the use of these compounds in post-harvest disease management depends on their characterization, dose and their mode of action. These mVOCs need to be applied in very low concentration and they are completely degradable. Compared to synthetic fungicides, they are less harmful due to no residual toxicity. To harness the use of mVOCs, a prerequisite is the availability of adequate in vitro test systems to generate the data to facilitate the legal and regulatory authorities in giving permission for their use in agriculture. A review by Ceremi et al. [106] give the list of in vitro systems for the evaluation of mVOCs on human being. They reviewed the submerged cultivation, air-liquid-interface (ALI), spheroids and organoids as well as their advantages and disadvantages. As these mVOCs are suspected to be allergenic, the methods to study the effects on human respiratory tract need to be updated. In our work too, (Pooja et al. unpublished data) though styrene was abundantly produced by *Hanseniaspora opuntiae* and *B. amyloliquefaciens*, but we did not use it due its ill effect on human.

8. Conclusion

Microbes are a remarkable source of active chemicals due to their diverse chemical makeup. VOCs in particular, when compared to traditional products, can offer evident environmental benefits due to their nil residual effect, renewability, biodegradability, and low toxicity. This makes them an effective aspect of an eco-chemical approach in the management of postharvest diseases. As the world is moving towards a green economy, with new production chains that begin in agriculture and end by returning back to agriculture. Products and by-products come together to establish a sustainable economic system that uses renewable resources. Locally and in many agricultural exporting nations, laws to limit chemicals are currently being adopted. In order to counteract the negative effects of microorganism infection during storage, these non-toxic and GRAS-recognized compounds will be added to the postharvest chain. This will improve the protection of human health and the environment. The use of mVOCs in disease management is evolving and its beneficial effect without harming human health need demonstration after in depth molecular studies to confirm their potential use in agriculture. The future thrust areas as suggested by Kanchiswamy [54] include application of nanotechnology in delivering these mVOCs, expanding the database of mVOCs, studies on mode of action of individual compounds and synergistic effect of cocktail of compounds, non-volatile biodegradable precursors of mVOCs and molecular basis of their mode of action.

Conflict of interest

The authors declare no conflict of interest.

Author details

Subbaraman Sriram[1*], Pooja Shekar Patel[2], Darisi Venkata Sudhakar Rao[1] and Kodthalu Seetharamaiah Shivashankara[1]

1 ICAR – Indian Institute of Horticultural Research, Bengaluru, India

2 Department of Plant Pathology, GKVK, University of Agricultural Sciences, Bengaluru, India

*Address all correspondence to: subbaraman.sriram@icar.gov.in

References

[1] Hanke MD, Flachowsky H. Fruit Crops. Berlin Heidelberg: Springer-Verlag; 2010. pp. 305-348

[2] Food and Agriculture Organization of the United Nations (FAO). Sustainability pathway: Food wastage foot print. 2019. Available from: http://www.fao.org/nr/sustainability/food-loss-and-waste

[3] Kruijssen F, Tedesco I, Ward A, Pincus L, Love D, Thorne-Lyman AL. Loss and waste in fish value chains: A review of the evidence from low and middle-income countries. Global Food Security. 2020;**26**:100434

[4] Block G, Patterson B, Subar A. Fruit, vegetables and cancer prevention: A review of the epidemiological literature. Nutrition and Cancer. 1992;**18**:1-29

[5] Ferro-Luzzi A, Cialfa E, Leclerq C, Toti E. The Mediteranean diet revisted: Focus on fruit and vegetables. International Journal of Food Sciences and Nutrition. 1994;**45**:291-300

[6] Morris DM, Kritchevsky SB, Davis CE. Serum carotenoids and coronary heart disease: The lipid research clinics coronary primary prevention trial and fellow-up study. Journal of the American Medical Association. 1994;**274**:1439-1441

[7] World Health Organization. Diet, Nutrition and the Prevention of Cronic Diseases: Report of WHO Study Group. WHO technical series report 797. Geneva: World Health Organization; 1990

[8] Coates L, Johnson G. Postharvest diseases of fruit and vegetables. In: Brown JF, Ogle HJ, editors. Plant Pathogens and Plant Diseases. Armidale NSW: Rockvale Publications; 1997:533-548

[9] Eckert JW, Eaks IL. Postharvest disorders and diseases of citrus fruits. In: Reuter W, Calavan EC, Carman GE, editors. The Citrus Industry. DANR. Berkeley, CA, USA: University of California Press; 1989. pp. 179-260

[10] Smilanick JL, Brown GE, Eckert JW. The biology and control of postharvest diseases. In: Wardowski WF, Miller WM, Hall DJ, Grierson W, editors. Fresh Citrus Fruits. Second ed. Longboat Key, FL, USA: Florida Science Source, Inc.; 2006a. pp. 339-396

[11] Tuset JJ. Podredumbres de los Frutos Cítricos. Generalitat, Valenciana, Valencia, Spain: Conselleria d'Agricultura i Pesca; 1988

[12] Sharma RR, Singh D, Singh R. Biological control of postharvest diseases of fruits and vegetable by microbial antagonist: A review. Biological Control. 2009;**50**:205-221

[13] Mari M, Neri F, Bertolini P. Novel approaches to prevent and control postharvest diseases of fruit. In: Stewart Postharvest Review, 3(6): Article 4. London, UK: Stewart Postharvest Solutions Ltd.; 2007

[14] Droby S, Chalutz E, Wilson CL, Wisniewski ME. Biological control of postharvest diseases: A promising alternative to the use of synthetic fungicides. Phytoparasitica. 1992;**20**:1495-1503

[15] Wilson CL, Wisniewski ME, Droby E, Chalutz E. A selection strategy for microbial antagonists to control postharvest diseases of fruit and vegetables. Scientia Horticulturae. 1993;**53**:183-189

[16] Filonow AB, Vishniac HS, Anderson JA, Janisiewicz WJ. Biological control of *Botrytis cinerea* in apple by yeasts from various habitats and their putative mechanism of antagonism. Biological Control. 1996;**17**(2):212-220

[17] Ippolito A, El-Ghaouth A, Wilson CL, Wisniewski MA. Control of postharvest decay of apple fruit by *Aureobasidium pullulans* and induction of defense responses. Postharvest Biology and Technology. 2000;**19**:265-272

[18] Jijakli MH, Grevesse C, Lepoivre P. Modes of action of biocontrol agents of postharvest diseases: Challenges and difficulties. Bulletin-OILB/SROP. 2001;**24**(3):317-318

[19] Janisiewicz WJ, Tworkoski TJ, Sharer C. Characterizing the mechanism of biological control of postharvest diseases on fruit with a simple method to study competition for nutrients. Phytopathology. 2000;**90**(11):1196-1200

[20] Barkai-Golan R. Postharvest Diseases of Fruit and Vegetables: Development and Control. Amsterdam, The Netherlands: Elsevier Sciences; 2001

[21] El-Ghaouth A, Wilson CL, Wisniewski ME. Biologically based alternatives to synthetic fungicides for the postharvest diseases of fruit and vegetables. In: Naqvi SAMH, editor. Diseases of Fruit and Vegetables. Vol. 2. The Netherlands: Kluwer Academic Publishers; 2004. pp. 511-535

[22] Ahmad F, Ahmad I, Khan MS. Screening of free-living rhizospheric bacteria for their multiple plant growth promoting activities. Microbiological Research. 2008;**163**:173-181

[23] Janisiewicz WJ, Korsten L. Biological control of postharvest diseases of fruit. Annual Review of Phytopathology. 2002;**40**:411-441

[24] Veselova MA, Plyuta VA, Khmel IA. Volatile compounds of bacterial origin: Structure, biosynthesis, and biological activity. Microbiology. 2019;**88**:261-274

[25] Poveda J. Beneficial effects of microbial volatile organic compounds (MVOCs) in plants. Applied Soil Ecology. 2021;**168**:104-118

[26] Stinson M, Ezra D, Hess WM, Sears J, Strobel G. An endophytic *Gliocladium* sp. of *Eucryphia cordifolia* producing selective volatile antimicrobial compounds. Plant Science. 2003;**165**:913-922

[27] Koitabashi M. New biocontrol method for parsley powdery mildew by the antifungal volatiles-producing fungus Kyu-W63. Journal of General Plant Pathology. 2005;**71**:280-284

[28] Chen H, Xiao X, Wang J, Wu LJ, Zheng ZM, Yu ZL. Antagonistic effects of volatiles generated by *Bacillus subtilis* on spore germination and hyphal growth of the plant pathogen, *Botrytis cinerea*. Biotechnology Letters. 2008;**30**:919-923

[29] Gu YQ, Mo MH, Zhou JP, Zou CS, Zhang KQ. Evaluation and identification of potential organic nematicidal volatiles from soil bacteria. Soil Biology and Biochemistry. 2007;**39**:2567-2575

[30] Fialho MB, Toffano L, Pedroso MP, Augusto F, Pascholati SF. Volatile organic compounds produced by *Saccharomyces cerevisiae* inhibit the *in vitro* development of *Guignardia citricarpa*, the causal agent of citrus black spot. World Journal of Microbiology and Biotechnology. 2010;**26**:925-932

[31] Bruce A, Verrall S, Hackett CA, Wheatley RE. Identification of volatile organic compounds (VOCs) from bacteria and yeast causing growth inhibition of sapstain fungi. Holzforschung. 2004;**58**:193-198

[32] Wan MG, Li GQ, Zhang JB, Jiang DH, Huang HC. Effect of volatile substances of *Streptomyces platensis* F-1 on control of plant fungal diseases. Biological Control. 2008;**46**:552-559

[33] Li QL, Ning P, Zheng L, Huang JB, Li GQ, Hsiang T. Fumigant activity of volatiles of *Streptomyces globisporus* JK-1 against *Penicillium italicum* on *Citrus microcarpa*. Postharvest Biology and Technology. 2010;**58**:157-165

[34] Kulakiotu EK, Thanassoulopoulos CC, Sfakiotakis EM. Postharvest biological control of *Botrytis cinerea* on kiwifruit by volatiles of 'Isabella' grapes. Phytopathology. 2004;**94**:1280-1285

[35] Schulz S, Dickschat J. Bacterial volatiles: The smell of small organisms. Natural Product Reports. 2007;**24**:814-842

[36] Korpi A, Pasanen AL, Pasanen P. Volatile compounds originating from mixed microbial cultures on building materials under various humidity conditions. Applied and Environmental Microbiology. 1998;**64**:2914-2919

[37] Strobel GA, Dirksie E, Sears J, Markworth C. Volatile antimicrobials from *Muscodor albus*, a novel endophytic fungus. Microbiology. 2001;**147**:2943-2950

[38] Inamdar AA, Morath S, Bennett JW. Fungal volatile organic compounds: More than just a funky smell? Annual Review of Microbiology. 2020;**74**:101-116

[39] Garcia-Alcega S, Nasir ZA, Ferguson R, Whitby C, Dumbrell AJ. Fingerprinting outdoor air environment using microbial volatile organic compounds (MVOCs)—A review. Trends in Analytical Chemistry. 2017;**86**:75-83

[40] Kaminski E, Libbey L, Stawicki S, Wasowicz E. Identification of the predominant volatile compounds produced by *Aspergillus flavus*. Applied Microbiology. 1972;**24**:721-726

[41] Seifert R, King A. Identification of some volatile constituents of *Aspergillus clavatus*. Journal of Agricultural and Food Chemistry. 1982;**30**:786-790

[42] Zhao L, Ni Y, Su M, Li H, Dong F. High throughput and quantitative measurement of microbial metabolome by gas chromatography/mass spectrometry using automated alkyl chloroformate derivatization. Analytical Chemistry. 2017;**89**:5565-5577

[43] Orban AM, Rühl M. Identification of volatile producing enzymes in higher fungi: Combining analytical and bioinformatic methods. Methods in Enzymology. 2022;**664**:221-242

[44] Wilson AD, Baietto M. Advances in electronic-nose technologies developed for biomedical applications. Sensors. 2011;**11**:1105-1176

[45] Kaddes A, Fauconnier ML, Sassi K, Nasraoui B, Jijakli MH. Endophytic fungal volatile compounds as solution for sustainable agriculture. Molecules. 2019;**24**:1065

[46] Bentley R, Meganathan R. Geosmin and methylisoborneol biosynthesis in streptomycetes: Evidence for an isoprenoid pathway and its absence in non-differentiating isolates. FEBS Letters. 1981;**125**:220-222

[47] Dudareva N, Klempien A, Muhlemann JK, Kaplan I. Biosynthesis, function and metabolic engineering of plant volatile organic compounds. New Phytologist. 2013;**198**(1):16-32

[48] Guevara-Avendano E, Bejarano-Bolivar AA, Kiel-Martinez AL, Ramirez-Vazquez M, Mendez-Bravo A,

Von Wobeser EA. Avocado rhizobacteria emit volatile organic compounds with antifungal activity against *Fusarium solani*, *Fusarium* sp. associated with Kuroshio shot hole borer, and *Colletotrichum gloeosporioides*. Microbiological Research. 2019;**219**:74-83

[49] Asari S, Matzen S, Petersen MA, Bejai S, Meijer J. Multiple effects of *Bacillus amyloliquefaciens* volatile compounds: Plant growth promotion and growth inhibition of phytopathogens. FEMS Microbiology Ecology. 2016;**92**:fiw070

[50] Gotor-Vila A, Teixido N, Di Francesco A, Usall J, Ugolini L, Torres R, et al. Antifungal effect of volatile organic compounds produced by *Bacillus amyloliquefaciens* CPA-8 against fruit pathogen decays of cherry. Food Microbiology. 2017;**64**:219-225

[51] Archana TJ, Gogoi R, Kaur C, Varghese E, Sharma RR, Srivastav M, et al. Bacterial volatile mediated suppression of postharvest anthracnose and quality enhancement in mango. Postharvest Biology and Technology. 2021;**177**:111525

[52] Rajaofera MJN, Wang Y, Dahar GY, Jin P, Fan L, Xu L. Volatile organic compounds of *Bacillus atrophaeus* HAB-5 inhibit the growth of *Colletotrichum gloeosporioides*. Pesticide Biochemistry and Physiology. 2019;**156**:170-176

[53] Giorgio A, de Stradis A, Cantore P, Iacobellis NS. Biocide effects of volatile organic compounds produced by potential biocontrol rhizobacteria on *Sclerotinia sclerotiorum*. Frontiers in Microbiology. 2015;**6**:1056

[54] Kanchiswamy CN, Malnoy M, Maffei ME. Chemical diversity of microbial volatiles and their potential for plant growth and productivity. Frontiers in Plant Science. 2015;**6**:151. DOI: 10.3389/fpls.2015.00151

[55] Gao H, Li P, Xu X, Zeng Q, Guan W. Research on volatile organic compounds from *Bacillus subtilis* CF-3: Biocontrol effects on fruit fungal pathogens and dynamic changes during fermentation. Frontiers in Microbiology. 2018;**9**:456

[56] Gao Z, Zhang B, Liu H, Han J, Zhang Y. Identification of endophytic *Bacillus velezensis* ZSY-1 strain and antifungal activity of its volatile compounds against *Alternaria solani* and *Botrytis cinerea*. Biological Control. 2017;**105**:27-39

[57] Wang L, Dou G, Guo H, Zhang Q, Qin X, Yu W. Volatile organic compounds of *Hanseniaspora uvarum* increase strawberry fruit flavor and defense during cold storage. Food Science & Nutrition. 2019;**7**:2625-2635

[58] Ezra D, Hess WM, Strobel GA. New endophytic isolates of *Muscodor albus*, a volatile-antibiotic-producing fungus. Microbiology. 2004;**150**:4023-4031

[59] Pena LC, Jungklaus GH, Savi DC, Ferreira-Maba L, Servienski A, Maia BH. *Muscodor brasiliensis* sp. nov. produces volatile organic compounds with activity against Penicillium digitatum. Microbiological Research. 2019;**221**:28-35

[60] Jiang CH, Xie YS, Zhu K, Wang N, Li ZJ, Yu GJ, et al. Volatile organic compounds emitted by *Bacillus* sp. JC03 promote plant growth through the action of auxin and strigolactone. Plant Growth Regulation. 2019;**87**:317-328

[61] Ryu CM, Farag MA, Hu CH, Reddy MS, Wei HX, Paré PW, et al. Bacterial volatiles promote growth in *Arabidopsis*. Proceedings of the National Academy of Sciences of the United States of America. 2003;**100**:4927-4932

[62] Rath M, Mitchell TR, Gold SE. Volatiles produced by *Bacillus mojavensis* RRC101 act as plant growth modulators and are strongly culture-dependent. Microbiological Research. 2018;**208**:76-84

[63] Contreras-Cornejo HA, Macías-Rodríguez L, Herrera-Estrella A, Lopez-Bucio J. The 4-phosphopantetheinyl transferase of *Trichoderma virens* plays a role in plant protection against *Botrytis cinerea* through volatile organic compound emission. Plant and Soil. 2014;**379**:261-274

[64] Hung R, Lee S, Bennett JW. *Arabidopsis thaliana* as a model system for testing the effect of *Trichoderma* volatile organic compounds. Fungal Ecology. 2013;**6**:19-26

[65] Jalali F, Zafari D, Salari H. Volatile organic compounds of some *Trichoderma* spp. increase growth and induce salt tolerance in *Arabidopsis thaliana*. Fungal Ecology. 2017;**29**:67-75

[66] Vaishnav A, Kumari S, Jain S, Varma A, Choudhary DK. Putative bacterial volatile-mediated growth in soybean (*Glycine max* L. Merrill) and expression of induced proteins under salt stress. Journal of Applied Microbiology. 2015;**119**:539-551

[67] del Rosario-Cappellari L, Banchio E. Microbial volatile organic compounds produced by *Bacillus amyloliquefaciens* GB03 ameliorate the effects of salt stress in *Mentha piperita* principally through acetoin emission. Journal of Plant Growth Regulation. 2019;**39**:764-775

[68] Naznin HA, Kiyohara D, Kimura M, Miyazawa M, Shimizu M, Hyakumachi M. Systemic resistance induced by volatile organic compounds emitted by plant growth-promoting fungi in *Arabidopsis thaliana*. PLoS One. 2014;**9**:e86882

[69] Li N, Kang S. Do volatile compounds produced by *Fusarium oxysporum* and *Verticillium dahliae* affect stress tolerance in plants? Mycology. 2018;**9**:166-175

[70] Ryu CM, Farag MA, Hu CH, Reddy MS, Kloepper JW, Paré PW. Bacterial volatiles induce systemic resistance in *Arabidopsis*. Plant Physiology. 2004;**134**:1017-1026

[71] D'Alessandro M, Erb M, Ton J, Brandenburg A, Karlen D, Zopfi J, et al. Volatiles produced by soil-borne endophytic bacteria increase plant pathogen resistance and affect tritrophic interactions. Plant, Cell & Environment. 2014;**37**:813-826

[72] Raza W, Ling N, Liu D, Wei Z, Huang Q, Shen Q. Volatile organic compounds produced by *Pseudomonas fluorescens* WR-1 restrict the growth and virulence traits of *Ralstonia solanacearum*. Microbiological Research. 2016b;**192**:103-113

[73] Velazquez-BecerraC,Macías-RodríguezLI, Lopez-Bucio J, Flores-Cortez I, Santoyo G, Hernandez-Soberano C, et al. The rhizobacterium *Arthrobacter agilis* produces dimethyl hexadecylamine, a compound that inhibits growth of phytopathogenic fungi in vitro. Protoplasma. 2013;**250**:1251-1262

[74] Raza W, Yousaf S, Rajer FU. Plant growth promoting activity of volatile organic compounds produced by biocontrol strains. Science Letters. 2016a;**4**:40-43

[75] Davis TS, Boundy-Mills K, Landolt PJ. Volatile emissions from an epiphytic fungus are semiochemicals for eusocial wasps. Microbial Ecology. 2012;**64**:1056-1063

[76] Usseglio VL, Pizzolitto RP, Rodriguez C, Zunino MP, Zygadlo JA,

Areco VA, et al. Volatile organic compounds from the interaction between *Fusarium verticillioides* and maize kernels as a natural repellents of *Sitophilus zeamais*. Journal of Stored Products Research. 2017;**73**:109-114

[77] Strobel GA. *Muscodor* species-endophytes with biological promise. Phytochemistry Reviews. 2011;**10**:163-172

[78] Mercier J, Jimenez B. Control of fungal decay of apples and peaches by the biofumigant fungus *Muscodor albus*. Postharvest Biology and Technology. 2004;**31**:1-8

[79] Mercier J, Smilanick JL. Control of green mold and sour rot of stored lemon by biofumigation with *Muscodor albus*. Biological Control. 2005;**32**:401-407

[80] Mlikota Gabler F, Fassel R, Mercier J, Smilanick JL. Influence of temperature, inoculation interval, and dosage on biofumigation with *Muscodor albus* to control postharvest gray mold on table grapes. Plant Disease. 2006;**90**:1019-1025

[81] Kim HY, Choi GJ, Lee HB, Lee S-W, Lim HK, Jang KS, et al. Some fungal endophytes from vegetable crops and their anti-oomycete activities against tomato late blight. Letters in Applied Microbiology. 2007;**44**:332-337

[82] Lee SO, Kim HY, Choi GJ, Lee HB, Jang KS, Choi YH, et al. Mycofumigation with *Oxyporus latemarginatus* EF069 for control of postharvest apple decay and Rhizoctonia root rot on moth orchid. Journal of Applied Microbiology. 2009;**106**:1213-1219

[83] Warner G. Pear rot emerging as postharvest problem. Good-Fruit-Grower. 1993;**44**:25

[84] Rouissi W, Ugolini L, Martini C, Lazzeri L, Mari M. Control of postharvest fungal pathogens by antifungal compounds from *Penicillium expansum*. Journal of Food Protection. 2013;**76**(11):1879-1886

[85] Mari M, Bautista-Banos S, Sivakumar D. Decay control in the postharvest system: Role of microbial and plant volatile organic compounds. Postharvest Biology and Technology. 2016;**122**:70-81

[86] Li Q, Wu L, Hao J, Luo L, Cao Y, Li J. Biofumigation on post-Harvest diseases of fruits using a new volatile-producing fungus of *Ceratocystis fimbriata*. PLoS One. 2015;**10**(7):e0132009

[87] Ling L, Zhao Y, Tu Y, Yang C, Ma C, Feng S, et al. The inhibitory effect of volatile organic compounds produced by *Bacillus subtilis* CL2 on pathogenic fungi of wolfberry. Journal of Basic Microbiology. 2021;**61**:110-121

[88] Calvo H, Mendiara I, Arias E, Gracia AP, Blanco D, Venturini ME. Antifungal activity of the volatile organic compounds produced by *Bacillus velezensis* strains against postharvest fungal pathogens. Postharvest Biology and Technology. 2020;**166**:111208

[89] Zheng L, Situ JJ, Zhu QF, Xi PG, Zheng Y, Liu HX, et al. Identification of volatile organic compounds for the biocontrol of postharvest litchi fruit pathogen *Peronophythora litchii*. Postharvest Biology and Technology. 2019;**155**:37-46

[90] Zheng M, Shi J, Shi J, Wang Q, Li Y. Antimicrobial effects of volatiles produced by two antagonistic Bacillus strains on the anthracnose pathogen in postharvest mangos. Biological Control. 2013;**65**:200-206

[91] Arrebola E, Sivakumar D, Korsten L. Effect of volatile compounds produced by Bacillus strains on postharvest decay. Biological Control. 2010;**53**:122-128

[92] Li Q, Ning P, Zheng L, Huang J, Li G, Hsiang T. Effects of volatile substances of *Streptomyces globisporus* JK-1 on control of *Botrytis cinerea* on tomato fruit. Biological Control. 2012;**61**:113-120

[93] AL-Taweel AM, El-Deeb KS, AL-Muhtadi FJ. Chemical composition and antimicrobial activity of the essential oil of *Kleina odora*. Saudi Pharmaceutical Journal. 2004;**J.12**:47-50

[94] Santos A, Sanchez A, Marquina D. Yeasts as biological agents to control *Botrytis cinerea*. Microbiological Research. 2004;**159**:331-338

[95] Liu J, Sui Y, Wisniewski M, Droby S, Liu Y. Review: Utilization of antagonistic yeasts to manage postharvest fungal diseases of fruit. International Journal of Food Microbiology. 2013;**167**:153-160

[96] Chanchaichaovivat A, Ruenwongsa P, Panijpan B. Screening and identification of yeast strains from fruits and vegetables: Potential for biological control of postharvest chilli anthracnose (*Colletotrichum capsici*). Biological Control. 2007;**42**:326-335

[97] El-Tarabily KA, Sivasithamparam K. Potential of yeasts as biocontrol agents of soil-borne fungal plant pathogens and as plant growth promoters. Mycoscience. 2006;**47**:25-35

[98] Nunes CA. Biological control of postharvest diseases of fruit. European Journal of Plant Pathology. 2012;**133**:181-196

[99] Huang R, Li GQ, Zhang J, Yang L, Che HJ, Jiang DH, et al. Control of postharvest Botrytis fruit rot of strawberry by volatile organic compounds of *Candida intermedia*. Phytopathology. 2011;**101**(7):859-869

[100] Parafati L, Vitale A, Restuccia C, Cirvilleri G. Biocontrol ability and action mechanism of food-isolated yeast strains against *Botrytis cinerea* causing post-harvest bunch rot of table grape. Food Microbiology. 2015;**47**:85-92

[101] Toffano L, Fialho MB, Pascholati SF. Potential of fumigation of orange fruits with volatile organic compounds produced by *Saccharomyces cerevisiae* to control citrus black spot disease at postharvest. Biological Control. 2017;**108**:77-82

[102] Arrarte E, Garmendia G, Rossini C, Wisniewski M, Vero S. Volatile organic compounds produced by Antarctic strains of Candida sake play a role in the control of postharvest pathogens of apples. Biological Control. 2017;**109**:14-20

[103] Zhang Y, Li T, Liu Y, Li X, Zhang C, Feng Z, et al. Volatile organic compounds produced by *Pseudomonas chlororaphis* subsp. *aureofaciens* SPS-41 as biological fumigants to control *Ceratocystis fimbriata* in postharvest sweet potatoes. Journal of Agricultural and Food Chemistry. 2019;**67**:3702-3710. DOI: 10.1021/acs.jafc.9b00289

[104] Araki A, Kanazawa A, Kawai T, Eitaki Y, Morimoto K, Nakayama K, et al. The relationship between exposure to microbial volatile organic compound and allergy prevalence in single-family homes. Science of the Total Environment. 2012;**423**:18-26. DOI: 10.1016/j.scitotenv.2012.02.026

[105] Piechulla B, Degenhardt J. The emerging importance of micro-biolvolatile organic compounds. Plant, Cell & Environment. 2014;**37**:811-812. DOI: 10.1111/pce.12254

[106] Cerimi K, Jackel U, Meyer V, Daher U, Reinert J, Klar S. In vitro systems for toxicity evaluation of microbial volatile organic compounds on humans: Current status and trends. Journal of Fungi. 2022;**8**:75. DOI: 10.3390/jof8010075

Chapter 5

Application of Phytohormones, Growth Regulators, and Calcium to Preserve Fruit Quality in Pre- and Post-Harvest

Ismael Aguilar-Ayala and Diana Herrera-Rojas

Abstract

The technological levels used to reduce fruit losses in post-harvest are closely linked to those used in pre-harvest. Applications of phytohormones, growth regulators, and calcium to fruit in pre- and post-harvest are viable alternatives to increase and preserve quality attributes. Knowledge of the action and response of fruits to the exogenous application of different phytohormones, growth regulators, and calcium in pre-harvest are fundamental when considering that fruit quality is acquired at this stage and that the purpose of post-harvest technology is to preserve fruit quality. This chapter describes research carried out to evaluate the response of different fruits to the application of phytohormones, phytoregulators, and calcium, which showed favorable responses in increasing fruit quality in pre-harvest and preserving quality in post-harvest.

Keywords: auxins, gibberellins, cytokinins, ethylene, calcium, fruit, pre-harvest, post-harvest, application, preservation, quality

1. Introduction

Currently, the generation, application, and dissemination of new techniques and technologies in the conservation of fruits and vegetables in post-harvest are of vital importance when considering that 20–50% of the agricultural production obtained is lost [1–4]. The losses of fruits and vegetables are different according to the economic development of the countries, in nations considered developed, from 5 to 35% is lost, and in developing countries from 20 to 50%. In these regions, it is considered that losses of fruits and vegetables occur mainly in the retail and consumption stages, while in countries considered poor or low-income, losses of fruits and vegetables are related to poor pre-harvest handling where low technological levels are applied, harvesting techniques that damage the fruits, poor transport from the orchard to the warehouse, packaging without technical design, and the lack and inadequate infrastructure for storage [1, 2, 5]. On the other hand, fruit and vegetable losses are generally classified into five stages: Stage 1: pre-harvest handling. Stage 2: packing process. Stage 3: post-harvest handling and storage. Stage 4: distribution from the warehouse to the retail market. Stage 5: consumption [6]. Fruits in post-harvest and without being subjected to any industrial process are living organisms with their own metabolism and enzymatic action, with high percentages of water content and susceptibility to attack by microorganisms, factors that determine the perishability of the fruit, i.e., the shelf life under storage conditions that can last from a few days [6] to years. To prolong the shelf life of fruits, different technologies have been implemented: use of low and high temperatures, refrigeration and freezing, immersion in chemical additives [7], controlled and modified atmospheres, vacuum packaging, and edible coatings [8]. The purpose of post-harvest handling technologies is to preserve fruit quality and minimize losses caused by transport and handling in storage. The quality of harvested fruit is not increased, only preserved; quality attributes are obtained pre-harvest, so it is very important to apply optimal technological levels both pre-harvest and post-harvest [8]. The application of phytohormones, growth regulators, and various chemical calcium compounds in the pre- and post-harvest handling of the fruit constitutes some alternative technologies to increase and preserve fruit quality. In this chapter, pioneering work and recent research on the application of phytohormones, growth regulators, and calcium to pre- and post-harvest fruit will be described in general terms.

2. Application of phytohormones and growth regulators to pre- and post-harvest fruits

Phytohormones are simple organic chemical compounds, without associated protein groups, which induce all morphogenic responses during plant ontogeny, and participate in most morphogenic responses in growth and development; they are synthesized in different plant structures such as shoot and root meristems, leaves, fruits, and seeds. Some plant hormones act at the same site of synthesis or have an effect on other plant structures and whose physiological response may last for a short or medium time. When phytohormones act in conjugation with others, they induce different morphogenic expression responses, for example, the formation of shoots and alternatively of roots and/or the production of amorphous cell masses [9]. Growth regulators are synthetic compounds of a different chemical nature from phytohormones, and they can generate similar or different effects to those expected and can even elicit more intense responses than plant hormones at the same molar concentration [9]. The main phytohormones and growth regulators used in pre- and post-harvest fruit management are auxins, gibberellins, cytokinins, and ethylene.

2.1 Auxins

In 1933, Kögl and collaborators synthesized indole-3-acetic acid from the amino acid tryptophan, a low-molecular-weight organic acid found in plants, and named it auxin, the word auxin comes from the Greek ¨auxein” meaning ¨to grow¨ [9]. Auxins synthesized in the vegetative structures of the plant are translocated through the phloem in a basipetal and cell-to-cell manner and have been found in both algae and fungi [10]. Indole-3-acetic acid elicits plant physiological responses when it binds to a receptor and undergoes signal transduction to generate diverse responses in plant growth and development (see **Table 1**). Auxins induce growth by cell division and elongation, and cell differentiation of apical, root, and cambium meristematic zones. Endogenous and exogenous auxins generate adventitious roots, and promote apical dominance and fruit growth and development [11].

In plant growth and development, auxins act in conjugation with gibberellins, cytokinins, and ethylene [12].

Plant growth regulators with auxin effect such as indole-3-propionic acid, naphthaleneacetic acid, 2,4-dichlorophenoxyacetic acid, and 2,4,5-trichlorophenoxyacetic acid are synthetic organic substances considered agrochemicals, and they are used to promote seed germination, rooting of cuttings, sprouting of differentiated buds, fruit setting, delaying the physiological maturity of fruits, and preserving the quality of some fruits in post-harvest [9, 11]. Exogenous application of auxins to plants in high concentrations induces vegetative growth by inhibiting flowering, growth, and fruit development [13].

The use of indole-3-butyric acid (IBA) by Solis et al. [14] in red pomegranate (*Punica granatum* L.) 'Apaseo tardía' had the purpose of evaluating the effect of indole-3-butyric acid on the preservation of fruit quality in post-harvest. The treatments applied were 0, 20, 40, and 80 mg/L^{-1} of IBA by immersion for 10 minutes, and the fruits were stored for 60 days under refrigeration at a temperature of 4 ± 1°C. The response variables evaluated were mass (g), size (cm) in diameter and length of the fruit, concentration of total anthocyanins, percentage of titratable acidity, ascorbic acid concentration, chlorophylls "a," "b," and total. The results indicate that the 20 mg/L^{-1} AIB treatment significantly preserved fruit size in length, while, in the mass variables, ascorbic acid concentration and anthocyanin concentration did not present significant statistical differences compared with the 0 mg/L^{-1} AIB treatment. However, the mass variables, ascorbic acid concentration, and total anthocyanins showed statistical trends for the preservation of fruit quality significantly to the application of AIB, so it is suggested to increase the immersion time of the fruit.

In another research conducted by Bustamante and Gómez [15], they applied auxins and calcium in commercial chemicals products to banana bunches to improve fruit quality, considering that the mass and size are important for their commercialization. The purpose of the research was to generate technology to increase fruit quality in pre-harvest with the application of commercial auxin and calcium chemicals. The treatments applied were as follows: Basfoliar Kelp (Auxins) + Basfoliar Calcium (Ca) [75 cc + 75 cc /L^{-1} water]; Basfoliar Kelp (Auxins) + Basfoliar Calcium (Ca) [100 cc +100 cc /L^{-1} water]; and Basfoliar Kelp (Auxins) + products containing Basfoliar Calcium (Ca) [12.5 cc + 12.5 cc/L^{-1} water] and the control and two applications were made, the first at 4 days after the beginning of flowering and the second at 10 days after flowering. The results found indicate that all treatments significantly increased fruit mass and size in length and diameter.

With the purpose of delaying fruit senescence and thus preserving the quality for a longer time, Da silva et al. [16] applied indole-3-acetic acid to fruits of Spondias (*Spondias dulcis*). The treatments evaluated were 0, 50, 100, 150, and 200 mg L^{-1} of AIA applied by immersion for 20 min, the evaluations were carried out at 5 and 10 days after application of AIA, the response variables were pulp firmness, titratable acidity, soluble solids, ascorbic acid content, and pulp and peel color. Univariate analysis was applied for each variable, and regression analysis was applied to determine the effect of auxin concentration. There was no correlation between the concentration of auxins applied on the concentration of total soluble solids and flesh color. There were no significant differences in the variables of titratable acidity, pH, and skin color. The concentrations of 50 and 100 mg L^{-1} of AIA showed the best results in reducing the ripening of Spondia fruits.

2.2 Gibberellins

Gibberellins (GAs) are phytohormones of plant growth and development involved in various physiological processes of plants, the fungus was isolated by Eichi

Kurosawa in 1926 from cultures of *Gibberella fujikoroi*, a fungus parasite of rice plants causing the ¨bakanae¨ disease, and the attack of the fungus induces excessive growth by elongation of stems and shoots and a strong decrease in productivity [17, 18].
In 1955, gibberellic acid (Ga3) was isolated from the secreted filtrate of *Gibberella fujikoroi* [19]. The synthesis of gibberellins occurs in various plant structures: buds, leaves and fruits, and their movements are basipetal. Endogenous gibberellins mainly induce height growth [20], and promote inflorescence development and flowering in long-day plants. In association with phytochromes, they induce the differentiation of vegetative meristems to reproductive meristems promoting flowering [21]. They induce parthenocarpy, and growth and development in many fruits [22]; likewise, gibberellins and other phytohormones that synthesize seeds within the fruit largely regulate their size. It has also been shown that the number and size of fruits attached to the plant correlate with the number of seeds contained in them and the absence of pollination causes seed-bearing fruits to drop. Genetically parthenocarpic fruits contain normal gibberellin levels [7], seedless fruits with relatively high gibberellin concentrations show adequate growth and development [19]. Commercial gibberellic acid obtained from extracts of the fungus *Gibberella* applied exogenously affects the growth and development stages of the plant passing from vegetative to intermediate stage and vice versa [23]. In long-day plants, it induces flowering [24, 25] and can affect sex determination by modifying the floral structure and leading to female or male flowers [21]. Gibberellins generate multiple responses when applied at different stages of fruit growth and development: It induces bud differentiation, and promotes fruit growth and development and parthenocarpy (see **Table 1**). In apples, it promot fruit development after pollination; in citrus, it delays senescence [26]. It is used to increase quality in seedless grape fruits, in apples, peaches, apricots, cherries, and in citrus fruits, it increases fruit yellowing; in the latter, GAs delay the coloration from green to orange and prevent alterations in the rind [27].

On the other hand, there are compounds of different chemical constitution that inhibit plant growth known as growth inhibitors or retardants: Chlormequat-Cl, Cycocel, AMO-1618, Phosphon-D, Ancymidol, Paclobutrazol, Uniconazole-P that reduce meristematic growth mainly by inhibiting cell division by preventing the development of the gibberellin synthesis cycle. The application of these chemical compounds reduces plant size but does not affect productivity. Plant growth arrest generated by the application of these chemicals is reversed if gibberellins are applied, even if they are mixed with the inhibitory chemical [28]. Some investigations conducted with applications of gibberellins to different fruits are described below.

The concentration of Mexican plum (*Spondias purpurea* L.) production causes this fruit to have unattractive prices for producers, so research was done [29] to phase production. Gibberellic acid was applied to pre-harvest fruit. The treatments were: 0.0, o.1, 0.2, and 0.4 mM Activol, and the application was made by spraying. The response variables evaluated were: blossom to harvest period in days, % bud burst, % flower buds, mass (g), size in length and diameter (cm), starch concentration, total soluble solids, ascorbic acid, and titratable acidity. The results indicate that production was advanced by 36 days with the 0.4 mM treatment, and this same treatment significantly induced fruit set with 51.92% with respect to the control treatment that achieved 2.67%. The 0.2 mM treatment showed a significant growth of fruit size in length with 4.32 cm compared with 3.97 cm of the control treatment, and fruit size in diameter was 3.44 cm significantly different from the control treatment with 3.17 cm; fruit mass(g) was significantly different with the 0.2 mM Activol treatment, which was 29.37 (g), and the control treatment was 23.39 (g). In the other evaluated variables such as starch, total soluble solids, total sugars, ascorbic acid, and titratable acidity, no significant statistical differences were found between treatments.

The fruit of red pomegranate cv. Apaseo tardía is very susceptible to fruit cracking. In consideration of the above, an investigation was carried out [30] where gibberellic acid was applied to fruits of red pomegranate cv, Apaseo tardía in pre-harvest. Gibberellic acid was used in treatments of 0, 50, 100, and 200 mg/l in three direct applications to the fruit at intervals of 15 days each. The results showed that the treatments of 50, 100, and 200 mg/l of acigigib showed highly significant differences in avoiding 100% of fruit cracking compared with the control treatment, which showed 62% of cracked fruit. In the quality variables, the 0 mg/l gibberellic acid treatment showed significant statistical differences in the concentration of total sugars compared to the other treatments. The concentration of chlorophyll "a" in the pericarp of red pomegranate showed significant differences between treatments. As for the concentration of chlorophyll "b" in the pericarp of the red pomegranate fruit, the control treatment was the only one in which chlorophyll "b" was observed. No significant statistical differences were found between treatments in the following quality variables: total soluble solids, ascorbic acid, titratable acidity, anthocyanins, hydrogen potential, fruit mass (g), size (cm) in length and diameter.

In order to evaluate the effectiveness of a pre-harvest application of gibberellic acid to fruits of late red pomegranate cv. Apaseo tardía to prevent fruit cracking and conserve fruit quality, treatments of 0 and 50 mg/l of gibberellic acid applied once were evaluated. The variables evaluated were the number of uncracked and cracked fruit, fruit size in length and diameter, mass (g), total soluble solids, ascorbic acid content, total sugars, hydrogen potential, anthocyanin concentration, titratable acidity, and fruit maturity index. The results showed that, out of a total of 250 fruits evaluated per treatment, the 50 mg/l gibberellic acid treatment showed a statistically significant decrease of only 30 cracked fruits against 77 of the control treatment. The fruit quality variables, size in length and diameter, mass(g), total soluble solids, ascorbic acid, total sugars, anthocyanins, hydrogen potential, titratable acidity and maturity index, did not show significant differences between treatments [31].

2.3 Cytokinins

Plant cytokinins were discovered in 1963 by Miller and Letham in corn kernels (*Zea mays*), and the first cytokinin was identified as 6-(4-hydroxy-3-methylbut-trans-2-enyl-amino) purine called zeatin. Naturally synthesized cytokinins of importance as zeatins are benzyladenine and kinetin, and natural plant cytokinins are synthesized at sites of continuous cell division such as the root and shoot meristems of the plant. Synthetic cytokinins such as benzyladenine (BA) or furfurylaminopurine and thidiazuron (TDZ) are more efficient than endogenous hormones because they are not degraded or metabolized by the tissue. Cytokinins promote cell division and differentiation, sprouting of axillary buds, delay leaf senescence, induce sprouting of dormant buds and fruit growth, stimulate nutrient mobilization, synthesis and degradation of pigments such as chlorophyll and protein degradation, and increase the movement of sugars, amino acids, and trace elements to developing organs and generate protein synthesis [32, 33].

The application of growth regulators such as cytokinins in the plant has been a very important practice for producers, because new production technologies are applied in fruit trees and thus ensure the production yield in flowers and fruits, inducing flowering and decreasing the number of aborted flowers, being factors that affect the production yield of any fruit tree.

Barbosa [34] used the application of cytokinins to ensure the yield in the production of papaya because it is a fruit of great economic importance, and it has a high percentage of flower drop during the flowering stage affecting the percentage of fruit set. The purpose of the research was to determine the effect of the combination of three doses of cytokinins (X-Cyte) and three doses of gibberellins (RyzUp) on fruit set, growth and development in a commercial papaya plantation. The treatments applied were 2, 7, and 12 ppm concentrations of cytokinins and 5, 10, and 15 ppm concentrations of gibberellins. As results, cytokinins (X-Cyte) at a dose of 12 ppm obtained the highest papaya production with 13.02 tm ha^{-1}, and gibberellins (Ryz-Up) had no effect on papaya production with the exception of the number of fruits harvested. Statistical significance was found in the interaction C x G: fruit mass, number of fruits tied, and polar diameter, which implies that both factors act jointly on these characteristics, stimulating fruit growth. The yield in papaya production was in relation to the number of fruits harvested, there were significant differences in relation to the control in the number of buds and flowers, and the application of gibberellins had no effect on the yield of papaya production.

In order to evaluate the effect of cytokinins on blueberry fruit production, Cano [35] applied Agrocimax Plus (cytokinin) to blueberry (*Vaccinium corimbosum* L.) Biloxi variety. The treatments consisted of the addition of Agrocimax Plus at concentrations of 1.25ml/L, 2.5ml/L, and the control without Agrocimax Plus. Applications were made from pre-flowering to fruit ripening. The variables evaluated were productivity, yield, and fruit size. The results obtained were that the treatment at a concentration of 1.25 ml/L obtained a yield of 3.19 kg per plant and a production of 15950 kg per hectare; for the 10–17 mm size, the yield was 1.54 kg per plant and a production of 7700 kg per hectare. In the 18–28 mm size a yield of 1.65 kg per plant and a production of 8250 kg per hectare were obtained. The economic evaluation of the use of Agrocimax Plus on the production of blueberry fruit (*Vaccinium corimbosum* L.) was with the Agrocimax Plus treatment (1.25ml/L) with a profitability index of 187% of the production cost.

In the research conducted by Tamalá [36], the inductor effect of cytokinins on flowering and fruit setting was determined in a soursop (*Annona muricata*) crop of approximately 2 years and 6 months of age, the treatments were cytokinin 1.5 ml/L of water, 1.25 ml/L of water, and the control, with a completely randomized block design, and the study variables were days of flowering, total number of flowers, aborted flowers, fruit set, and fruit circumference. The results obtained were that the concentration of 1.5 ml/L of water accelerated flower production and fruit formation compared to the control, because cytokinins increased the number of inflorescences, decreased flower drop compared to the other treatments, increased fruit size in length and diameter, and increased fruit production.

2.4 Ethylene

Ethylene is a gaseous type phytohormone that is synthesized from the amino acid methionine to form S-adenosyl-L-methionine (SAM) catalyzed by the enzyme SAM synthetase, to form 1-aminocyclopropane-1-carboxylic acid (ACC) from SAM through the participation of the enzyme ACC synthase (ACS) and finally the conversion of ACC to ethylene [37–39]. It is of great importance in physiology and pre- and post-harvest technology; it is responsible for various physiological responses in the plant and fruit in pre-harvest and in post-harvest it has a decisive influence on fruit preservation. Neljubow observed in etiolated plants of Pisum sativum the triple response induced by this gas: reduction of growth by elongation, thickening of the hypocotyl, and change in growth orientation. Ethylene has been studied in biological models such as Arabidopsis thaliana to know and understand its mode of action and effect on growth and development of flowering plants [13, 37].

The application of ethylene known as the plant hormone of fruit ripening provariety of different effects; in plants (see **Table 1**), it induces growth by cell division, causes leaf abscission, promotes the sprouting of differentiated buds, and increases fruit quality; and when ethylene is applied post-harvest, it modifies the quality of climacteric and non-climacteric fruits [40]. Climacteric fruits continue with the ripening process after being harvested, and increase their respiration rate and synthesize ethylene; non-climacteric fruits once harvested do not continue with the ripening process [41]. Ethylene, as a growth regulator, is used to modify the ripening of many non-climacteric fruits, since it induces the destruction of chlorophylls and facilitates the appearance of pigments that give the typical color of these fruits. In general, it causes softening of the epicarp and pulp [41]. In climacteric fruits, it increases color, total soluble solids, odor, flavor, and sugars, and decreases firmness and acidity, among other quality attributes [42].

"Ethephon" or "Ethrel" (2-chloroethyl phosphonic acid) is a commercial chemical that is applied by spraying or dipping. Treated fruits release ethylene in response. In an investigation, 2-chloroethyl phosphonic acid (¨Ethrel¨) was applied at doses of 0, 50, 100, 200, and 400 mg/l to fruits of 2 phenotypes of pink and white Ilama (*Annona diversfolia Saff*) in post-harvest. The treatments were by immersion for 5 minutes and subsequent storage for 5 days at room temperature at an average of 17 °C. Total soluble solids, fruit mass (g), % malic acid, total soluble solids ratio, and total sugar concentration were evaluated. The results showed the responses of fruits subjected to different Ethrel concentrations: loss of mass, increase in soluble solids, decrease in malic acid, increase in Brix [42].

In saladette tomato (*Solanum lycopersicum* L.), pre-harvest application of Ethephon (Ethrel 240) has been used by Martinez-Damian [43], to determine the physicochemical quality of the fruit. The variables evaluated were color, weight, equatorial and polar diameter, roundness index, firmness, total soluble solids, titratable acidity, and lycopene concentration. The results of the individual application of Ethephon in combination with iodine and/or sodium selenite increased fruit weight, firmness, and citric acid with respect to the control.

Currently, there is a need to look for possible solutions that contribute to homogenize the ripening of coffee fruits, and for this Alvarado and Vera [44] worked on the proposal to evaluate growth regulators and sugar mobilizers in the ripening of Arabica coffee fruits. They applied 10 treatments with three replicates per treatment, and the phyto-regulators applied were Ethefon, Mepiquat Chloride, and the combination of both. The treatments consisted of the application of the regulators to pre-harvest fruit and foliage, the variables measured were the weight of ripe fruit per plant, % of ripe fruit per plant, % of empty fruit, fruit harvested per plant, and number of harvests. The results showed significant differences between the treatments in relation to the control.

3. Calcium application to fruits in pre- and post-harvest

Calcium is an essential nutrient for plant growth and development, participates in the formation and structuring of the cell wall, provides selective permeability and preserves cell membranes, and generates cell signaling responses [45]. Calcium moves in the plant by xylem, and in phloem is very little mobile and accumulates during the period of fruit growth and development [46, 47]. Calcium deficiency directly affects cell wall formation and development, and by causing the cell wall to lose its integrity, physiological disorders are induced by modifying extra- and intracellular processes, decreasing fruit firmness, and increasing senescence [48]. In post-harvest, calcium applied to fruit promotes tissue firmness and cell turgor, extending the shelf life of the fruit [49]; it also minimizes pathogen attack, reduces physiological disorders, and reduces mass loss by increasing the stability of the cell wall by presenting greater resistance to water leakage [50] (see **Table 1**).

The calcium application technique in pre-harvest consists of spraying the calcium solution directly to the fruits of the plant, the most commonly used calcium salts are calcium chloride and calcium nitrate, and in post-harvest the technique of immersing the fruits in a calcium solution for a previously established period of time is used; in this technique, the fruits must be in constant movement within the solution to avoid and prevent oxidation reactions that could generate alterations in the quality of the fruit. Different calcium salts are used, such as calcium lactate, calcium chloride, calcium phosphate, calcium propionate, and calcium gluconate. The factors to be considered for the application of this technique in post-harvest are pH of the solution, immersion time, temperature, and concentration [51].

Regarding the results obtained in different investigations with the application of calcium in pre- and post-harvest, the one is carried out on Logan cultivar blackberry, a very perishable fruit that preserves its post-harvest quality attributes for a maximum period of 24 hours. The research consisted of 2 experiments in which the treatments applied were obtained from the combination of 1000, 3000, and 5000 mg/l of $CaCl_2$ with 0.03% Twinn-20, 1% urea, and 1% starch plus a control treatment: In the first experiment, the application of the treatments was carried out in three periods, the first consisted of three applications of the treatments to the same fruit, at intervals of 10 days each, starting 25 days after flower opening (anthesis); in the second period, the treatments were applied twice to the same fruit at 35 and 45 days after anthesis; the third period consisted of a single application at 45 days after anthesis. In the second experiment, four epochs were considered, and in the first three epochs, the treatments were applied only once at 25, 35, and 45 days after anthesis: The fourth epoch consisted of three applications of the treatments mentioned above at 38 days after anthesis and two subsequent applications with an interval of 3 days each. The evaluation of the quality variables of 'Logan' blackberry fruit in post-harvest was carried out at 0, 24 and 48 hours after harvest. The results found indicate that the variables such as firmness, color, titratable acidity, ascorbic acid, brix/acid ratio, calcium content in pulp, and respiration showed significant statistical differences, while the variables of total soluble solids and % dry matter did not show significant statistical differences [52].

Among other research [53] where calcium was applied to apple fruit in pre-harvest, it was found that the application of calcium significantly reduced fruit rot in post-harvest for 6 months, likewise in calcium nitrate was applied to mango cv. Haden in pre-harvest, evaluating treatments of 0, 5, 10, 10, 15, and 29 g/l with 5 sprays of calcium nitrate foliar and to the fruit. The results indicate that there was a significant increase in the production of the trees that received treatment compared with the control treatment, and the mango fruits that were kept at a temperature of 20 ± 2°C and 74 ± 4% relative humidity presented significant differences in the following response variables: respiration, weight loss, acidity, and total soluble solids as for the variables that did not present significant statistical differences were: firmness, color, and enzymatic activity of pectin methyl esterase. Other studies showed significant inhibition of the presence of fungi and spores and consequently a decrease in fungal infections [54].

4. Conclusion

The application of phytohormones, growth regulators, and calcium to fruit in pre- and post-harvest is an important alternative to preserve the shelf life of fruit due to the diversity of physiological responses that they induce in the fruit (see **Table 1**).

Phytohormone, growth regulator, calcium	Physiological response in pre-harvest	Physiological response in post-harvest
Auxin	• Promotes fruit growth and development [11]. • Delays physiological maturity [9, 11]. • Increases mass and size [16].	• Preserves quality [14].
Gibberellin	• Induces parthenocarpy [22]. • Promotes growth and development [22]. • Promotes bud sprouting [26]. • Increases quality [27]. • Delays senescence in citrus [26]. • Advances production by reducing the period of physiological maturity [28]. • Prevents cracking [29, 30].	
Cytokinins	• Induce sprouting of axillary buds [32, 33]. • Induce fruit growth and development [32, 33]. • Increases yield [34, 35].	
Ethylene	• Increases quality [41]. • Increases mass and size [41]. • Increases color [41].	• Induces respiration in climacteric fruits [41]. • Destroys chlorophylls in non-climacteric fruit [41]. • Causes softening of epicarp and pulp in non-climacteric fruits [41]. • Promotes quality [42]. • Induces loss of mass and size [42]. • Decreases firmness [41]. • Decreases shelf life [41].
Calcium		• Promotes firmness [48]. • Preserves quality [48, 52, 53]. • Minimizes pathogen attack [50]. • Decreases physiological disorders [50]. • Reduces loss of mass [52]. • Decreases respiration [52, 53].

Table 1.
Physiological responses of fruit to the application of phytohormones, growth regulators, and calcium in pre- and post-harvest.

Author details

Ismael Aguilar-Ayala* and Diana Herrera-Rojas
Universidad Nacional Autónoma de México, Facultad de Estudios Superiores Iztacala, Mexico

*Address all correspondence to: biofrut94@gmail.com

References

[1] Kitinoja L, Saran SS, Roy SK, Kader AA. Postharvest technology for developing countries: Challenges and opportunities in research, outreach and advocacy. Journal of the Science of Food and Agriculture. 2011;**91**:597-603

[2] Ahmad MS, Siddiqui MW. Postharvest Quality Assurance of Fruits. Suiza: Springer; 2015

[3] Yang Q, Wang H, Zhang H, Zhang X, Apaliya MT, Zheng X, et al. Effect of *Varrowia lipolytica* decay of grape caused by *Talaromyces rugulosus* and the protein expression profile of *T. rugulosus*. Postcosecha Biology and Technology. 2017;**26**:15-22

[4] Anaya-Esparza LM, Pérez-Larios A, Ruvalcaba-Gómez JM, Sánchez-Burgos JA, Romero-Toledo R, Montalvo-González E. Funcionalización de los recubrimientos a base de quitosano para la conservación postcosecha de frutas y hortalizas. Revista especializada en Ciencias Químico-Biológicas. 2020;**23**:1-4

[5] Kitinoja L, Kader AA. Medición de las pérdidas poscosecha de frutas y hortalizas frescas en los países en desarrollo. Documento de la Fundación de Educación Poscosecha. 2015. Available from: http://postharvest.org/PEF_White_Paper_15-02_PHFVmeasurement.pdf

[6] Porat R, Lichter A, Terry LA, Harker R, Buzby J. Postharvest losses of fruits and vegetables during retail and in consumers homes: Quantifications, causes and means of prevention. Postharvest Biology and Technology. 2018;**139**:135-149

[7] Lurie S. Temperature management. In: Knee M, editor. Fruit Quality and Its Biological Basis. USA: Shefffield Academic Press; 2002. pp. 106-121

[8] Mir N, Beaudry R. Atmosphere control using oxygen and carbón dioxide. In: Knee M, editor. Fruit Quality and Its Biological Basis. USA: Shefffield Academic Press; 2002. pp. 121-153

[9] Jordán M, Casaretto J. Hormonas y reguladores del crecimiento: Auxinas, Giberelinas y Citocininas. In: Squeo FA, Cardemil L, editors. Fisiología vegetal. La serena Chile: Ediciones Universidad de la Serena; 2006. pp. 1-27

[10] Acosta EM, Sánchez BJ, Bañón AM. Auxinas. In: Azcón-Bieto J, Talón M, editors. Fundamentos de Fisiología Vegetal. 2nd Edición ed. Barcelona: McGraw-Hill–Interamericana; 2013. pp. 377-399

[11] Munguía RAG, Martínez TM. Las auxinas: síntesis, transporte y señalización. Biológicas. 2018;**20**(1):1-7

[12] Garay-Arroyo A, Sánchez MP, García-Ponce B, Álvarez-Buylla RE, Gutiérrez C. La homeostasis de las auxinas y su importancia en el desarrollo de *Arabidopsis thaliana*. Rebrape. 2014;**33**(1):13-22

[13] Alcántara CJS, Acero GJ, Alcántara CJD, Sánchez MRM. Principales reguladores hormonales y sus interacciones en el crecimiento vegetal. NOVA. 2019;**17**(32):109-129

[14] Solís-Guzmán CE, Aguilar-Ayala I, Molina GMG. Efecto del ácido indol-3-butírico sobre la calidad poscosecha de granada roja (*Punica granatum* L.) 'Apaseo tardía'. BIOCYT. 2019;**12**(46):855-868

[15] Bustamante GV, Gómez TD. Efecto de la aplicación de auxinas y calcio a las últimas manos del racimo del banano

para mejorar calibración y largo de dedos de la fruta. Economía y Desarrollo. 2019;**20**(3):48-53

[16] da Silva FC, Amaral SH, De MJO S, Carvalho TM, Martins RD, SAP S. Uso de ácido indol-3-ácetico (IAA) en postcosecha de cajá-manga (*Spondias dulcis*). Colombian Journal of Horticultural Science. 2020;**14**(2):201-208

[17] Paque S, Weijers D. Auxin: The plant molecule that influences almost anything. BMC Biology. 2016;**14**(67):1-5

[18] Tamura S. In: Takahasshi N, Phinney BO, Macmillan J, editors. Historical Aspects of Gibberelliins. New York: Springer-Verlag; 1990. pp. 1-8

[19] Malonek S, Bomke C, Bornberg-Bauer E, Rojas MC, Hedden P, Hopkins P, et al. Distribution of gibberellin biosynthetic genes and gibberellin production in the *Gibberella fujikuroi* species complex. Phytochemistry. 2005;**66**:1296-1311

[20] Domingo J, Giberelinas TM. Fundamentos de Fisiología Vegetal. 2nd ed. Barcelona: McGraw-Hill-Interamericana; 2013. pp. 399-417

[21] Hedden P, Kamilla Y. Gibberellin biosynthesis: Enzymes, genes and their regulation. Annual Review in Plant Physiology. 1997;**48**:431-460

[22] Yu H, Ito T, Zhao Y, Peng J, Kumar P, Meyerowitz EM. Floral homeotic genes are targets of gibberellin signaling in flower development. PNAS. 2004;**101**(20):7827-7832

[23] Kato K, Ohara H, Takahashi E, Matsui H, Nakayama M. Endogenous gibberellin-induced parthenocarpy in grape berries. Acta Horticulture. 2000;**514**:233-239

[24] Stuart NW, Cathey HM. Applied Aspects of gibberellins. Annual Review of Plant Physiology. 1961;**12**:369-394

[25] Gocal GF, Poole AT, Gubler F, Watts RJ, Blundell C, King RW. Long day up-regulation of a GAMYB gene during *Lolium temulentum* inflorescence formation. Plant Physiology. 1999;**119**:1271-1278

[26] King RW, Evans LT. Gibberellins and flowering of grasses and cereals: Prizing open the lid of the ¨Florigen¨ black box. Annual Review of Plant Biology. 2003;**54**:307-328

[27] García-Martínez JL, Hedden P. Gibberellins, and fruit development. In: Tomas-Barberan FA, Robins RJ, editors. Phytochemistry of Fruit and Vegetables. Oxford, UK.: Clarendon Press; 1997. pp. 263-285

[28] Rademacher W. Growth retardants: Effects on gibberellin biosynthesis and other metabolic pathways. Annual Review of Plant Physiology and Plant Molecular Biology. 2000;**50**:502-531

[29] Lozano-Trenado L. Efecto del ácido 2-cloroetil fosfónico (ethephon) y ácido giberélico en el desfasamiento de la producción y la calidad de la Ciruela Mexicana (*Spondias purpurea* L.). [Tesis de Licenciatura]. Universidad Nacional Autónoma de México. Facultad de Estudios Superiores Iztacala. 2008

[30] Escudero MYE. Efecto del acigigib sobre el agrietamiento y la calidad del fruto de la granada roja cv. Apaseo tardía (*Punica granatum*). [Tesis de licenciatura] Universidad Nacional Autónoma de México. Facultad de Estudios Superiores Iztacala. 2016

[31] Velázquez PMA. Evaluación del efecto del nitrato de potasio en la brotación de yemas y efecto del ácido giberélico en la calidad del fruto de granada roja (*Punica granatum* L.) 'Apaseo tardía'. [Tesis de Licenciatura]. Universidad Nacional Autónoma de México. Facultad de Estudios Superiores Iztacala. 2017

[32] Segura J. Fundamentos de Fisiología Vegetal. 2nd Edición ed. Barcelona: McGraw-Hill–Interamericana; 2013. pp. 421-441

[33] Díaz D. Las Hormonas Vegetales en las Plantas. México: Artículos Técnicos de INTAGRI; 2017. p. 2022

[34] Barboza IME. "Efecto de la combinación de diferentes dosis de citoquininas y giberelinas sobre el cuajado, retención y crecimiento de frutos en el cultivo de papaya (*Carica papaya* L.) en cieneguillo sur- sullana, 2016." [Tesis de Licenciatura]. Universidad Nacional de Piura. Facultad de Agronomía. Perú. 2018

[35] Cano CEG. "Efecto de aplicación de diferentes dosis de Agrocimax plus (citoquinina), sobre el rendimiento de fruta en arándano (*Vaccinium corimbosum* L.) variedad biloxi en la provincia de Huaylas-Ancash". [Tesis de Licenciatura]. Universidad Nacional Santiago Antúnez de Malayo. Perú. 2018

[36] Tamalá POS. Efecto de inductor de floración sobre la formación de frutos en el cultivo de guanábana (*Annona muricata*) en la comuna Bambil Collao, Santa Elena. [Tesis de Licenciatura]. Universidad Estatal Península de Santa Elena. Facultad de Ciencias Agraria. Carrera Agropecuaria. Ecuador. 2020

[37] Zacarías L. Fundamentos de Fisiología Vegetal. 2nd Edición ed. Barcelona: McGraw-Hill–Interamericana; 2013. pp. 445-453

[38] Balaguera-López HE, Salamanca-Gutiérrez FA, Camilo GJ, Herrera-Arévalo C. Etileno y retardantes de la maduración en la poscosecha de productos agrícolas. una revisión. Colombian Journal of Horticultural Science. 2014;**8**(2):302-313

[39] Depaepe VDS. Regulators of Growth: Ethylene. 2nd ed. pp. 403-410

[40] Watkins CB. Ethylene synthesis, mode of action, consequences and control. In: Knee M, editor. Fruit Quality and Its Biological Basis. USA: Shefffield Academic Press; 2002. pp. 181-224

[41] Martínez-González ME, Balois-Morales R, Alia-Tejacal I, Cortes-CruzMA,Palomino-HermosilloYA, López-Gúzman GG. Poscosecha de frutos: maduración y cambios bioquímicos. Rev. de Cien. Ağrı. 2017;**19**:4075-4087

[42] Piña-Soto L. Fenología de Ilama (*Annona diversifolia Saff.)* y efecto del ácido 2-cloroetil fosfónico en la maduración del fruto [Tesis de Licenciatura]. México D. F. Universidad Nacional Autónoma de México. Facultad de estudios Superiores Iztacala; 2006

[43] Martínez-Damián MT, Cano-Hernández R, Moreno-Pérez CEC, Sánchez-del Castillo F, Cruz-Álvarez O. Efecto de biorreguladores del crecimiento en precosecha sobre la calidad fisicoquímica de tomate saladette. Rev. Chapingo Serie Horticultura. 2019;**25**(1):29-43

[44] Alvarado INM, Vera VRS Efecto de fitorreguladores y movilizadores de azúcares sobre la maduración homogénea de frutos de café (*Coffea arabica L*). [Tesis de Licenciatura]. Escuela Superior Politécnica Agropecuaria de Manabí "Manuel Félix López" Carrera de Ingeniería Agrícola. Calceta, Ecuador. 2021

[45] Morteza SA, Hassanpouraghdam PMB, Farmani B. The language of calcium in postharvest life of fruits, vegetables and flowers. Scientia Horticulturae. 2012;**144**:102-115

[46] Rogiers SY, Keller M, Holzapfel BP, Virgona JM. Accumulation of potassium and calcium by ripening berries on field vines of *Vitis vinifera* (L) cv. Shiraz. Australian Journal of grape Wine Research. 2000;**6**:240-243

[47] Penter MG, Stassen PJ. The effect of pre-and postharvest calcium applications on the postharvest quality of Pinkenton avocados. South African Avocado Grower Association yerabppk. 2000;**23**:1-7

[48] Ciccarese A, Stellacci AM, Gentilesco G, Rubino P. Effectiveness of pre- and post-veraison calcium applications to control decay and maintain table grape fruit quality during storage. Biología y Tecnología de Postcosecha. 2013;**75**:135-141

[49] Angeletti P, Castagnasso H, Miceli E, Terminiello L, Concellon A, Chaves A, et al. Vicente. Effect of preharvest calcium applications on postharvest quality, softening and cell wall degradation of two blueberries (*Vaccinium corymbosum*) varieties. Biología y Tecnología de Postcosecha. 2010;**58**(2):98-103

[50] Martín-Diana R, Frías JM, Barat JM, GTM H, Barry-Ryan C. Calcio para extender la vida útil de todo fresco y frutas y verduras mínimamente procesadas: una revisión. Trends in Food Science & Technology. 2007;**18**(4):210-218

[51] Imre JH, Ball B, Vámos A, Gáll JM. Influence of Preharvest calcium applications fruit injury, and storage atmospheres on postharvest brown rot of Apple. Postharvest Biology and Technology. 2012;**67**:29-36

[52] Aguilar AI. Influencia del cloruro de calcio ($CaCL_2$) sobre la calidad de frutos de zarzamora 'logan'. [Tesis de Maestría] Montecillo, México Colegio de postgraduados. 1989

[53] Romero-Gomezcaña NR, Sanchez-Garcia P, Rodríguez-Alcazar J, Saucedo-Veloz C. Calcium foliar sprinkling and its relation with fruit quality of mangoes cv. Haden. Agric. Téc. Méx. 2006;**32**(1):5-15

[54] Elmer PAG, Spiers TM, Wood PN. Effects of pre-harvest foliar calcium spray on fruit calcium levels and Brown rot of peaches. Crop Protection. 2007;**26**:11-18

Section 2

Recent Developments in Handling Practices

Chapter 6

New Advances in Postharvest Technology: An Overview for Feed Production from Postharvest Wastes and By-Products

Kian Sadeghi, Farhad Parnian-khajehdizaj, Mahdi Ganjkhanlou, Reza Faraji and Zahra Abdollahi

Abstract

Globally agricultural production system generates a substantial proportion of postharvest waste that causes environmental pollution resulting in economic losses and human health-related problems. It is therefore important to make an assessment of this loss and turn it back to the consumption cycle. Processing and conversion of by-products, residues, and agricultural wastes and their reuse in the production cycle is a suitable solution for the economic use of these types of postharvest waste, especially in feeding livestock animals or in related industries. This chapter provides an overview of the assessment of the postharvest wastes that are generated in the field or on the farm at the time of harvest or processing industry. After introducing the potential use of technologies to upgrade postharvest waste for animal feed purposes and briefly discussing livestock performance, this review presents the latest and most interesting research on the use of postharvest wastes as feed.

Keywords: plant by-product, postharvest waste, technology, animal feed, livestock performance

1. Introduction

Livestock production is rapidly increasing due to population growth, changing lifestyles, and dietary habits in developed industrialized countries. It accounts for approximately 40% of the gross value of agricultural products. Projections indicate that total demand for animal products in developing countries will double by 2030 [1]. Considering this issue and with the aim of dealing with food insecurity and strengthening sustainable agriculture, it is possible to use feeding strategies and feedstuffs that are able to increase livestock productivity and have fewer environmental effects compared to conventional livestock production [2]. In addition to reducing the environmental impacts associated with animal feed production, valorizing plant by-products for feed formulation can maximize resource efficiencies and helps the competitiveness of feed manufacturers by making available of more sustainable raw materials that could reduce dependence on current raw materials [3]. Plant by-products (PBP) include a wide range of secondary residues produced from the industrial processing of plants into valuable commercial products [4]. These by-products are considered safe and widely accepted as animal feed [5]. These by-products include residues from food factories, fruit and vegetable wastes, and grain harvesting by-products. However, the use of plant waste as animal feed has limitations arising from the processes of agricultural products transformation industries, which can affect the possibility of evaluating its nutritional value. For example, its high water content, which is frequently greater than 80%, makes it more difficult and can hasten the spread of microbiological contamination [6]. On the other hand, the use of PBP in animal nutrition is limited due to restrictions such as diversity in nutrient composition and technical requirements for storage, which are necessary to stabilize the product and reduce seasonal availability of resources. Furthermore, preservation methods such as thermal processing can be expensive and diminish the environmental sustainability of PBP feed [5]. Also, the lack of efficient storage strategies is a fundamental barrier that limits the use of PBP in animal feeding, as their intrinsic instability causes quick quality deterioration and severe changes in nutritional composition [7]. This chapter provides information on some agricultural postharvest residues, their chemical composition, processing methods, nutritional value, and guidelines for including PBP in livestock diets. It also covers aspects related to the use of such post-harvest waste as a substrate for the production of value-added products. It is expected that this chapter will promote conversion of agricultural waste into valuable resources and help create opportunities for development. The recycling of these resources saves livestock feed and also reduces the environmental pollution associated with the disposal of PBP.

2. Grape pomace

Grapes are one of the most widely grown crops in the world, with an expected production of more than 79 million tons in 2018 [8]. Grape seeds have a significant amount of oil, which contains large amounts of unsaturated fatty acids, more than 80% of which include linoleic acid (**Figure 1**) [9].

Due to the presence of antioxidant substances such as flavonoids, grape pomace plays an important role in preventing the phenomenon of oxidation by removing free radicals produced from the environment under heat stress [10]. Proanthocyanidins of grape seed extract act as antioxidants in poultry feed, improve the performance of broiler chickens, and treat clinical symptoms caused by oxidative stress caused by Eimeria tenella infection [11]. Grape pomace can be used in animal nutrition, so that the use of its extract in amounts of 62 and 92 mg in the diet of broiler chickens from the age of 3 to 6 weeks prevented the lipid oxidation of the chicken carcasses during their storage in the refrigerator [12]. When grape pomace is fed to cattle as a dietary supplement, it can potentially increase the oxidative stability of beef products by increasing intestinal absorption and transfer of polyphenolic compounds to meat [13]. It has been reported that the use of grape pomace up to 10% of the ration of fattening lambs has no negative effects on the growth performance of lambs (**Figures 2** and **3**) [15].

Figure 1.
Grape plant.

Figure 2.
Grape pomace.

The effects of grape pomace on the quality and characteristics of fatty acids in meat samples obtained from lactating lambs were investigated in the ewe's diet. The results revealed that grape pomace incorporation did not have any negative effect on the carcasses but improved the water holding capacity [16]. Compared to diets lacking naturally occurring antioxidants, adding antioxidants such polyphenol-rich grape by-products to animal diets enhances meat quality while also preventing oxidation [17]. The appropriate use of grape pomace could increase the growth and production rate and reduce the feed conversion ratio (FCR) in Afshari fattening lambs [18]. Also, in another report, the use of grape pomace resulted in an increase in voluntary feed intake, growth rate, and a decrease in FCR in fattening lambs [19].

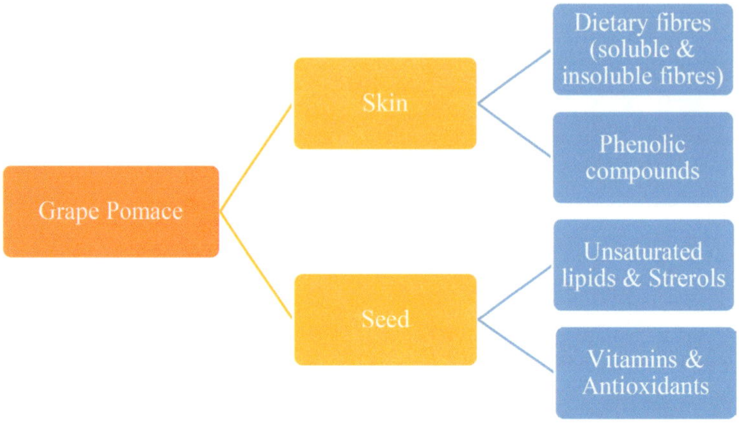

Figure 3.
Components of grape pomace [14].

Figure 4.
Pomegranate plant.

3. Pomegranate pomace

Pomegranate (*Punica granatum* L.) is a member of the Punicaceae family and an important native product of subtropical Asia (**Figure 4**) [20, 21].

Pomegranate fruit has approximately 48% peel and 52% fruit (w/w), which is the edible part of fruit and consists of 78% juice and 22% seeds [22, 23]. Pomegranate is rich in vitamins A, B, C, and E, minerals potassium, and iron, which have healing properties. It is also a rich source of folic acid and antioxidants. This fruit contains a significant amount of phenolic compounds such as anthocyanins, punicalin, and various flavonols, which have antimicrobial, antibacterial, and anti-inflammatory properties and increase the immune system in vitro and in vivo [24]. Pomegranate peel contains 55.4% non-fibrous carbohydrates (NFC), 8.4% crude protein (CP),

16.7% lignin, 34.5% neutral detergent fiber (NDF), and 0.84–1.0% total condensed tannins [25]. Pomegranate pomace is a by-product of the pomegranate juice industry, which has strong antioxidant power, anti-inflammatory compounds, vitamin E, sterols, phenols, and natural estrogens [26]. Pomegranate by-products offer excellent nutritional value as ruminant feed and can be utilized efficiently to substitute grains in ruminant diets. Feeding cattle calves and a fresh pomegranate peel diet improved feed intake and plasma alpha-tocopherol amount [27]. The antioxidants in pomegranate peel prevent diseases in lambs and have been useful in improving the quality of meat [28]. Substitution of a portion of grain in the diet with dry pomegranate seed pulp had no effect on growth performance, carcass traits, and nutrient digestibility, but it reduced the cost of meat production and increased the antioxidant capacity of lambs. As the pomegranate seed pulp was increased in the diet, it caused a decrease in kidney fat and a tendency to increase the apparent digestion of crude fat [29]. In fattening lambs, it has also been shown that pomegranate pomace silage can significantly replace a part of the fodder, which will result in a reduction in production costs and saving environment from the waste pollution coming from pomegranate processing industries. In another study, dairy goats fed a diet supplemented with dry pomace pomegranate seeds at 14% instead of cereal grain with no detrimental effects on animals. Therefore, it was suggested that pomegranate seed pomace, as a cheap by-product, can be replaced in the diet as an energy source [30]. Dairy cows fed diets supplementing with concentrated pomegranate extract at 1 and 2% based on dry matter, revealed increased the antioxidant activity of milk by 15 and 17.2%, respectively, and compared to the control group [31]. As a new feed for beef cattle, the antioxidant potential and nutrients of fresh and ensiled pomegranate by-products (pomegranate peel) were investigated. The results of this experiment showed that fresh pomegranate peel caused a significant increase in feed intake and alpha-tocopherol concentration in plasma [31]. Inclusion of wet fresh pomegranate peels in diets of bull calves promoted an increase in feed intake, with a tendency to increased weight gain [27]. In contrast, Oliveira et al. [32] found that feeding a pomegranate extract to young calves for the first 70 d of life did not change the digestibility of dry matter, organic matter, or starch, but it suppressed the intake of grain and whole tract digestibility of fat and crude protein, likely because of its high tannin content.

4. Tomato pomace

Tomato (*Lycopersicon esculentum*) has an annual global production of 170 million tons, of which 127.5 million tons are used for fresh consumption and 42.5 million tons for industrial processing [33] (**Figure 5**). Asia produces 61.1% of the world's tomatoes, whereas Europe, the Americas, and Africa provide 13.5, 13.4, and 11.8% of total tomato yield, respectively [34].

Tomato pomace is a by-product of tomato processing that refers to the skin (peel) and seeds of tomatoes and accounts for 10–40% of all processed tomatoes (**Figure 6**) [35]. Tomato pomace contains approximately 33% seeds, 27% peel, and 40% pulp, while dry pulp contains approximately 44% seeds and 56% peel and pulp [36] which average protein is 21.9% in tomato pomace and 38.7% in fat-free tomato seeds [37]. Tomato pulp is a good source of lycopene, carotene, vitamin E, vitamin C

Figure 5.
Tomato plant (fruit).

Figure 6.
Tomato pomace.

and nucleosides [38], carotenoids, lycopene, flavonoids, and soluble dietary fiber (**Figure** 7) [40, 41].

Tomato pomace is used as an ingredient in small ruminant diets due to its chemical composition and good animal acceptability [42]. In feeding ruminants, tomato pomace is sometimes considered as a concentrate due to the high content of nutrients and sometimes as forage because of its high content of the cell wall [43].

Recently, studies have been conducted using tomato residues in the form of silage and considering tomatoes along with other industrial residues in goat diets [44]. Silage of tomato pomace with 10% of wheat straw can be a good quality forage source for sheep when the forage is not available [45]. In a research, dietary replacing 10% of corn silage with ensiled tomato pomace had a positive effect on the vitamin content of

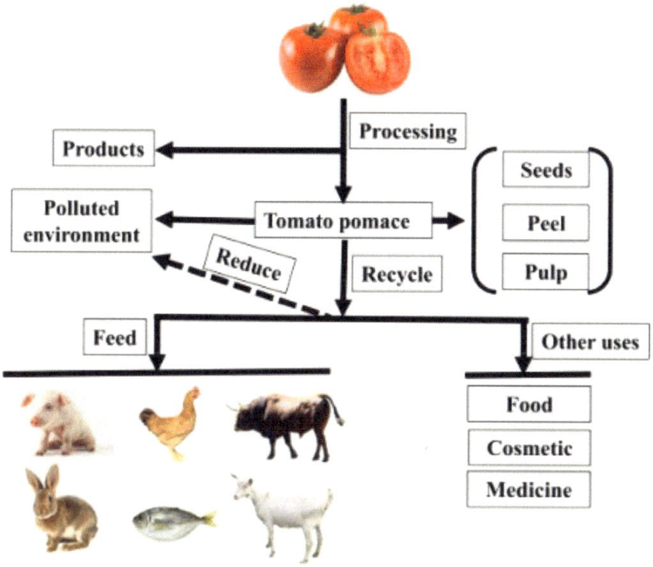

Figure 7.
Processing and uses of tomato pomace [39].

milk, antioxidant function, and immunity in early lactating cows [46]. Ruminants fed diets containing tomato pomace showed higher nutritional intake as well as apparent digestibility, organic matter, dry matter (DM), and crude protein indices [47]. Tomato pomace, if properly preserved, can be included in a significant portion of animal rations for a longer period of time and can also be used as a protein and energy supplement in feeding ruminants [48]. Up to 15% of the diets supplied to sheep can be substituted with dried tomato pomace without having any negative impact on growth [49]. The substitution of barley grain diet with tomato and cucumber waste was studied on rumen fermentation and microbial communities in goats and found that up to 250 grams per kilogram of tomato waste can replace the barley grain diet [50]. Supplement consumption increased milk quality and fat content by 20 and 40%, respectively. This demonstrated that, despite the fact that TP decreased the body weight of breastfeeding goats, it might enhance the quality and fat content of their milk, which may be related to the TP's own energy content and fatty acid composition [39]. In a study, supplementation of different levels of tomato pomace had no effect on the body weight of goats and sheep, carcass length, blood sugar, total protein, urea, or cholesterol [51].

Tomato pomace has also been used in poultry diets, so feeding 5% tomato pomace to chickens at the age of 1–28 days can increase body weight and production index, also increase the activity of superoxide dismutase (SOD) and glutathione peroxidase (GSH-Px), and reduce HDL cholesterol and serum triglyceride concentrations [52]. The results of another study revealed that dietary inclusion of tomato pomace at 12% can significantly improve the immune system function, antioxidant enzymes, and digestive enzymes of Japanese quails [53]. The concentrate mixture in the feed of male buffaloes can be substituted with sun-dried tomato pomace without having any negative effects on urinary purine derivatives, DM intake, nutritional digestibility, microbial protein synthesis, or production of volatile fatty acids in the rumen [54].

Figure 8.
Fodder beet plant.

5. Fodder beet pomace

Fodder beetroot belongs to the Chenopdiaceae family, which is native to the temperate region of Europe and was obtained from a cross between white and red garden beet [55] (**Figure 8**). Fodder beet is a plant with tuberous roots and broad leaves in some countries and is considered one of the most important winter fodder sources for feeding livestock, especially dairy cows [56].

Fodder beet has a high digestibility, so its organic matter digestibility is reported to be 87–90%. It has been reported that the net energy of lactation is 7.9 to 8.2 MJ/kg of dry matter, which is equal to or even more than cereal grains [57]. In an experiment with the substitution of 24% beet pomace with barley in diets based on straw fodder in Merino fattening lambs, it was shown that the consumption of concentrate and daily weight gain in the control diet (100% barley) was more than the diet containing beet pomace [58]. Also, dairy cows fed alfalfa or silage grass had lower milk urea and lower urea or creatinine concentrations in their urine when fodder beet was added to their diet [59]. During storage, beets continue to decrease in quality and are prone to spoilage and decay due to high humidity and sugar. However, in some countries, beets are ensiled with other forages to preserve their nutritional value for cattle [60].

6. Olive pomace

Olive (Olea europaea L.) is a small tree of the Oleaceae family, which is widely cultivated in the Mediterranean region. After extracting oil from olives, that is used in food preparation [61], a significant amount of olive pomace is obtained, which includes the peel, fruit, core, fleshy part of the fruit, and the shell of the wooden core (about 25% of the weight of the olive product) (**Figure 9**) [62].

Olive pomace, that is obtained by extracting oil from olive fruit, has a lot of fat and moisture, so with the drying process, its moisture content decreases [63]. Olive

Figure 9.
Olive plant.

Figure 10.
Olive pomace.

pomace also has chemical properties in terms of nutrition, with a high percentage of cellulose, hemicellulose, and lignin. It is also a good source of fatty acids, minerals, and phenolic compounds [64] (**Figure 10**). Despite having positive impacts, using olive waste as animal feed is not commercially viable because of concerns with digestion, taste, and safety [65].

Non-starch polysaccharides (NSP) are among the anti-nutritional factors found in olives that negatively affect the digestion and absorption of other nutrients. Soluble non-starch polysaccharides increase the viscosity of the contents of the gastrointestinal tract and reduce the digestibility of nutrients, while insoluble NSPs limit enzyme access to the substrate by trapping nutrients [66]. Olive pomace is less used by ruminants due to its high lignin, low crude protein, and poor digestibility compared to forage. However, inclusion of olive pomace to the diet had a positive effect on the production and fat percentage of cow and sheep milk [67]. Several studies have been carried out on the solid-state fermentation of olive pomace using fungal strains, which allows its use as a feed additive for ruminants (cows, sheep, goats, and camels) and poultry [68, 69].

Studies have shown that the use of olive pomace in the diet of laying hens increased egg weight and shell weight, but had no effect on other functional parameters and egg quality characteristics [70, 71]. In a research study, olive pulp was used at 2.5, 5, 7.5, and 10% in preliminary and final diets of broilers, and the results showed a significant reduction in feed intake and conversion rate, which could be due to the presence of inhibitory substances in olive pulp [72]. Because of their antibacterial effect against pathogenic intestinal bacteria, phenolic compounds derived from olive leaves may be useful to broilers [73]. Additionally, broilers' digestive enzymes that alter nutrient digestibility can be stimulated or inhibited by phenolic substances [74].

7. Apple pomace

Apple (Malus domestica Borkh) is one of the earliest known fruits and is commonly grown in temperate climates [75] (**Figure 11**). Apple pomace is a heterogeneous mixture consisting of skin, core, seed, calyx, stem, and soft tissue.

Apple pomace is rich in pectin, fermentable carbohydrates, minerals, and crude fiber, which increases its use for animal feed [76]. It also contains large amounts of sugar and is rich in various sources of carbon [77] (**Figure 12**).

Aside from its antioxidant characteristics, apple pomace also contains antibacterial, antiviral, and anti-inflammatory capabilities [78, 79]. **Figure 13** shows apple processing and apple pulp production. When fermented apple pomace is added to sheep diets, meat oxidation was decreased during storage without altering other aspects of meat quality [81]. Apple pomace dietary supplement improved milk production, apparent digestibility of crude protein and neutral detergent fiber, decreased rumen pH, and improved milk quality and serum biochemical parameters

Figure 11.
Apple plant (fruit).

Figure 12.
Apple pomace.

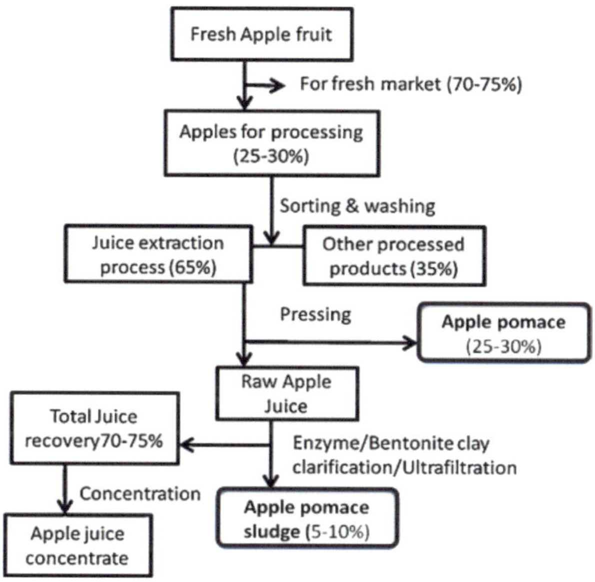

Figure 13.
Processing of apple and generation of apple pomace [80].

in Guanzhong dairy goats [82]. Feeding fermented apple pomace with basic rations to dairy cows leads to an average increase in milk production (1.90–1.89 kg per cow per day), milk fat, milk protein, and milk solid content, with a decrease in the incidence of disease in cows [83].

Moreover, the alfalfa forage in the ration of fattening lambs was replaced with ensiled apple pomace in proportions of 20, 40, and 60% and result of the experiment revealed that a diet containing 20% apple pomace silage increased the daily feed intake and increased the body weight of the lambs [84]. In another experiment, replacing of alfalfa in the diet of Arabian sheep with 30% dry apple pomace improved the activity of rumen microbes in digestion and fermentation of diet nutrients by reducing the duration of rumination and chewing time [85]. The reason for this was the probability of the cell wall and the size of the smaller particles in the diet containing apple pomace. In other words, it has been discovered that increasing the amount of cell wall in the diet or the size of the forage particles increases chewing activity [86]. However, apple pomace has disadvantages. The main concern of apple pomace is related to the environmental pollution caused by its waste accumulation, and it is not considered as a high-quality feed for animals due to its low protein content [87].

8. Fruit and vegetable waste

In recent years, the use of agricultural by-products in feeding animals has proven to be a successful solution for reducing feed costs, reducing environmental pollution, and ensuring the quick and inexpensive return of these materials to the nature cycle [88]. Nutrition is thought to be the most important determinant of increasing animal output. However, in agriculturally dependent nations, poor nutrition and pricey feedstuffs, particularly concentrate feed, are the main obstacles [89]. Dry citrus pomace is the best nutritional citrus product for livestock, and it is prepared for feeding all year round. Fresh citrus pomace has a moisture content of 85–88%, so adding moisture-absorbing materials in ensiling it can increase the quality of silage production [90]. The chemical and physical composition of citrus by-products is different depending on the type of fruit and the type of processing in the processing factories [91]. Vitamins, polyphenols (particularly anthocyanins), dietary fiber, and important unsaturated fatty acids are among the bioactive components found in fruit pulp [92]. Citrus pomace, tomato pomace, apple pomace, sugarcane pomace, and pistachio peel are among the by-products of agricultural transformation industries that are used as potential sources for animal feed [93]. Fruit and vegetable waste (FVW), waste from the food industry, vegetable industry, and general markets can be added to animal feed without adverse effects due to the presence of nutrients, minerals, fiber, vitamins, and bioactive compounds [94]. Considering that the cost of animal feed is increasing due to the increase in the cost of fertilizer and unsuitable climate for agriculture, therefore, food waste is an alternative source of feed ingredients. It can reduce feed and disposal costs and reduce environmental pollution [95]. Fruit and vegetable processing industries produce a large amount of waste, which is an excellent source of nutrients for livestock [7]. The use of FVW as an animal feed material has the potential to assist meet the increasing demand for animal protein as the world's population grows through 2050 [54]. In addition, transferring FVW to livestock feed can help sustain livestock production and reduce competition for land and water use [96]. In order to grow livestock populations and combat the feed crisis while lowering environmental risks and addressing the problems provided by diverse biophysical elements, the interaction between waste management and sustainable livestock feed production can be crucial [97]. Dairy cows' diets supplemented with 18% fruit and vegetable residues as part of the concentrate resulted in milk with a higher percentage of beneficial fatty acids without reducing daily milk production [98]. A mixture of waste juices from various fruits and vegetables, such as carrots, apples, mangoes, avocados, and oranges, can make up to 20 percent of a broiler's diet [99]. The main limitation of using agricultural waste and products from transformation industries as animal feed is the abundance of secondary compounds such as saponins, tannins, and essential oils in these products, which can limit the widespread use of these co-products in animal feed [100].

Carrot (*Ducus carota*) is one of the root vegetables that are damaged during harvest or discarded due to low quality, which can be a good feed for ruminants. Carrot pomace spoils very quickly due to high water activity. Drying carrot pomace is considered a suitable solution for maximizing the use of abundant nutrient sources in carrot pomace as well as increasing its shelf life. Carrot pomace is rich in insoluble fiber such as lignocellulose, which is a combination of pectin polysaccharides, hemicellulose, and cellulose, and these components have favorable physiological properties. Carrot pomace based on dry matter contains 2.7% crude protein, 24% insoluble fiber in neutral detergent, 15% insoluble fiber in acidic detergent, and 4.9% phenolic compounds, as well as carotenoids and soluble sugars such as sucrose, fructose, and xylose, which can be used in animal feeding [101].

Potato peel is one of the agricultural wastes that can be utilized as an alternative feed for animals due to its natural sources of energy and fiber with low protein levels [102]. Potato peel as a by-product of the food industry is a completely cheap, valuable, and cost-effective raw material for the production of economically important materials, added value, and product extraction, including biopolymers, natural antioxidants, dietary fiber, and natural food additives [103]. Potato peel contains polyphenols and various phenolic acids that are responsible for its antioxidant activities. Moreover, chemical composition of this by-product consists of 25% starch, 30% non-starch polysaccharide, 20% acid-soluble and acid-insoluble lignin, 18% protein, 6% ash, and 1% fat in dry form [104, 105]. When potato peel with high starch concentrations was fed, milk fat content was higher in dairy cows. It seems that the slow breakdown of starch from potato peels in the rumen can increase the higher transport of precursors of milk fat synthesis in the udder [106]. Also, starch, as one of the side products of potatoes, is the most abundant source of energy for the most livestock [107].

Another study was conducted on adult rams to determine the chemical composition, nutrient digestibility, and mineral content of potato compared to alfalfa as a forage in ruminants. The findings revealed that potato had much higher mineral, DM, and NDF digestibility than alfalfa. It can be stated that potato leaves are a nutritious alternative to other types of forages for ruminants because of their high nutritional value [108]. In one study, a significant increase in milk production was observed after supplementation with 6 kg of potato waste per day [109].

9. Waste and by-products of grain post-harvest

Plant residues are a post-harvest by-product, and the quantity harvested is directly related to all the factors that normally affect crop yield. Rice bran is a rice processing by-product which accounts for tons of food waste per year. In comparison with other grains, rice bran is rich in terms of nutrient density, amino acid, and fatty acid characteristics, including 74% unsaturated fatty acids and tocopherol content. Both protein and fat in rice bran have relatively high biological values [110]. Rice bran can be used up to 10% without any adverse effect on laying performance, digestive organs, and egg quality [111]. The use of fine-grain straw has also become common in the diet of dairy cows. There are three primary reasons why straw should be included in diets provided to dry and lactating dairy cows or dairy heifers [112]: 1) to reduce the density of nutrients (primarily energy) in the diet. For dairy heifer diets, straw is usually added to the diet to dilute the energy content. 2) For "drying" wet diets. Straw can be added to diets formulated with wet ingredients to increase the amount of dry matter in the ration and make it more suitable for dairy cows. 3) Changing the ratio of cation to anion in the diet of dry cows. Straws are often low in potassium, and low-potassium forages can help prevent milk fever in dairy cows [112].

Grain processing techniques are classified into two groups: physical and chemical processing. Physical processing, includes rolling, crimping, and grinding, breaks the outer tissues of the grain and provides access to rumen microorganisms and digestive enzymes. Chemical treatment with alkalis, for example, sodium hydroxide or ammonia, has a similar effect to rolling or crushing on access to rumen microbes and digestive enzymes [113]. Attempts have been made to replace concentrate feed in traditional animal diets with fermented wheat straw, especially to reduce the cost of animal feed. The use of wheat straw as an additive causes the least decomposition of dry matter compared to other additives [114]. Using symbiotic lignocellulose-decomposing bacteria from termite guts to process agricultural by products can improve their nutritional value by degrading lignin, a component resistant to rumen fermentation, and boosting plant cell wall digestibility [115]. For 6 weeks, lignocellulosic biomass from wheat straw (LBWS) and palm leaf (Phoenix dactylifera) (LBDL) were incubated with lignocellulose-degrading bacteria isolated from termite gut, which altered their chemical composition and boosted nutrient digestibility [116]. Barley straw has a better nutritional value than wheat straw, with an average of 90.9% dry matter, 3.8% crude protein, and 6 mega joules of metabolizable energy per kilogram of dry matter. However, it is high in lignocellulose and low in calcium and phosphorus. Ruminant animals can be fed with barley straw because rumen microorganisms can ferment the cell walls [117]. In addition to the physical and biological methods used in straw processing, chemicals can be used to break the interpolymeric bonds in the cell wall to release carbohydrates (such as hemicelluloses) that are readily fermented by ruminal microorganisms. The use of alkaline substances such as NaOH, KOH, $Ca(OH)_2$, wood ash, or urea is mainly associated with improved digestibility and may improve the value of low-quality feeds [118–120] Corn residue has been used for decades for grazing, bedding, or harvesting as supplemental feed for beef and dairy cattle [121]. Corn processing methods that economically increase digestibility and acceptability without adversely affecting rumen pH or disrupting digestive function include: (1) particle size reduction, which results in dry rolled or dry ground grain with or without moisture addition (tempering); 2) ensiling with inherent moisture before the grain has dried in the field to create high-moisture maize, or reconstituting maize before ensiling and/or feeding.; 3) steaming flaking [122] and microwave irradiation [118, 123, 124]. Sugarcane bagasse is a fibrous residue from the process of extracting water from sugarcane stalks that is available in large quantities and can be used as an alternative source of forage for ruminant feed. Due to the conversion of agricultural and industrial waste into animal feed, sugarcane bagasse is regarded as a tool for achieving value-added and ecologically responsible activities [125]. The effect of sugarcane bagasse on beef cattle has been investigated with the aim of maximizing their performance, and it was reported that sugarcane bagasse can be used as an exclusive source of forage for beef cattle [126]. However, it has low nutritional value and high indigestible fiber content as dry matter (DM), such as ether extract (0.31%), protein (2.67%), hemicellulose (51.5%), cellulose (54.61%), and lignin (14.29%), which leads to low digestibility (26.7%) and consequently poor animal performance [127]. The results of several studies have shown that the pistachio by-product has a high nutritional value and has good potential to be used in ruminant diets [128].

The pistachio by-product is high in non-fiber carbohydrates (36.40–9.4%), neutral detergent fiber (30.9%), and crude protein (11.4%) [129]. Feeding 21% pistachio by-products and palm waste silage to lambs increases their lean meat yield compared with control [130].

10. Food waste

Food waste can be used to replace some of the grains and vegetable protein sources used in animal feed, reducing food competition between humans and animals [131]. Food waste continues to be a global problem with negative environmental, economic, and social consequences [132]. Recently, FAO [34] defined food waste using two indicators: 1) Food that is lost through production or the supply chain before it reaches the retail level 2) Food that is then thrown out by consumers or retailers [133]. As "up-cyclers," livestock may turn inedible items into high-quality protein in the form of meat, eggs, and milk, decreasing food loss and waste [134].

The use of food waste in feeding animals has the potential to increase food security, reduce the environmental effects of the agro-food system, and reduce the costs of producing animal products [135]. **Figure 14** illustrates the significance of food waste and food loss as ecosystem services connected with animal production.

Some food waste can be used directly as livestock feed, while others must be processed further. Food waste, which primarily consists of rice, pasta, and vegetables, comprises a high percentage of volatile substances and a high moisture level of 74–90%. It is mainly composed of degradable carbohydrates (41–62%), fats (13–30%), and proteins (15–25%) [137, 138]. In different feeding experiments, the proportion of food waste used in diets varied from 10–100%. Animal weight increase and/or feed efficiency responses differed according to animal species and physiological stage, period of experimental feeding, and type of food waste [134].

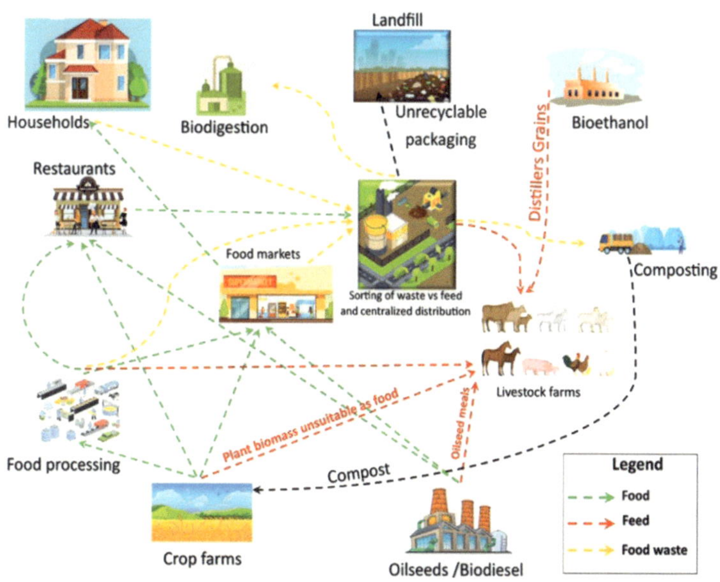

Figure 14.
Food waste is generated in food processing plants, restaurants, household, and food markets [136].

A variety of technologies for processing food waste have been documented, which can be divided into three groups: Wet-based, dry-based, and ensiling/fermentation treatments. Wet-based methods usually involve a simple heating step to sterilize the raw material, making it safe for animals. Wet-based feed products are high in moisture content (70–80%) with a relatively short storage life. For example, García et al. [6] sorted food waste out of municipal solid waste and heated it to 65–80°C for 10–60 min and then analyzed for nutrients, microbes, and toxins as potential feed. Westendorf et al. [139] heated food waste and food processing by-products at 100°C for 4 h to be used in pig feeding trials. Dry-based treatment combined with heating (sterilization) can produce long shelf life feeds (80–95% DM) that are easier to handle. Paek et al. [140] processed household food waste by rinsing, grinding, dewatering, and vacuum dehydration. Kim and Kim [141] described conversion of residential and restaurant food waste to dry feed by shredding and dewatering, heat-sterilizing, further dewatering, and drying. The ensiling/fermentation operation usually involves a heating sterilization process followed by the addition of prescribed microbial/yeast agents [134]. Procedures and conditions of ensiling/fermentation varied depending on individual studies. For example, Moon et al. [142] ground household food waste, heated it to 140°C, then aerobically fermented it for 24 h at 30–40°C with a probiotic microbial mix containing yeast, lactic acid bacteria, and *E. coli*. In another study, Kwak and Kang [143] aerobically fermented ground restaurant food waste with a microbial culture and poultry litter at 55–60°C for 4 h, then vacuum-dried it. Ensiling/fermentation treatment helps prolong the storage of the end product. For instance, Murray Martinez et al. [144] reported that feed produced from cafeteria food waste after fermentation was stable for up to 30 days. The primary or industrial processing of food intended for human and animal use has produced a significant amount of wastes that, despite their potential to cause pollution, have nutritional value and can be used to create monogastric meals [145].

11. Animal waste as a source of protein

Animal by-products from slaughtered animals that are not directly consumed by humans are commonly used as feed ingredients, for example, meat meal, bone meal, feather meal, blood meal, skin, slaughterhouse waste such as rumen content [146]. There are two sources of dietary protein, namely animal-based proteins and plant-based proteins. Plant proteins are usually low in lysine and methionine and have less biological value [147]. In the diet of broiler chickens, feather meal, fish meal, poultry by-products, and meat and bone meal are mainly used [148]. One of these slaughterhouse wastes is the rumen content, and it is considered as a potential alternative protein source [146]. The rumen content is relatively rich in crude protein and other microflora such as fungi, protozoa, and bacteria, so they are dried and crushed and mixed into animal and poultry diets. Using it as animal feed also increases the economic efficiency of slaughterhouse by-products [149].

A good supply of animal fat, protein, calcium, and readily available phosphorus is meat and bone meal. However, the results of laboratory tests from different rendering plants show a wide variation in crude protein (67.7–38.5%), ash (13–56.5%), crude fat (4.3–15.3%), and gross energy showed 9.4 22.3 MJ/kg [150]. Due to the diversity in the composition of raw materials and rendering processes, meat and bone meal probably has the high diversity in nutrient quality [151]. Poultry waste is high in minerals, total digestible nutrients (TDN), and protein (approximately 25% protein equivalent).

Feeding ruminants with poultry waste lowers feeding costs and lessens the impact of pollution on the environment in places where poultry farming occurs [152]. The use of chicken manure as a protein source in feeding ruminants not only reduces environmental problems, but can also replace part of the protein sources as a valuable food material and reduce the total price of the diet [153]. Dried poultry manure is utilized as ruminant feed and can greatly enhance dairy and meat production [154]. Poultry manure begins to breakdown quickly after disposal and emits ammonia, which in excessive amounts can harm the health and production of animals as well as farm employees' health [155]. Feather meal is rich in amino acids such as serine, proline, glycine, arginine, phenylalanine and threonine [156] which is considered a suitable protein source in the case of proper processing and can be used as a substitute for part of the protein sources in the diet, especially in monogastric animals [157]. Adding fully hydrolyzed feather meal the diet of lactating cows was evaluated to investigate its effects on whole-body protein digestibility and energy utilization. This experiment showed that fully

	DM	CP	NDF	ADF	EE	Ash	Ca	P	Reference
Soybean meal	86.51	50.19	15.13	9.98	1.78	6.82	—	—	[161]
Cotton seed meal	94.2	29.0	50.0	36.5	6.1	4.0	—	—	[162]
Guar meal	95.5	49.6	15.1	6.62	4.7	5.1	—	—	[163]
Sunflower meal	88.3	36.5	56.6	35.9	4.1	7.2	—	—	[164]
Olive meal	91.37	7.61	61.37	52.34	9.11	4.06	0.20	0.16	[165]
Canola meal	94.9	44.0	27.5	20.8	2.4	6.2	—	—	[162]
Tomato pomace	91.54	12.37	55.25	48.18	0.25	4.35	0.34	0.20	[166]
Apple pomace	30.70	5.60	45.30	38.0	4.70	2.60	0.11	0.12	[167]
Grape pomace	—	13.8	24.3	19.3	3.17	18.9	—	—	[168]
Lemon pomace	32.26	17.52	31.01	25.96	8.11	10.42	—	—	[169]
Pomegranate pomace	33.02	9.2	35.35	30.61	6.33	3.2	—	—	[170]
Sugarcane bagasse	96.1	2.18	85.5	60.0	0.84	2.70	—	—	[171]
Barley straw	93.1	4.4	77.3	—	—	7.7	0.45	0.24	[172]
Oat straw	93.3	4.8	77.0	—	—	8.8	0.32	0.22	[172]
Wheat straw	93.6	4.6	77.0	—	—	7.7	0.23	0.14	[172]
Rice bran	92.0	15.0	24.0	—	—	10.7	1.72	—	[172]
Wheat bran	84.82	17.58	49.83	15.13	5.02	6.31	—	—	[161]

Table 1.
Chemical compositions (%DM) of some agricultural post-harvesting by-products used as animal feed.

hydrolyzed feather meal in the diet of dairy cows can replace blood meal and even lead to more energy in the diet and the efficiency of energy use for milk production [158]. In the pre-starter (1 to 7 days old) and starter (8 to 21 days old) stages, broiler diets can contain up to 6% feather and blood meal (FBM) [159]. Feathers have a high crude protein content, mainly composed of keratins, simple proteins resistant to proteolytic enzymes in the animal's stomach and intestines [160]. **Table 1** provides chemical composition of some agricultural post-harvesting by-products.

Using 4% feather meal along with 4% poultry slaughterhouse waste powder in the diet of laying hens increased the feed conversion ratio compared to when the diet contained 5% feather meal and poultry slaughterhouse waste alone [173]. Fish meal is widely used to increase the content of DHA and n-3 unsaturated fatty acids in animal products such as chicken meat and eggs [174]. The traditional processing of fish meal, including cooking, pressing, drying, and grinding, is expensive and has a complex process, and the heat used to dry fish meal leads to a decrease in the digestibility of fish meal [175]. Supplementation with 100 grams of fish meal per cow per day resulted in the highest milk production compared to the control [176]. In another experiment, the use of fish meal at the level of 5% of the diet of cows with frequent estrus can improve the pregnancy condition and increase milk production [177]. Ruminal degradable protein supplement, especially fish powder, has a better effect than urea in increasing the flow of non-ammonia-N (NAN) from the rumen [178].

12. Conclusion

Animal feeding studies identified in the present chapter suggest that agricultural post-harvesting by-products are generally nutritious that can be converted into safe animal feed ingredients using advanced technologies. These by-products can be incorporated into animal diets without compromising animal health and performance. However, animal nutritionists, experts, and farm advisors have the opportunity to assist producers in lowering feed costs in certain regions of the country by incorporating agricultural post-harvest by-products into safe and effective rations. A basic understanding of the availability of the waste stream in certain regions and seasons, along with the knowledge of production and conversion, sanitary aspects in the storage of these by-products are necessary to ensure the absence of mold and bacterial contamination and pathogens.

Conflict of interest

The authors declare no conflict of interest.

Author details

Kian Sadeghi[1*], Farhad Parnian-khajehdizaj[2*], Mahdi Ganjkhanlou[1], Reza Faraji[1] and Zahra Abdollahi[1]

1 Department of Animal Science, Campus of Agriculture and Natural Resources, University of Tehran, Karaj, Alborz, Iran

2 Animal Science Research Department, East Azarbaijan Agricultural and Natural Resources Research and Education Center, AREEO, Tabriz, Iran

*Address all correspondence to: kian.sadeghi@ut.ac.ir and farhad.parnian@yahoo.com

References

[1] Bruinsma J, editor. World Agriculture: Towards 2015/2030: An FAO Perspective. London: Earthscan Publications Ltd; 2003. pp. 158-176

[2] FAO. Animal Nutrition in FAO and Sustainable Development Goals. Rome, Italy: FAO; 2017

[3] Myer RO, Brendemuhl JH, Johnson DD. Evaluation of dehydrated restaurant food waste products as feedstuffs for finishing pigs. Journal of Animal Science. 1999;**77**:685-692. DOI: 10.2527/1999.773685x

[4] Santana-Méridas O, González-Coloma A, Sánchez-Vioque R. Agricultural residues as a source of bioactivenaturalproducts.Phytochemistry Reviews. 2012;**11**:447-466. DOI: 10.1007/s11101-012-9266-0

[5] Bremer VR, Watson AK, Liska AJ, Erickson G, Cassman K, Hanford KJ, et al. Effect of distillers grains moisture and inclusion level in livestock diets on greenhouse gas emissions in the corn-ethanol-livestock life cycle. The Professional Animal Scientist. 2011;**27**:449-455. DOI: 10.15232/S1080-7446(15)30517-9

[6] García AJ, Esteban MB, Marquez MC, Ramos P. Biodegradable municipal solid waste: Characterization and potential use as animal feedstuffs. Waste Management. 2005;**25**:780-787. DOI: 10.1016/j.wasman.2005.01.006

[7] Wadhwa M, Bakshi SP. In: Makkar HPS, editor. Utilization of Fruit and Vegetable Wastes as Livestock Feed and as Substrates for Generation of Other Value-Added Products. Rome, Italy: FAO; 2013

[8] FAOSTAT. Food and Agriculture Organization of the United Nations. 2018. Available from: http://www.fao.org/faostat/en/#data/QC

[9] Lau DW, King AJ. Pre-and post-mortem use of grape seed extract in dark poultry meat to inhibit development of thiobarbituric acid reactive substances. Journal of Agricultural and Food Chemistry. 2003;**51**:1602-1607. DOI: 10.1021/jf020740m

[10] Brenes A, Viveros A, Goni I, Centeno C, Sáyago-Ayerdy S, Arija I, et al. Effect of grape pomace concentrate and vitamin E on digestibility of polyphenols and antioxidant activity in chickens. Poultry Science. 2008;**87**:307-316. DOI: 10.3382/ps.2007-00297

[11] Wang ML, Suo X, Gu JH, Zhang WW, Fang Q, Wang X. In fluence of grape seed proanthocyanidin extract in broiler chickens: Effect on chicken coccidiosis and antioxidant status. Poultry Science. 2008;**87**:2273-2280. DOI: 10.3382/ps.2008-00077

[12] Sáyago-Ayerdi S, Brenes A, Viveros A, Goñi I. Antioxidative effect of dietary grape pomace concentrate on lipid oxidation of chilled and long-term frozen stored chicken patties. Meat Science. 2009;**83**:528-533. DOI: 10.1016/j.meatsci.2009.06.038

[13] Sharma SD, Meeran SM, Katiyar SK. Dietary grape seed proanthocyanidins inhibit UVB-induced oxidative stress and activation of mitogen-activated protein kinases and nuclear factor-κB signaling in in vivo SKH-1 hairless mice. Molecular Cancer Therapeutics. 2007;**6**:995-1005. DOI: 10.1158/1535-7163.MCT-06-0661

[14] Kalli E, Lappa I, Bouchagier P, Tarantilis PA, Skotti E. Novel application and industrial exploitation of winery

by-products. Bioresources and Bioprocessing. 2018;**5**(1):1-21

[15] Bahrami Y, Chekani-Azar S. Some blood biochemical parameters and yield of lambs fed ration contained dried grape pomace. Global Veterinary. 2010;**4**:571-575

[16] Gómez-Cortés P, Guerra-Rivas C, Gallardo B, Lavin P, Mantecón A, De La Fuente M, et al. Grape pomace in ewes diet: Effects on meat quality and the fatty acid profile of their suckling lambs. Food Research International. 2018;**113**:36-42. DOI: 10.1016/j.foodres.2018.06.052

[17] Brenes A, Viveros A, Chamorro S, Arija I. Use of polyphenol-rich grape by-products in monogastric nutrition. A review. Animal Feed Science and Technology. 2016;**211**:1-17. DOI: 10.1016/j.anifeedsci.2015.09.016

[18] Safaei AR, Torbatinejad NM, Mansouri H, Zerehdaran S. Biological effects of grape pomace on performance parameters and blood metabolites of fattening lambs. Animal Science Journal. 2015;**28**:61-74. DOI: 10.22092/ASJ.2015.102338

[19] Baumgartel T, Kluth H, Epperelein K. A note on digestibility and energy value for sheep of different grape pomace. Small Ruminant Research. 2007;**67**:302-306. DOI: 10.1016/j.smallrumres.2005.11.002

[20] Gil MI, Tomas-Barberan FA, Hess-Pierce B, Holcroft DM, Kader AA. Antioxidant activity of pomegranate juice and its relationship with phenolic composition and processing. Journal of Agricultural and Food Chemistry. 2000;**48**:4581-4589. DOI: 10.1021/jf000404a

[21] Wadhwa M, Bakshi MPS. Utilization of fruit and vegetable wastes as livestock feed and as substrates for generation of other value-added products. Bangkok: Rap Publication; 2013;**4**:67

[22] Sarica S. Using possibilities of pomegranate juice by-products in animal nutrition. Gazi Osman Paşa Üniversitesi Ziraat Fakültesi Dergisi. 2011;**28**:97-101

[23] ZareiMM,ZeinolabedinBS.Evaluation of physicochemical characteristics of pomegranate (Punica granatum L.) fruit during ripening. Fruits. 2011;**66**:121-129. DOI: 10.1051/fruits/2011021

[24] Adams LS, Seeram NP, Aggarwal BB, Takada Y, Sand D, Heber D. Pomegranate juice, total pomegranate ellagitannins, and punicalagin suppress inflammatory cell signaling in colon cancer cells. Agricultural Food and Chemistry. 2006;**54**:980-985. DOI: 10.1021/jf052005r

[25] Kara K. Effect of dietary fibre and condensed tannins concentration from various fibrous feedstuffs on in vitro gas production kinetics with rabbit faecal inoculum. Journal of Animal and Feed Sciences. 2016;**25**:266-272. DOI: 10.22358/jafs/65563/2016

[26] Fadavi A, Barzegar M, Azizi MH. Determination of fatty acids and total lipid content in oilseed of 25 pomegranates varieties grown in Iran. Journal of Food Composition and Analysis. 2006;**19**:676-680. DOI: 10.1016/j.jfca.2004.09.002

[27] Shabtay A, Eitam H, Tadmor Y, Orlov A, Meir A, Weinberg P, et al. Nutritive and antioxidative potential of fresh and stored pomegranate industrial byproduct as a novel beef cattle feed. Journal of Agricultural and Food Chemistry. 2008;**56**:10063-10070. DOI: 10.1021/jf8016095

[28] Viuda-Martos M, Ruiz-Navajas Y, Fernández-López J, PérezÁlvarez JA. Effect of adding citrus waste water, thyme and oregano essential oil on the chemical, physical and sensory

characteristics of a bologna sausage. Innovative Food Science & Emerging Technologies. 2009;**10**:655-660. DOI: 10.1016/j.ifset.2009.06.001

[29] Emami A, Ganjkhanlou M, Fathi Nasri MH, Zali A, Rashidi L. Pomegranate seed pulp as a novel replacement of dietary cereal grains for kids. Small Ruminant Research. 2015;**123**:238-245. DOI: 10.1016/j.smallrumres.2014.12.001

[30] Modaresi SJ, Fathi Nasri MH, Rashidi L, Dayani O, Kebreab E. Short communication: Effects of supplementation with pomegranate seed pulp on concentrations of conjugated linoleic acid and punicic acid in goat milk. Journal of Dairy Science. 2011;**94**:4075-4080. DOI: 10.3168/jds.2010-4069

[31] Shabtay A, Nikbachat M, Zenou A, Yosef E, Arkin O, Sneer O, et al. Effects of adding a concentrated pomegranate extract to the ration of lactating cows on performance and udder health parameters. Animal Feed Science and Technology. 2012;**175**:24-32. DOI: 10.1016/j.anifeedsci.2012.04.004

[32] Oliveira RA, Narciso CD, Bisinotto RS, Perdomo MC, Ballou MA, Dreher M, et al. Effects of feeding polyphenols from pomegranate extract on health, growth, nutrient digestion, and immunocompetence of calves. Journal of Dairy Science. 2010;**9**:4280-4291. DOI: 0.3168/jds.2010-3314

[33] Coelho M, Pereira R, Rodrigues AS, Teixeira JA, Pintado ME. Extraction of tomato by-products' bioactive compounds using ohmic technology. Food and Bioproducts Processing. 2019;**117**:329-339. DOI: 10.1016/j.fbp.2019.08.005

[34] FAOSTAT. 2019. Available from: http://www.fao.org/faostat/en/#home

[35] Nagarajan J, Ramanan RN, Raghunandan ME, Galanakis CM, Krishnamurthy NP. Carotenoids. In: Galanakis CM, editor. Nutraceutical and Functional Food Components: Effects of Innovative Processing Techniques. Cambridge, Massachusetts: Academic Press; 2017. pp. 259-296. ISBN 9780128052570

[36] Poojary MM, Passamonti P. Extraction of lycopene from tomato processing waste: Kinetics and modelling. Food Chemistry. 2015;**173**:943-950. DOI: 10.1016/j.foodchem.2014.10.127

[37] Mildner-Szkudlarz S, Bajerska J, Górnaś P, Segliņa D, Pilarska A, Jesionowski T. Physical and bioactive properties of muffins enriched with raspberry and cranberry pomace powder: A promising application of fruit by-products rich in biocompounds. Plant Foods for Human Nutrition. 2016;**71**:165-173. DOI: 10.1007/s11130-016-0539-4

[38] Albanese D, Adiletta G, D'Acunto M, Cinquanta L, Di Matteo M. Tomato peel drying and carotenoids stability of the extracts. International Journal of Food Science and Technology. 2014;**49**:2458-2463. DOI: 10.1111/ijfs.12602

[39] Lu S, Chen S, Li H, Paengkoum S, Taethaisong N, Meethip W, et al. Sustainable valorization of tomato pomace (Lycopersicon esculentum) in animal nutrition: A review. Animals. 2022;**12**:3294. DOI: 10.3390/ani12233294

[40] Pinela J, Montoya C, Carvalho AM, Martins V, Rocha F, Barata AM, et al. Phenolic composition and antioxidant properties of ex-situ conserved tomato (Solanum lycopersicum L.) germplasm. Food Research International. 2019;**125**:108545. DOI: 10.1016/j.foodres.2019.108545

[41] Belovi CM, Torbica A, Lijakovic IP, Tomic J, Loncarevic I, Petrovic J. Tomato pomace powder as a raw material for ketchup production. Food. Bioscience. 2018;**26**:193-199. DOI: 10.1016/j.fbio.2018.10.013

[42] Álvarez S, Méndez P, Martínez-Fernández A. Fermentative and nutritive quality of banana by-product silage for goats. Journal of Applied Animal Research. 2015;**43**:396-401. DOI: 10.1080/09712119.2014.978782

[43] Yuangklang C, Vasupen K, Wongsuthavas S, Panyakaew P, Alhaidary A, Mohamed HE, et al. Growth performance in beef cattle fed rations containing dried tomato pomace. Journal of Animal and Veterinary Advances. 2010;**9**:2261-2264

[44] Pardo G, Martin-Garcia I, Arco A, Yañez-Ruiz DR, Moral R, Del Prado A. Greenhouse-gas mitigation potential of agro-industrial by-products in the diet of dairy goats in Spain: A life-cycle perspective. Animal Production Science. 2016;**56**:646-654. DOI: 10.1071/AN15620

[45] Denek N, Can A. Feeding value of wet tomato pomace ensiled with wheat straw and wheat grain for Awassi sheep. Small Ruminant Research. 2006;**65**:260-265. DOI: 10.1016/j.smallrumres.2005.06.024

[46] Tuoxunjiang H, Yimamu A, Li XQ, Maimaiti R, Wang YL. Effect of ensiled tomato pomace on performance and antioxidant status in the peripartum dairy cow. Journal of Animal and Feed Sciences. 2020;**29**:105-114. DOI: 10.22358/jafs/124049/2020

[47] So S, Cherdthong A, Uriyapongson S. Potential use of tomato pomace as ruminants diet – a review. Bioscience Research. 2019;**16**:1538-1548

[48] Ziaei N, Molaei S. Evaluation of nutrient digestibility of wet tomato pomace ensiled with wheat straw compared to alfalfa hay in Kermani sheep. Journal of Animal and Veterinary Advances. 2010;**9**:771-773

[49] Omar HAA, Abdel-Magid SS. Incorporation of dried tomato pomace in growing sheep rations. Global Veterinaria. 2015;**14**:1-16. DOI: 10.5829/idosi.gv.2015.14.01.9256

[50] Soto E, Khelil H, Carro M, Yanez-Ruiz D, Molina-Alcaide E. Use of tomato and cucumber waste fruits in goat diets: Effects on rumen fermentation and microbial communities in batch and continuous cultures. The Journal of Agricultural Science. 2015;**153**:343-352. DOI: 10.1017/S0021859614000380

[51] Abdullahzadeh F. The effect of tomato pomace on carcass traits, blood metabolites and fleece characteristic of growing Markhoz goat. Journal of Animal Science. 2012;**8**:848-852

[52] Hosseini-Vashan S, Golian A, Yaghobfar A. Growth, immune, antioxidant, and bone responses of heat stress-exposed broilers fed diets supplemented with tomato pomace. International Journal of Biometeorology. 2016;**60**:1183-1192. DOI: 10.1007/s00484-015-1112-9

[53] Reda FM, Madkour M, Abd El-Azeem N, Aboelazab O, Ahmed SY, Alagawany M. Tomato pomace as a nontraditional feedstuff: Productive and reproductive performance, digestive enzymes, blood metabolites, and the deposition of carotenoids into egg yolk in quail breeders. Poultry Science. 2022;**101**:101730. DOI: 10.1016/j.psj.2022.101730

[54] Bakshi MP, Kaur J, Wadhwa M. Nutritional evaluation of sun dried

tomato pomace as livestock feed. Indian Journal of Animal Nutrition. 2012;**29**:6-19

[55] McDonald P, Henderson AR, Heron SJ. The Biochemistry of Silage. Shedfield, UK: Chalcombe Publications; 1991. Available from: https://www.chalcombe.co.uk/contact.htm

[56] Solhjoo A, Amiri GZ. The effect of using different levels fodder beet silage on performance and carcass characteristics of gray Shirazi fattening lambs. Research on Animal Production (Scientific and Research). 2021;**2**:74-81. DOI: 20.1001.1.22518622.1400.12.33.3.8

[57] Donosă RE. The nutritive value of fodder beet in some dairy cow farms from Verşeni village (Iaşi County). Lucrări Ştiinţifice, Universitatea de Stiinte Agricole Şi Medicină Veterinară" Ion Ionescu de la Brad" Iaşi. Seria Agronomie. 2010;**53**:207-210

[58] Bodas R, Giráldez FJ, López S, Rodríguez AB, Mantecón AR. Inclusion of sugar beet pulp in cereal-based diets for fattening lambs. Small Ruminant Research. 2007;**71**:250-254. DOI: 10.1016/j.smallrumres.2006.07.006

[59] Eriksson T, Ciszuk P, Burstedt E. Proportions of potatoes and fodder beets selected by dairy cows and the effects of feed choice on nitrogen metabolism. Livestock Science. 2009;**126**:168-175. DOI: 10.1016/j.livsci.2009.06.018

[60] Schmidt S, Steingass H, Jungbluth T, Drochner W. Sugar beet mash silage as a component of a total-mixedration for dairy cows – Effects on parameters of digestion and animal performance. Archives of Animal Nutrition. 2001;**54**:47-59. DOI: 10.1080/17450390109381965

[61] Boskou D, Blekas G, Tsimidou M. History and characteristics of the olive tree. In: Olive Oil Chemistry and Technology. England: Taylor & Francis; 1996

[62] Dal Bosco A, Castellini C, Cardinali R, Mourvaki E, Moscati L, Battistacci L, et al. Olive cake dietary supplementation in rabbit: Immune and oxidative status. Italian Journal of Animal Science. 2007;**6**:713-715. DOI: 10.4081/ijas.2007.1s.761

[63] Al-Mughrabi KI, Shahrour W, Anfoka GH, Aburaj TA. Antifungal activity of olive cake extracts. Phytopathologia Mediterranea. 2001;**40**:1000-1005

[64] Serafini A, Tonetto GM. Production of fatty acid methyl esters from an olive oil industry waste. Brazilian Journal of Chemical Engineering. 2019;**36**:285-297. DOI: 10.1590/0104-6632.20190361s20170535

[65] Serra A, Conte G, Giovannetti M, Casarosa L, Agnolucci M, Ciucci F, et al. Olive pomace in diet limits lipid peroxidation of sausages from cinta senese swine. European Journal of Lipid Science and Technology. 2018;**120**:1700236

[66] Bedford MR. The use of NSPases for improving efficiency of nutrient extraction from corn for poultry. Poultry Bulletin. 2009;**11**:91-114

[67] Alcaide EM, Ruiz DY, Moumen A, Garcıa IM. Chemical composition and nitrogen availability for goats and sheep of some olive by-products. Small Ruminant Research. 2003;**49**:329-336. DOI: 10.1016/S0921-4488(03)00148-2

[68] Neifar M, Jaouani A, Ayari A, Abid O, Salem HB, Boudabous A, et al. Improving the nutritive value of olive cake by solid state cultivation of the medicinal mushroom Fomes

fomentarius. Chemosphere. 2013;**91**(1):110-114. DOI: 10.1016/j.chemosphere.2012.12.015

[69] Fadel M, El-Ghonemy DH. Biological fungal treatment of olive cake for better utilization in ruminants nutrition in Egypt. International Journal of Recycling of Organic Waste in Agriculture. 2015;**4**:261-271. DOI: 10.1007/s40093-015-0105-3

[70] Zangiabadi H, Torki M. The effect of a β-mannanase-based enzyme on growth performance and humoral immune response of broiler chickens fed diets containing graded levels of whole dates. Tropical Animal Health and Production. 2010;**42**:1209-1217. DOI: 10.1007/s11250-010-9550-1

[71] Taklimi SM, Ghahri H, Pour-Reza J, Fazaeli H, Lotfollahian H. Investigation into the possible use of olive pulp in commercial layer diets. British Poultry Science. 1999;**40**:40-41. DOI: 10.1080/00071669986765

[72] Rabayaa E, Abo Omar JM, Othman RA. Utilization of olive pulp in broiler rations. An-Najah University Journal for Research-Natural Sciences. 2001;**15**:133-144

[73] Sarica S, Urkmez D. The use of grape seed-, olive leaf-and pomegranate peel-extracts as alternative natural antimicrobial feed additives in broiler diets. European Journal of Poultry Science. 2016;**80**:1-13. DOI: 10.1399/ eps.2016.121

[74] Leskovec J, Levart A, Žgur S, Jordan D, Pirman T, Salobir J, et al. Effects of olive leaf and marigold extracts on the utilization of nutrients and on bone mineralization using two different oil sources in broilers. The Journal of Poultry Science. 2018;**55**:17-27. DOI: 10.2141/jpsa.0170059

[75] Musacchi S, Serra S. Apple fruit quality: Overview on pre-harvest factors. Scientia Horticulturae. 2018;**234**:409-430. DOI: 10.1016/j.scienta.2017.12.057

[76] Pirmohammadi R, Rouzbehan Y, Rezayazdi K, Zahedifar M. Chemical composition, digestibility and in situ degradability of dried and ensiled apple pomace and maize silage. Small Ruminant Research. 2006;**66**:150-155. DOI: 10.1016/j.smallrumres.2005.07.054

[77] Oreopoulou V, Tzia C. Utilization of plant by-products for the recovery of proteins, dietary fibers, antioxidants, and colorants. In: Utilization of by-Products and Treatment of Waste in the Food Industry. US: Springer; 2007. pp. 209-232. DOI: 10.1007/978-0-387-35766-9_11

[78] Yuan Y, Yue T, Bai X, Zhang H. Fractionation and anti-inflammatory effects of polyphenolenriched extracts from apple pomace. Bangladesh Journal of Pharmacology. 2012;7:28-32. DOI: 10.3329/bjp.v7i1.10194

[79] Guil-Guerrero JL, Ramos L, Moreno C, Zúñiga-Paredes JC, Carlosama-Yépez M, Ruales P. Plant-food by-products to improve farm-animal health. Animal Feed Science and Technology. 2016;**220**:121-135. DOI: 10.1016/j.anifeedsci.2016.07.016

[80] Beigh YA, Ganai AM, Ahmad HA. Utilisation of apple pomace as livetock feed: A review. Indian Journal of Small Ruminants (The). 2015;**21**:165-179. DOI: 10.5958/0973-9718.2015.00054.9

[81] Alarcon-Rojo AD, Lucero V, Carrillo-Lopez L, Janacua H. Use of apple pomace in animal feed as an antioxidant of meat. South African Journal of Animal Science. 2019;**49**:131-139. DOI: 10.4314/sajas.v49i1.15

[82] Xiong ML, Wu XJ, Zhu XF, Zhang WJ. Effects of different apple

pomace levels on lactation performance, nutrient apparent digestibility, serum biochemical indices and the rumen pH of Guanzhong dairy goats. Acta Prataculturae Sinica. 2021;**30**:81-88. DOI: 10.11686/cyxb2020162

[83] Diao QY, Tu Y, Gao F, Cao B, Zhang X, Xi X, et al. Effects of fermented apple residues on milk performance and immunity of dairy cows. China Dairy Cattle. 2003;**5**:21-24

[84] Kafilzadeh F, Taasoli G. Effects of dried and ensiled apple pomace from puree making on performance of finishing lambs. Pakistan Journal of Biological Sciences. 2008;**11**:294-297

[85] Boldaji ZT, Mohammadabadi T, Sari M, Chaji M. Study of the effect of dried apple pomace on digestibility, rumination microbiol fermentation characteristic and blood glucose and urea of Arabi sheep. Journal of Veterinary Research. 2016;**71**:255-262

[86] Allen DM, Grant RJ. Interactions between forage and wet corn gluten feed as sources of fiber in diets for lactating dairy cows. Journal of Dairy Science. 2000;**83**:322-331

[87] Suárez B, Álvarez ÁL, García YD, del Barrio G, Lobo AP, Parra F. Phenolic profiles, antioxidant activity and in vitro antiviral properties of apple pomace. Food Chemistry. 2010;**120**:339-342. DOI: 10.1016/j.foodchem.2009.09.073

[88] Vasta V, Nudda A, Cannas A, Lanza M, Priolo A. Alternative feed resources and their effects on the quality of meat and milk from small ruminants. Animal Feed Science and Technology. 2008;**147**:223-246. DOI: 10.1016/j.anifeedsci.2007.09.020

[89] Vasta V, Makkar HP, Mele M, Priolo A. Ruminal biohydrogenation as affected by tannins in vitro. The British Journal of Nutrition. 2008;**102**:82-92. DOI: 10.1017/S0007114508137898

[90] Migwi PK, Gallagher JR, Van Barneveld RJ. The nutritive value of citrus pulp ensiled with wheat straw and poultry litter for sheep. Australian Journal of Experimental Agriculture. 2001;**41**:1143-1148. DOI: 10.1071/EA00080

[91] Vatandoost M. Investigation of different levels of sugar beet pulp and citrus residue on productive performance of Baluchi lambs. Research on Animal Production (Scientific and Research). 2019;**10**:16-24. DOI: 10.29252/rap.10.25.16

[92] Tańska M, Roszkowska B, Czaplicki S, Borowska EJ, Bojarska J, Dąbrowska A. Effect of fruit pomace addition on shortbread cookies to improve their physical and nutritional values. Plant Foods for Human Nutrition. 2016;**71**:307-313. DOI: 10.1007/s11130-016-0561-6

[93] Besharati M, Palangi V, Nemati Z, Safari R, Salem AZM. Valorization of dietary lemon pomace waste to enhance lucerne silage composition and quality characteristics, and ruminal biogas production and fermentation. Research Square. 2021;1-12. DOI: 10.21203/rs.3.rs-392203/v1

[94] Conti C, Bacenetti J, Tedesco DE. Earthworms for feed production from vegetable waste: Environmental impact assessment. Environmental Engineering and Management Journal. 2019;**18**:2117-2122

[95] Rivin J, Miller Z, Matel O. Using Food Waste as Livestock Feed. United States: University of Wisconsin System Board of Regents and University of Wisconsin-Extension, Cooperative Extension; 2012

[96] Das NG, Huque KS, Amanullah SM, Makkar HP. Feeding of processed vegetable wastes to bulls and its potential environmental benefit. Animal Nutrition. 2019;**5**:87-94. DOI: 10.1016/j.aninu.2018.04.002

[97] Sahoo A, Chaturvedi OH, Thirumurgan P, Naqvi SMK. Cactus: Ensuring Round the Year Feed Supply. National Innovation in Climate Resilient Agriculture. Avikanagar: Central Sheep and Wool Research Institute; 2017. p. 26

[98] Angulo J, Mahecha L, Yepes SA, Yepes AM, Bustamante G, Jaramillo H, et al. Nutritional evaluation of fruit and vegetable waste as feedstuff for diets of lactating Holstein cows. Journal of Environmental Management. 2012;**95**:S210-S214. DOI: 10.1016/j.jenvman.2011.06.050

[99] Rizal Y, Mahata ME, Andriani M, Wu G. Utilization of juice wastes as corn replacement in the broiler diet. International Journal of Poultry Science. 2010;**9**:886-889

[100] Min BR, Attwood GT, Reilly K, Sun W, Peters JS, Barry TN, et al. Lotus corniculatus condensed tannins decrease in vivo populations of proteolytic bacteria and affect nitrogen metabolism in the rumen of sheep. Canadian Journal of Microbiology. 2002;**48**:911-921

[101] Nawirska A, Kwaśniewska M. Dietary fibre fractions from fruit and vegetable processing waste. Food Chemistry. 2005;**91**:221-225. DOI: 10.1016/j.foodchem.2003.10.005

[102] Ncobela CN, Kanengoni AT, Hlatini VA, Thomas RS, Chimonyo M. A review of the utility of potato by-products as a feed resource for smallholder pig production. Animal Feed Science and Technology. 2017;**227**:107-117. DOI: 10.1016/j.anifeedsci.2017.02.008

[103] Chiellini E, Cinelli P, Chiellini F, Imam SH. Environmentally degradable bio-based polymeric blends and composites. Macromolecular Bioscience. 2004;**4**:218-231. DOI: 10.1002/mabi.200300126

[104] Nelson M. Utilization and application of wet potato processing coproducts for finishing cattle. Journal of Animal Science. 2010;**88**:E133-E142. DOI: 10.2527/jas.2009-2502

[105] Liang S, McDonald AG, Coats ER. Lactic acid production with undefined mixed culture fermentation of potato peel waste. Waste Management. 2014;**34**:2022-2027. DOI: 10.1016/j.wasman.2014.07.009

[106] Jurjanz S, Colin-Schoellen O, Gardeur JN, Laurent F. Alteration of milk fat by variation in the source and amount of starch in a total mixed diet fed to dairy cows. Journal of Dairy Science. 1998;**81**:2924-2933

[107] Svihus B, Uhlen AK, Harstad OM. Effect of starch granule structure, associated components and processing on nutritive value of cereal starch: A review. Animal Feed Science and Technology. 2005;**122**:303-320. DOI: 10.1016/j.anifeedsci.2005.02.025

[108] Salehi SA, Lashkari SA, Abbasi RE, Kamangar H. Nutrient digestibility and chemical composition of potato (Solanum tuberosum L.) vine as alternative forage in ruminant diets. Agricultural communications. 2014;**2**:63-66

[109] Montoya NF, Pino ID, Correa HJ. Evaluación de la suplementación con papa (Solanum tuberosum) durante la lactancia en vacas Holstein. Revista

Colombiana de Ciencias Pecuarias. 2004;**17**:241-249

[110] Khan AD. Making rice bran a cereals alternative. Feed International June. 2004;June:18-19

[111] Samli HE. Using rice bran in laying hen diets. Journal of Central European Agriculture. 2006;7:137-140. DOI: 10.5513/jcea.v7i1.365

[112] Anderson T, Hoffman P. Nutrient composition of straw used in dairy cattle diets. Focus Forage. 2006;**1**:1-3

[113] Campling RC. Processing cereal grains for cattle—A review. Livestock Production Science. 1991;**28**:223-234. DOI: 10.1016/0301-6226(91)90144-F

[114] Kakkar VK, Dhanda S. Comparative evaluation of wheat and paddy straws for mushroom production and feeding residual straws to ruminants. Bioresource Technology. 1998;**66**:175-177. DOI: 10.1016/S0960-8524(97)00098-9

[115] Okano K, Ohkoshi N, Nishiyama A, Usagawa T, Kitagawa M. Improving the nutritive value of madake bamboo, Phyllostachys bambusoides, for ruminants by culturing with the white-rot fungus Ceriporiopsis subvermispora. Animal Feed Science and Technology. 2009;**152**:278-285. DOI: 10.1016/j.anifeedsci.2009.04.021

[116] Azizi-Shotorkhoft A, Mohammadabadi T, Motamedi H, Chaji M, Fazaeli H. Isolation and identification of termite gut symbiotic bacteria with lignocellulose-degrading potential, and their effects on the nutritive value for ruminants of some by-products. Animal Feed Science and Technology. 2016;**221**:234-242. DOI: 10.1016/j.anifeedsci.2016.04.016

[117] McDonald P, Edwards R, Greenhalgh J, Morgan C, Sinclair L, Wilkinson R. Animal Nutrition. 7th ed. London, England, UK: Longman Group UK Ltd.; 2010. p. 693

[118] Parnian F, Taghizadeh A. Evaluation of microwave irradiation effects on nutritive value of broom sorghum grain using an in vitro gas production technique. In: Proceedings of the British Society of Animal Science. Vol. 2009. Cambridge, United Kingdom: Cambridge University Press; 2009. pp. 182-182. DOI: 10.1017/S1752756200030210

[119] Adesogan AT, Arriola KG, Jiang Y, Oyebade A, Paula EM, Pech-Cervantes AA, et al. Symposium review: Technologies for improving fiber utilization. Journal of Dairy Science. 2019;**102**:5726-5755. DOI: 10.3168/jds.2018-15334

[120] Bachmann M, Martens SD, Le Brech Y, Kervern G, Bayreuther R, Steinhöfel O, et al. Physicochemical characterisation of barley straw treated with sodium hydroxide or urea and its digestibility and in vitro fermentability in ruminants. Scientific Reports. 2022;**12**:20530. DOI: 10.1038/s41598-022-24738-w

[121] Sindelar AJ, Coulter JA, Lamb JA, Vetsch JA. Agronomic responses of continuous corn to Stover, tillage, and nitrogen management. Agronomy Journal. 2013;**105**:1498-1506. DOI: 10.2134/agronj2013.0181

[122] Zinn RA, Barreras A, Corona L, Owens FN, Plascencia A. Comparative effects of processing methods on the feeding value of maize in feedlot cattle. Nutrition Research Reviews. 2011;**24**:183-190. DOI: 10.1017/S0954422411000096

[123] Sadeghi AA, Shawrang P. Effects of microwave irradiation on ruminal protein and starch degradation of corn grain. Animal Feed Science and

Technology. 2006;**127**:113-123. DOI: 10.1016/j.anifeedsci.2005.07.004

[124] Parnian-khajehdizaj F, Taghizadeh A, Nobari BB. Effect of feeding microwave irradiated sorghum grain on nutrient utilization, rumen fermentation and serum metabolites in sheep. Livestock Science. 2014;**167**:161-170. DOI: 10.1016/j.livsci.2014.06.004

[125] Gunun N, Wanapat M, Gunun P, Cherdthong A, Khejornsart P, Kang S. Effect of treating sugarcane bagasse with urea and calcium hydroxide on feed intake, digestibility, and rumen fermentation in beef cattle. Tropical Animal Health and Production. 2016;**48**:1123-1128. DOI: 10.1007/s11250-016-1061-2

[126] Leme PR, Silva SI, Pereira AS, Putrino SM, Lanna DP, Nogueira Filho JC. Sugarcane bagasse utilization in diets with high-proportion of concentrated for Nelore cattle in the feedlot. Revista Brasileira de Zootecnia. 2003;**32**:1786-1791

[127] So S, Cherdthong A, Wanapat M. Improving sugarcane bagasse quality as ruminant feed with lactobacillus, cellulase, and molasses. Journal of Animal Science and Technology. 2020;**62**:648. DOI: 10.5187/jast.2020.62.5.648

[128] Shakeri P, Riasi A, Alikhani M, Fazaeli H, Ghorbani GR. Effects of feeding pistachio by-products silage on growth performance, serum metabolites and urine characteristics in Holstein male calves. Journal of Animal Physiology and Animal Nutrition. 2013;**97**:1022-1029

[129] Ghaffari MH, Tahmasbi AM, Khorvash M, Naserian AA, Ghaffari AH, Valizadeh H. Effects of pistachio by-products in replacement of alfalfa hay on populations of rumen bacteria involved in biohydrogenation and fermentative parameters in the rumen of sheep. Journal of Animal Physiology and Animal Nutrition. 2014;**98**:578-586. DOI: 10.1111/jpn.12120

[130] SoltaniNezhad B, Dayani O, Khezri A, Tahmasbi R. Performance and carcass characteristics in fattening lambs feed diets with different levels of pistachio by-products silage with wasted date. Small Ruminant Research. 2016;**137**:177-182. DOI: 10.1016/j.smallrumres.2016.03.015

[131] Ferguson JD. Food residue, loss and waste as animal feed. In: Encyclopaedia of Renewable and Sustainable Materials. Vol. 5. Amsterdam, The Netherlands: Elsevier; 2020. pp. 395-407

[132] Spang ES, Moreno LC, Pace SA, Achmon Y, Donis-Gonzalez I, Gosliner WA, et al. Food loss and waste: Measurement, drivers, and solutions. Annual Review of Environment and Resources. 2019;**44**:117-156. DOI: 10.1146/annurev-environ-101718-033228

[133] FAO. Global Food Losses and Food Waste–Extent, Causes and Prevention. Rome (Italy): The Food and Agriculture Organization; 2011

[134] Dou Z, Toth JD, Westendorf ML. Food waste for livestock feeding: Feasibility, safety, and sustainability implications. Global Food Security. 2018;**17**:154-161. DOI: 10.1016/j.gfs.2017.12.003

[135] Georganas A, Giamouri E, Pappas AC, Papadomichelakis G, Galliou F, Manios T, et al. Bioactive compounds in food waste: A review on the transformation of food waste to animal feed. Food. 2020;**9**:291. DOI: 10.3390/foods9030291

[136] Ominski K, McAllister T, Stanford K, Mengistu G,

Kebebe EG, Omonijo F, et al. Utilization of by-products and food waste in livestock production systems: A Canadian perspective. Animal Frontiers. 2021;**11**:55-63. DOI: 10.1093/af/vfab004

[137] Fisgativa H, Tremier A, Dabert P. Characterizing the variability of food waste quality: A need for efficient valorisation through anaerobic digestion. Waste Management. 2016;**50**:264-274. DOI: 10.1016/j.wasman.2016.01.041

[138] Braguglia CM, Gallipoli A, Gianico A, Pagliaccia P. Anaerobic bioconversion of food waste into energy: A critical review. Bioresource Technology. 2018;**248**:237-256. DOI: 10.1016/j.biortech.2017.06.145

[139] Westendorf ML, Dong ZC, Schoknecht PA. Recycled cafeteria food waste as a feed for swine: Nutrient content, digestibility, growth, and meat quality. Journal of Animal Science. 1998;**76**:2976-2983. DOI: 10.2527/1998.76122976x

[140] Paek BH, Kang SW, Cho YM, Cho WM, Yang CJ, Yun SG. Effects of substituting concentrate with dried leftover food on growth and carcass characteristics of Hanwoo steers. Asian-Australasian Journal of Animal Sciences. 2005;**18**:209-213

[141] Kim MH, Kim JW. Comparison through a LCA evaluation analysis of food waste disposal options from the perspective of global warming and resource recovery. The Science of the Total Environment. 2010;**408**:3998-4006. DOI: 10.1016/j.scitotenv.2010.04.049

[142] Moon JS, Kwon IK, Chae BJ. Effects of wet feeding of diets with or without food waste on growth performance and carcass characteristics in finishing pigs. Asian-Australians. Journal of Animal Science. 2004;**17**:504-510

[143] Kwak WS, Kang JS. Effect of feeding food waste-broiler litter and bakery byproduct mixture to pigs. Bioresource Technology. 2006;**97**:243-249. DOI: 10.1016/j.biortech.2005.02.008

[144] Murray Martínez R, Cerezal Mezquita P, Bermúdez P, Bugueño MR. Use of food wastes for the production of lactic silage. Brazilian Archives of Biology and Technology. 2012;**5**:118-126. DOI: 10.1590/S1516-89132012000100015

[145] Alagawany M, El-Saadony MT, El-Rayes TK, Madkour M, Loschi AR, Di Cerbo A, et al. Evaluation of dried tomato pomace as a non-conventional feed: Its effect on growth, nutrients digestibility, digestive enzyme, blood chemistry and intestinal microbiota of growing quails. Food and Energy Security. 2022;**11**:e373. DOI: 10.1002/fes3.373

[146] Makinde O, Sonaiya B, Adeyeye S. Conversion of abattoir wastes into livestock feed: Chemical composition of sun-dried rumen content blood meal and its effect on performance of broiler chickens. International Journal of Poultry Science. 2008;**6**:875-882

[147] Ahmad MH, Miah MY, Ali MA, Hossain MA. Effect of different protein concentrates replacement of fish meal on the performance of broiler. International Journal of Poultry Science. 2006;**5**:959-963

[148] Hasni MS, Sahito HA, Memon MA, Sanjrani MI, Gopang MA, Soomro NA. Effect of feeding various levels of feather meal as a replacement of fish meal on the growth of broiler. International Journal of Agriculture Innovations and Research. 2014;**3**:505-511

[149] Esonu BO, Ogbonna UD, Anyanwu GA, Emenalom OO, Uchegbu MC, Etuk EB, et al. Evaluation of performance, organ characteristics and economic analysis of broiler finisher fed dried rumen digesta. International Journal of Poultry Science. 2006;**5**:1116-1118

[150] Ravindran V, Hendriks WH, Camden BJ, Thomas DV, Morel PC, Butts CA. Amino acid digestibility of meat and bone meals for broiler chickens. Australian Journal of Agricultural Research. 2002;**53**:1257-1264. DOI: 10.1071/AR02055

[151] Davis TM, Parsons CM, Utterback PL, Kirstein D. Evaluation of the pepsin digestibility assay for predicting amino acid digestibility of meat and bone meals. Poultry Science. 2015;**94**:1003-1008

[152] Saleh HM, Elwan KM, El-Fouly HA, Ibrahim II Salama AM, Elashry MA. The use of poultry waste as a dietary supplement for ruminants. Egyptian Journal of Nutrition and Feeds. 2002;**3**:1-8

[153] Talib NH, Ahmed FA. Performance and carcass characteristics of intact zebu bulls fed different levels of deep stacked poultry litter. Journal of Animal and Veterinary Advances. 2008;**7**:1467-1473

[154] Alam MS, Khan MJ, Akbar MA, Kamruzzaman M. Broiler litter and layer manure in the diet of growing bull calves. Bangladesh Veterinarian. 2008;**25**:62-67. DOI: 10.3329/bvet.v25i2.4619

[155] Zhang W, Lau A. Reducing ammonia emission from poultry manure composting via struvite formation. Journal of Chemical Technology and Biotechnology: International Research in Process, Environmental & Clean Technology. 2007;**82**:598-602. DOI: 10.1002/jctb.1701

[156] Pfeuti G, Osborne V, Shoveller AK, Ignatz EH, Bureau DP. Development of a novel enzymatic pretreatment for improving the digestibility of protein in feather meal. Agricultural Engineering. 2019;**1**:475-484. DOI: 10.3390/agriengineering1040034

[157] Sarmwatanakul A, Bamrongtum B. Aquarium Fish Nutrition. Bangkok, Thailand: Ornament Fish Research and Public Aquarium; 2000. p. 1

[158] Morris DL, Judy JV, Kononoff PJ. Use of indirect calorimetry to evaluate utilization of energy in lactating Jersey dairy cattle consuming diets with increasing inclusion of hydrolyzed feather meal. Journal of Dairy Science. 2020;**103**:4206-4217. DOI: 10.3168/jds.2019-17762

[159] Xavier SA, Stringhini JH, Brito AB, Andrade MA, Café MB, Leandro NS. Feather and blood meal in pre-starter and starter diets for broilers. Revista Brasileira de Zootecnia. 2011;**40**:1745-1752. DOI: 10.1590/S1516- 35982011000800018

[160] da Silva Scapim MR, Loures EG, Rostagno H, Cecon PR, Scapim CR. Avaliação nutricional da farinha de penas e de sangue para frangos de corte submetida a diferentes tratamentos térmicos. Acta Scientiarum. Animal Sciences. 2003;**25**:91-98. DOI: 10.4025/actascianimsci.v25i1.2104

[161] Fang J, Xia G, Cao Y. Effects of replacing commercial material with apple pomace on the fermentation quality of total mixed ration silage and its digestibility, nitrogen balance and rumen fermentation in wethers. Grassland Science. 2020;**66**:124-131

[162] Maesoomi SM, Ghorbani GR, Alikhani M, Nikkhah A. Canola meal as a substitute for cottonseed meal in diet of midlactation Holsteins. Journal of Dairy

Science. 2006;**89**:1673-1677. DOI: 10.3168/jds.S0022-0302(06)72234-2

[163] Narimani S, Taghizadeh A, Sis NM, Parnian F, Nobari BB. Effects of compound treatment of exogenous feed enzymes and microwave irradiation on in vitro ruminal fermentation and intestinal digestion of guar meal. The Indian Journal of Animal Sciences. 2014;**84**:436-341

[164] Alcaide EM, Ruiz DY, Moumen A, Garcıa AM. Ruminal degradability and in vitro intestinal digestibility of sunflower meal and in vitro digestibility of olive by-products supplemented with urea or sunflower meal: Comparison between goats and sheep. Animal Feed Science and Technology. 2003;**110**:3-15. DOI: 10.1016/j.anifeedsci.2003.08.002

[165] Ashraf A, Sharma RK, Rastogi A. Effect of lime treatment of olive meal on in vitro utilization of total mixed ration containing olive meal as partial maize replacer. Veterinary World. 2013;**6**:440-443. DOI: 10.5455/vetworld.2013.440-443

[166] Yang P, Fan Y, Zhu M, Yang Y, Ma Y. Energy content, nutrient digestibility coefficient, growth performance and serum parameters of pigs fed diets containing tomato pomace. Journal of Applied Animal Research. 2018;**46**:1483-1489. DOI: 10.1080/09712119.2018.1546181

[167] Abdollahzadeh F, Pirmohammadi R, Fatehi F, Bernousi I. Effect of feeding ensiled mixed tomato and apple pomace on performance of Holstein dairy cows. Slovak Journal of Animal Science. 2010;**43**:31-35

[168] Guerra-Rivas C, Gallardo B, Mantecón ÁR, del Álamo-Sanza M, Manso T. Evaluation of grape pomace from red wine by-product as feed for sheep. Journal of the Science of Food and Agriculture. 2017;**97**:1885-1893. DOI: 10.1002/jsfa.7991

[169] Besharati M, Palangi V, Salem AZ, De Palo P, Lorenzo JM, Maggiolino A. Substitution of raw lucerne with raw citrus lemon by-product in silage: In vitro apparent digestibility and gas production. Frontiers in Veterinary Science. 2022;**9**:1672. DOI: 10.3389/fvets.2022.1006581

[170] Kara K, Guclu BK, Baytok E, Aktug E, Oguz FK, Kamalak A. Atalay AI. Investigation in terms of digestive values, silages quality and nutrient content of the using pomegranate pomace in the ensiling of apple pomace with high moisture contents. Journal of Applied Animal Research 2018; 46: 1233-1241.

[171] Ahmed MH, Babiker SA, Fadel Elseed AEMA, Mohammed AM. Effect of urea-treatment on nutritive value of sugarcane bagasse. ARPN Journal of Science and Technology. 2013;**3**:834-838

[172] Shaver R, Hoffman P. Use of straw in dairy cattle diets. Focus on Forage. 2010;**12**(2)

[173] Senkoylu N, Samli HE, Akyurek H, Agma A, Yasar S. Performance and egg characteristics of laying hens fed diets incorporated with poultry by-product and feather meals. Journal of Applied Poultry Research. 2005;**14**:542-247. DOI: 10.1093/japr/14.3.542

[174] Howe PR, Downing JA, Grenyer BF, Grigonis-Deane EM, Bryden WL. Tuna fishmeal as a source of DHA for n– 3 PUFA enrichment of pork, chicken, and eggs. Lipids 2002; 37: 1067-1076. DOI: 10.1007/s11745-002-1002-3

[175] Faid M, Zouiten A, Elmarrakchi A, Achkari-Begdouri A. Biotransformation of fish waste into a stable feed ingredient.

Food Chemistry. 1997;**60**:13-18. DOI: 10.1016/S0308-8146(96)00291-9

[176] Asghedom G, Austbo D, Kjos NP. Effect of fishmeal supplmementation on milk yield of Holstein-Friesian cows in Eritrea. Tanzania Journal of Agricultural Sciences. 2007;**8**:39-50

[177] Kamali Dehkordi F, Amanlou H, Sadeghi BN. The effect of replacement soybean meal with fish meal on milk production, milk composition and reproductive performance in repeat breeder Holstein cows. Iranian Journal of animal Science. 2022;**52**:271-283. DOI: 10.22059/ijas.2021.329251.653844

[178] Brito AF, Broderick GA, Reynal SM. Effects of different protein supplements on omasal nutrient flow and microbial protein synthesis in lactating dairy cows. Journal of Dairy Science. 2007;**90**:1828-1841. DOI: 10.3168/jds.2006-559

Chapter 7

Pre-Harvest and Postharvest Factors Affecting Quality and Shelf Life of Harvested Produce

Oluyinka Benedicta Adewoyin

Abstract

Food security and access to quality food are major challenges in the efforts against global hunger. Despite producing a large amount of food each year to boost the economy, a significant portion is lost due to pre-harvest and postharvest factors affecting produce's quality and shelf life. Numerous interventions have been implemented to address this to improve postharvest management, but there is still an urgent need to identify and manage the various factors contributing to postharvest losses. Factors contributing to postharvest losses include agents of food deterioration inherent in the produce before harvesting, inappropriate cultural practices, genetic composition, harvesting methods, quality of water for irrigation, microbial invasion, insect pest inoculum remnants and more. Postharvest handling involves interactive activities from harvest to consumer's final decision to eat or reject the food. Produce quality is determined by local conditions, policies, stakeholders' cultural practices, market demand, road condition, handling methods, packaging materials, transportation methods and level of knowledge and awareness in that environment. This study is to elucidate, through literature, pre-harvest and postharvest factors affecting quality of harvested produce. This study showed that understanding and appropriate management of pre-harvest and postharvest factors would reduce quality losses and increase the shelf life of produce.

Keywords: pre-harvest factors, postharvest factors, quality, shelf life, harvested produce

1. Introduction

One of the most urgent and essential needs of today is ensuring food security for the rapidly growing world population and, at the same time, ensuring long-term sustainable development in the reduction of food losses. Postharvest losses significantly increase food insecurity, reduce farmer's income and enhance inefficiency in the global food system. The essential elements of postharvest losses challenge include problem of multiple points of intervention, multiple technologies, complex value chain, and poorly developed food systems [1]. In accordance with projections by FAO, food production will need to grow by 70% to feed the world population, which will reach over 10 billion by the year 2050. As efforts are being geared towards increased production, there must be corresponding efforts for an integrated and innovative approach to the global efforts to ensure sustainable food production, consumption and loss reduction [2]. One major way of strengthening food security is by reducing losses. Postharvest loss can be defined as degradation in both quantity and quality of food from harvest to consumption. Reduction in these losses would increase the amount of food available for human consumption and enhance global food security. Food losses occur due to poor infrastructure, logistics issues, lack of technology, lack of prompt access to markets, insufficient skills, and inadequate knowledge and management capacity of supply chain actors. Losses also occur at the production, postharvest and processing stages in the food supply chain [3, 4]. Food waste refers to food appropriate for human consumption being discarded along the food chain due to consumers' behaviour [5, 6]. Damage restricts the use of a product, whereas loss makes its use impossible. These losses occur because harvested agricultural produce consists of living tissues that respire and undergo physiological changes caused by conditions such as high temperature, low atmospheric humidity, physical injury, biotic contamination and enzyme actions. Food losses reduce the food available for human consumption and incur costs of waste management; loss of scarce resources used in crop production generates about 6–10 per cent of human-greenhouse gas emissions in the land where food wastes decompose anaerobically [1]. Pre-harvest refers to every activity embarked on by the producer in the production of crops before harvest, and this includes site selection, land preparations, appropriate planting date, optimum seed rate, recommended spacing, appropriate tools and equipment used, proper tillage activities and seedbed preparation, pests, disease and weed management, irrigation, mulching, staking and use of hormones. An adequate supply of potassium nutrition in tomato production enhances titratable acidity and fruit colour quality and reduces the incidence of the yellow shoulder [7, 8], while the inadequate application of potassium in aqua-phonic tomato production results in ripening disorders [9]. An increase in nitrogen supply to tomatoes grown in a controlled environment may reduce fruit quality by decreasing the sugar content of the fruits [10]. A high nitrogen supply of about 250 kg/ha can impair some important quality traits of tomato fruits, such as total soluble solids [11], glucose, fructose, and pH [12]. The addition of ammonium in tomato production results in improved fruit flavours [13]. The quality of tomato fruit is also affected by the amount of boron used. Lower amounts of boron supply reduce fruit firmness [14]. The compositional quality of harvested produce is affected by maturity stage; Howard [15] observed that total vitamin C content of red pepper was about 30% higher compared to green pepper. Tomato fruits harvested green at table ripeness contain less vitamin C than those harvested at the full ripe stage. Tomato fruits at the 'breaker' stage contained only 69% of their vitamin C concentration. Quality refers to the state of excellence of a produce, which may be either good or bad. It refers to a property or group of properties that make a produce acceptable or desired by a consumer. It is subjective and changes according to culture, customs, environment, social status and mindset. These parameters change from one food commodity to the other. Several attributes have been used to describe quality: size, shape, colour, consistency, flavour or organoleptic properties like texture, smell and tastes. Other properties that are used to measure qualities include appearance as presentation, nutritive value, dependability and wholesomeness. Higher quality will translate to higher prices and more consumers' satisfaction. Quality can be described based on produce usages, such as edible quality, dessert quality, shipping quality, table quality, nutritional quality, internal quality and appearance quality. The quality of harvested produce is a combination of characteristics, attributes and properties that give the produce value to humans for food. Quality standards are usually sets to essentially meet specification and demand of the consumer. Every change in food that causes it to lose its desired quality and eventually become unpalatable is called food spoilage or rot. Food spoilage is also a metabolic process that causes foods to be undesirable or unacceptable for human consumption due to changes in sensory characteristics [16]. Quality losses include those that affect the nutrient composition, acceptability, and edibility of a given product, while quantity losses refer to those that result in the loss of the amount of a product [3]. Reducing food losses requires an understanding of pre-harvest and postharvest factors resulting in loses, cultural background and economic level of the people involved. It is essential because all food losses occur at a particular socio-cultural environment. Reduction of postharvest food losses is critical in ensuring future global food security. The issue of food losses is of high importance in the efforts to combat hunger, raise income and improve food security in the world today. It is very important to know the pattern and scale of these losses across the world, especially in developing countries and identify their causes and possible solutions.

2. Quality attributes

Appearance: This is the evaluation of quality by sense of sight, shape, wholesomeness and pattern. It includes quality traits such as:

Shape: The shape should not deviate from the typical accepted standard set for that produce. It must not have any damage or bruise, for example, cucumbers are supposed to be elongated in shape and also robust, not sickle or twisted in shape.

Size: The size of some harvested produce determines their market acceptability. Examples of such are plumb, robust banana finger, carrot, okra, plantain are more acceptable than a slim one; also in cauliflower, size and compactness of the head are quality parameters that determine market acceptability. Matured plants are harvested at about 15 cm in diameter, and protruding floral parts indicate over maturity.

Colour: Colour of harvested produce like fruits and vegetables is expressed through their various pigments and can be grouped into red: anthocyanin pigments, green: chlorophyll and yellow to orange: carotenoids. There are many factors that influence colour, such as genetic constitution, maturity, climate, environment, season, soil type, plant nutrition, plant density and postharvest treatment. Maturity is a very important factor in determining the colour of harvested produce. At early stages of development, colour is usually green and only attains its characteristic colour at full maturity. The chlorophyll colour is important in vegetables, and if this is lost, the vegetables are not acceptable. Examples of such green vegetables are celery, green bell peppers, chayote squash, cucumbers, collard greens, green beans, green onions, green peas, leeks, lettuce, mustard greens, endive, kale, kohlrabi, jalapenos, okra, snap peas, snow peas, Swiss chard, watercress and zucchini. It must show the true colour of the product in terms of lightness, transparency, turbidity, and glossy nature [17, 18].

Wholesomeness: This involves the sanitary factors of the product. It must be clean without impurities, extraneous matter, sediments or specks. Also, a product must be a whole product, not part of a whole.

Pattern: This also describes shape and size; it should follow the specific pattern peculiar to the product.

Firmness: For cauliflower, a firm and compact head of white to cream-white curds surrounded by a crown of well-trimmed, turgid green leaves are required quality indices with freedom from severe yellowing, defects due to handling and decay.

Texture: This is the hand and mouth feel, the assessment of quality by the sense of touch and taste. It also indicates coarseness or crispness, firmness, turgidity, density, viscosity, surface tension, juiciness or dryness, fibrousness or chewiness, softness, mildness or stickiness. The texture of tomato is majorly contributed by the insoluble solids derived from cell walls which determined the consistency, smoothness and juiciness of the fruit [24].

Consistency: This may also be considered by a sense of touch. This refers to visualisation, flow or spread proportion of the produce. Appearance and texture can be measured by a team of experts ranging in order of quality importance. It is, therefore, a very elusive factor to measure because it depends on individual judgement. This means of measurement is usually referred to as organoleptic test [19–22].

Flavour: This is the quality evaluation by the sense of taste or smell. It refers to terms such as odour, fragrance, acid, burnt or gutty; taste: sweet, sour, bitter, salty or bland, off-flavour, enzymatic reaction, physiological deterioration and chemical contamination, overcooked or stale. For example, apples should have a crunchy texture and a sweet flavour while pomegranates should have a juicy texture and a sweet-tart flavour while strawberries should have a juicy texture and a sweet flavour. Tomatoes are typically juicy and firm, but can range from soft and mushy to hard and crunchy feel. Cucumbers are typically crunchy and juicy; while carrots are typically crunchy and sweet. Beetroots are firm and sweet, and cabbage is crunchy and slightly bitter while lettuce is crisp and slightly bitter, and ginger and turmeric are fibrous and slightly spicy. Enzyme reactions also occur, resulting in desirable flavour. For example, hydroperoxide lyase catalyses the production of tomato flavours [23]. To enhance sales, these quality factors are essential for market acceptability and consumers' choice to buy or reject the product. Sometimes, initial sales occur based on appearance, but repeated purchases are driven by expected quality factors determined by flavour compounds and texture [24, 25].

Nutritional value: This quality attribute is very important in ascertaining produce composition. It includes: total protein content, amino acid composition, mineral and vitamins, juice content, total soluble solids and vitamin C. Vitamin C content varies in fruits and vegetables from one to 150 mg/100 g [26]. In berry fruits, it ranges from 14 to 103 mg/100 g [27]. Rosehip, jujube, guava, kiwifruit, peppers, citrus fruit, spinach, broccoli and cabbage are rich in ascorbic acid. Tomato is about 93–95% water and 5–7% total solids. The lipids constituent in grapes is 0.1%, in bananas is 0.2% and in apple is 0.06%. Lipids content of 35 to 70% of dry mass is obtained in avocados, olives and nuts [28]. Fat-rich fruits and nuts include avocado, cauliflower, broccoli, carrots, hazelnuts, almond, walnut, Brazil nuts and chestnut [29]. Hazelnuts and almonds have flavonoid content of 18 and 15 mg/100 g, respectively [9]. Walnuts and Brazil nuts have phenolic acid content of 36 and 11 mg/100 g, respectively [10–14]. Good examples of fibre-rich foods are mango, orange, papaya, sweet lime, watermelon and apple [30]. Grains, fruits and vegetables are good sources of fibre. Fruits and vegetables provide 37% of the fibre in the diet and grains (36%) while legumes supplied 13% [31, 32]. Pro-vitamin A refers to precursors of vitamin A, obtained from fruits and vegetables such as carrot, pumpkin, peach and mango. Riboflavin is the central component of flavour proteins. It can be obtained from beans, beetroot, pepper and spinach. Niacin is derived from almonds, avocado and cape gooseberries. Vitamin B5 or pantothenic is obtained from meats, potatoes, oat cereals, tomato products and

Vitamin A	Vitamin B	Vitamin C	Vitamin D	Vitamin E	Carotenoid	Phenolic
Carrot	Beans	Strawberry	Cod liver oil	Almond	Pineapple	Plum
Pumpkin	Sunflower seed	Black currants.	Salmon	Corn	Carrot	Blackberry
Mango	Avocado	Kiwifruit	Sardine	Broccoli	Tomato	Apple
Peach	melons,	Orange	Tuna Fish	Spinach	Melon	Strawberry
Broccoli	Broccoli	Pepper	Beef Liver	Peanut	Mango	Raspberry
Red bell pepper	Cabbage	Guava	Egg yolk	Avocado	Peach	Blueberry
Lettuce	Spinach	Rosehip	Natural Sun	Grains	Plum	Gooseberry
Spinach	Peas	Persimmon	Mushroom	Nuts	Lemon	Soursop
Cantelope	Citrus	Straw berries	Fortified food	Leafy vegetables	Acerola	Curry
Sweet potato	Banana	Broccoli	Oil from Sheep wool	Seeds	Amta	Guava

Source: Adewoyin O.B.

Table 1.
Food sources supplying vitamins, carotenoids and phenolic.

whole grains. Sources of vitamin B6 include beans, cabbage, cauliflower, spinach, sweet potato, grape, avocado and banana. It occurs in peas, beans, nuts, broccoli, mushrooms, potatoes, strawberries and sweet potatoes. Folic acid is essential for reproduction and normal growth. It is present in strawberry, tomato, avocado, spinach, cabbage and other green vegetables [33, 34]. **Table 1** showed food sources supplying vitamins, carotenoids and phenolic.

2.1 Quality deterioration

This is associated with advance spoilage; any change in food quality, when optimum quality is not obtained, is referred to as spoilage. The major causes of food deterioration in harvested crops are microorganisms, natural food enzymes, insects, rodents and parasites, heat and cold, moisture and dryness, air, for exmaple, O_2, light and time.

i. *Activities of microorganism:* Several thousands of microorganisms are useful in food processing and fermentation in brewing while many others have undesirable changes in produce (referred to as spoilage), thereby reducing the shelf life of harvested produce. They are found on plants, in the soil at growth stages on the field, distorting roots and leaves, resulting in reduced yield and loss of quality. After harvest, microorganism activities continued everywhere: on peels of harvested produce in the hands of handlers and in water [35].

ii. *Natural Food Enzymes:* This is also an important cause of food spoilage. Natural food enzymes like microorganism are controlled by heat, cold, dryness, certain chemicals and radiation. Treatments applied to inactive microorganism will partially or completely destroy the enzyme. Natural enzymes are present in harvested food crops, without which germination to fruit maturation would not have occurred. These enzymatic reactions persist after maturation. Some of these enzymes may be more active in fruit crops such as pepper, tomatoes, plantain and banana to cause fruit deterioration. It is not in all cases that low temperature is required to preserve most tropical produce. The colder, the worse for some produce. For example, if the avocado is stored in a fridge, it turns black without ripening because the respiratory pattern is upset and other biological processes take place, leading to deterioration. Specific conditions for preservation must be selected in accordance with the unique spoilage pattern of individual harvested produce [35–37].

iii. *Insects, rodents and parasites:* The effect of insects, rodents and parasites depend on the environment. These factors can then be controlled easily by ensuring an appropriate environment free from pest invasion compared to microorganisms. Insects and rodents do physical damage to harvested produce such as scars, holes which provide entry point for secondary infection by microbes [38–40]

iv. *Heat and cold:* Increase in temperature can double the rate of chemical and enzymatic reactions which lead to increase in microorganism invasion, and result in an increased loss of vitamins A, C and riboflavin. Under low temperature, banana, avocado and pear blacken in the freezer and during thawing the skin may crack and secondary invasion will set in bananas should be stored under conditions where the temperature range is 10–15°C [41].

v. *Moisture and dryness:* When a produce absorbs water, it losses quality and becomes moist. An example of such is caking and lumping in stored grain like soybean, rice, maize and beans. Quality changes occur in produce due to moisture pick up. Moisture picked up occurs during thawing at low temperatures but without control of relative humidity, which allows moisture to dense over the surfaces of the produce consequently resulting in the rapid multiplication of microorganism [37, 41–43]. **Table 2** showed various crops and their safe moisture content.

vi. *Air:* Oxygen in the air can be destructive due to its high oxidative power, and it supports multiplication and optimum living condition and reproductive cycles of insect pests, rodents and microorganisms. Off-colour can also occur due to oxygen, which is essential for microbial growth, especially mould, which grows on the surfaces of produce. Activities of pests and rodents increase rapidly when oxygen is adequately available in the storage system. Oxygen can be controlled from harvested produce by the waxing of fruits, packaging in O_2 impermeable skin, tight plastic, vacuum packaging with the removal of oxygen and supplementing with nitrogen.

vii. *Light:* Some vitamins, such as vitamins A, C, and riboflavin, are destroyed by light and the only way to control this is to use dark-coloured containers for storage. Vitamins A, C and riboflavin, under light, change the colour of food, and react with lipids, fats and oils.

viii. *Time:* All the above factors are affected by time. All the reactions proceed with time. Produce has to be preserved at the peak of quality, which is impossible due to time. It is always better to consume food immediately to avoid any loss,

Product	Safe moisture content (%)
Maize flour	11.5
Millet	16.0
Rice (milled)	13.0
Rice	15.0
Sorghum	13.5
Wheat	13.5
Wheat flour	12.0
Pulses	15.0
Lentil, pea	14.0
Carrot	12.7
Cabbage	21.6
Banana	9.95
Plantain	9.97
Mango	9.147
Lettuce	12
Spinach	4.7
Guava	6.94
Yellow Yam	59.3
White Yam	64.97
Bitter Yam	64.9

Source: Adewoyin O.B.

Table 2.
Crops and their safe moisture content.

but this is not often possible due to time. Microorganisms have plenty of time to act and cause deterioration. Time is needed to produce toxic substances, exposure of food to cold or heat takes time, and length of time affects deterioration. In some food industry, fermentation may be beneficial in producing the required flavour and aroma; this also takes time to develop. All deterioration takes time, and all processes leading to deterioration must be arrested on time [43, 44].

3. Factors influencing quality

3.1 Pre-harvest factors

Site selection: The soil properties of a given site for crop production will determine the ultimate compositional and physical quality of the harvested produce. Appropriate site selection, free from heavy metals, toxic materials and adequate fertility level is essential for maximum quality. The soil should be analysed, and the soil condition should be determined before planting.

Genetic constituent of produce: The potential quality of harvested produce is a factor in the genetic constitution of the plant. Varieties with shorter shelf lives have higher postharvest losses while those with thick peel, high firmness quality, low respiration rate and low ethylene production rates have longer shelf life. Their different quality traits characterise each variety; these peculiar genetic quality character traits make some varieties more desirable to producers and consumers. The choice of adequate-yielding crop variety with desired qualities and longer shelf life is a vital decision for producers and an important pre-harvest factor that determines the shelf life of harvested produce [10]. The varieties that have the potential traits to withstand the rigours of marketing and distribution will have reduced losses after harvest. Varieties with resistance to low-temperature disorders, pathogens can be stored efficiently for longer duration with minimum storage losses. To prolong shelf life, enhance sustainable food availability and maintain good quality, producers must choose varieties that have inherently good quality and extended storage potential in addition to the high yield and pest resistance potentials. Failure to select an appropriate cultivar may lead to lower yield, low-quality fruits or less market acceptability. Fruits of different cultivars differ in size, colour, texture and flavour as well as storage potential. Getinet *et al.* [45] showed the influence of tomato cultivars on some postharvest qualities of tomatoes stored under different conditions, they established that tomato cultivar Roma VF has higher sugar content and lower weight loss compared to other cultivars. The genetic constituent of a produce is, therefore, critical to the postharvest storage life and utilisation qualities of such produce [37].

Planting period: The quality of crops planted during the dry season differ in size, firmness, fibre content and nutritional composition compared to those cultivated in rainy season when there is adequate water availability for chemical processes necessary for plant growth and development [46].

Irrigation: Some crops are not drought resistant hence, yield decreases in terms of size and nutritional quality after short periods of water stress. Proper irrigation planning is crucial for optimum crop development and adequate nutritional composition. Efficient water management scheme is vital to maintaining quality crop and maximum yield [47]. It is observed that deficit irrigation reduced fruit water accumulation and fresh fruit yield but increased fruit total soluble solids in tomatoes [48]. Mitchell *et al.* observed that deficit irrigation reduced fruit water accumulation and fresh fruit yield but increased fruit total soluble solids in tomato [48]. A higher level of moisture stress affects both yield and quality by decreasing cell enlargement. Crops which have higher moisture content generally have poorer storage characteristics. Some hybrid onions give a high yield of bulbs with low dry matter content and short storage life. Fully matured banana harvested soon after rainfall or irrigation may easily split during handling operations resulting in microorganism infection and rotting. If orange is too turgid at harvest, gland in the skin can be ruptured during harvesting, releasing phenolic compounds and causing oleocellosis or oil spotting (green spot on the yellow/orange coloured citrus fruit after degreening). In green leafy vegetables, too much rain or irrigation can make leaves harder and brittle, making them more susceptible to damage and decay during handling and transportation. Generally, crops with higher moisture or low dry matter content have poorer storage characteristics. Keeping quality of bulb crops like onion and garlic will be poor if irrigation is not stopped before 3 weeks of harvesting [11].

Thinning and Pruning: Thinning is a post-planting operation that reduces plant population, and competition between plants, increase maximum exposure to light, supply adequate water and nutrient to plants, consequently promoting good balance in the vegetative stage during fruit production and also improves the quality of harvested produce. It affects fruit texture and size due to inadequate exposure to sunlight. Thinning also improves the textural characteristics of harvested produce, which consequently greatly affect firmness. Studies also revealed that fruit firmness positively correlated with fruit size, implying that larger fruits were slightly firmer at harvest than smaller ones [18]. Appropriate pruning enhances fruit texture characteristics by optimise light distribution to all fruit on the tree. Inappropriate pruning may result in fruit shading with consequent smaller undesirable under, ripe fruit with a hard and grainy texture. Pruning is done to control the number of flowers and fruits by reducing the competition between fruits. Pruning, therefore, ensures nutrients are channelled to fewer fruits which can lead to increased fruit size [49] and increased sugar content of fruits in some cases [50]. On the other hand, the effect of pruning on other quality traits of the fruit produced depends on many factors, including the sink developmental stage, fruit-to-leaf ratio, truss position, and genetic composition of the plant [51].

Maturity Stage: Fruit maturity stage influences the total antioxidant capacity of the produce. These changes are determined by crop type and stage of maturity. For example, in tomato, pepper, mango, and prunus species, total antioxidant capacity increases as carotenoids and vitamin C accumulate during ripening [52]. Adewoyin and Babatola observed that pepper fruits harvested at 10% ripe (breakers stage) retained firmness, and weight loss was minimal compared to those harvested at 100% ripe stage (fully ripe) in all storage medium studied [53]. Wang and Lin observed that the shelf life of all tomatoes is longest when harvested at green mature stage though fruit nutritional values and appearance may be affected when harvested green [54]. Delays between harvest and consumption or processing can result in losses of flavour and nutritional quality. The magnitude of these losses increases with exposure to inappropriate temperatures, relative humidity and concentrations of O_2, CO_2, and C_2H_4 [26]. During berry ripening, anthocyanins accumulate while phenolic acids decrease [55] in products in which anthocyanins or chlorophylls dominate, carotenoids decrease during development; in cherry, ascorbic acid accumulates during ripening [56].

Climatic condition: Many plants are very sensitive to environmental conditions, and thus quality will not be optimised when crop is produced under adverse conditions. Poor weather at harvesting time affects the operations and functionality of harvesting machines or human labour and usually increases the moisture content of the harvested products, consequently resulting in loss of quality and reduced shelf life [57].

Heat management: Physiological and biochemical processes involved in plant growth, yield and maturation is influenced by temperature. Higher temperature during field conditions decreases shelf life and quality of the produce. At high temperatures, plants respire at a faster rate, and stored carbohydrates in harvested produce are depleted rapidly during respiration. High temperature during the fruiting season of tomato leads to quick ripening of fruits. Orange grown in the tropics have higher sugar content and total soluble solids than those grown in the subtropics. Tropically grown oranges tend to be green in colour and peel less easily. This is due to the higher temperature that occurs in the tropics, which results in rapid maturation of fruit which halts the process of the typical temperate orange colour development [58].

Light: Light regulates several physiological processes like chlorophyll synthesis, phototropism, respiration and stomatal opening. The duration, intensity and quality of light affect the quality of fruits and vegetables at harvest. Most of the produce needs high light intensity. Absorption of red light through pigments, phytochrome, is essential for carbohydrate synthesis, which determines the shelf life of the produce.

Citrus and mango fruits produced in full sun generally had thinner skin, a lower weight, low juice content and lower acidity but a higher total soluble solid. Citrus fruits grown in the shade may be less susceptible to chilling injury when stored in cold storage. In tomatoes, leaf shading of fruits produced a deeper red colour during the ripening than in the case of those exposed to light. The side of the fruit that was exposed to the sun was firmer than those that were not exposed to sunlight, the lower the light intensity, the lower the ascorbic acid content of plant tissues. In leafy vegetables, leaves are larger for those exposed to adequate light and thinner under conditions of low light intensity.

Humidity: High humidity during the growing season results in thin rind and increased size in some horticultural produce, and this produce is more prone to a high incidence of disease during postharvest period. Humid atmosphere may cause the development of fungal and bacterial diseases, which damages produce during storage and transport. Damaged produce removes water very quickly and emits a larger ethylene concentration than healthy ones. Low humidity may cause browning of leaf edge on plants with thin leaves or leaflets. High humidity can maintain the water-borne pollutants in a condition so that they can be more easily absorbed through the cuticles or stomata. Reduced transpiration leads to calcium and other elemental deficiency [59].

Rainfall: Rainfall affects the water supply to the plant and influences the composition of the harvested plant part. This affects its susceptibility to mechanical damage and decay during subsequent harvesting and handling operations. Excess water supply to plants results in the cracking of fruits such as orange, cherries, plums and tomatoes. If root and bulb crops are harvested during heavy rainfall, the storage losses will be higher [60].

Seasons: Seasonal fluctuation and time of the day at harvest will greatly affect the postharvest quality of produce. Synthesis of higher amount of carbohydrates during the day and its utilisation through translocation and respiration at night is responsible for the variation in the longevity of some harvested produce. Roses and tuberose have been found to show longer keeping quality in the winter under ambient conditions than in the summer. Produce harvested early in the morning or in the evening hours exhibits longer postharvest life than produce harvested during hot time of the day. If long-day onion (temperate) is grown during short-day (tropics) conditions, it will result in very poor storage quality [61].

Fertiliser application: Poor fertiliser management will increase physiological disorders due to deficiencies of some minerals or increase of others, leading to toxicity. In both cases, quality will be negatively affected. The use of trace elements or the practice of soilless tomato production can be made possible during irrigation, where fertilisers are added to the irrigation water in a form of solution and administered. These trace elements are selected depending on the specific postharvest quality traits needed in the fruits. Nutrient balance is crucial for maintaining optimal fruit texture and size: fruits from nitrogen deficient trees are usually smaller with firmer texture, while excess nitrogen leads to rapid loss of firmness and decreased storability. Potassium deficiency also leads to textural changes resulting in small, poorly coloured fruit that may not ripen, leaving fruit hard and inedible. A lack of boron can result in fruits with a mealy texture [62–64].

Pest and Disease Management: Pathogens and insects have very negative effects on quality of harvested produce. The effect of insect is more pronounced on grains but can also cause a lot of damages in fruits and vegetables. Nematodes cause various injuries to fruits and vegetables and continue the deterioration during storage. Parasites are therefore seen to be important in damage to farm produce as well as food preservation. In the case of insects, produce attacked by them in agriculture may consume over 50% of the harvest. Insects at times lay eggs in the produce, making it almost impossible to eliminate all insects pest in the produce. Parasites like nematodes and amoeba may infect the produce, and the same is true when produce comes in contact with water; this is very common in Africa. Rodents contaminate food with their urine and droppings. They also produce large litter, for example, two rats can give up to 30,000 litres per year. Through their contamination, they spread diseases like salmonellosis, plague and typhoid fever. Various efforts to control the detrimental effects of these organisms have resulted in great hazard to human health due to misuse of the chemicals [65].

Weed management: Failure to control weeds will result in a lot of damage to crop quantity and quality. Weeds harbour diseases and pests that easily infest crops both on the field and in the store. Weeds also contaminate harvested produce by mixing with the seeds [66].

Presence of heavy metals: The site for crop production must not be just any site or dump site which is loaded with heavy metals. Appropriate soil tests should be done to ascertain the soil condition because some crops absorb heavy metals, which are easily assimilated by human system [66–69].

Harvesting methods and time: The time of the day at which harvesting is done must be considered to avoid excessive field heat, which can cause rapid deterioration of the harvested produce. Loss is also caused by improper harvesting methods such as rough handling, untimely harvesting, lack of appropriate and poorly-designed harvesting tools, equipment, and harvesting containers [70].

Method of Processing: Processing can decrease phenolic antioxidants [71]. Anthocyanin losses in processed berries are reduced by blanching, indicating enzymatic degradation [71]. A study conducted comparing manually tearing lettuce into strips to shredding with a sharp knife showed that the retention of ascorbate in lettuce sliced by a machine was 25–63% lower than in lettuce shredded by manual tearing [72]. Effects of slicing and shredding radishes on quality during storage at 1, 5, and 10°C were determined [73]; on the 10th day, intact radishes stored at 1°C had the lowest respiration rate, while sliced radishes stored at 10°C had the highest. Shredded radishes showed the most undesirably low levels soluble solids, higher weight loss, ascorbic acid, and lightness as compared to intact or sliced radishes [74].

4. Postharvest factors

Harvested produce are living tissues and subject to continual changes after harvest. Such changes cannot be stopped but can be controlled within certain limits by using various postharvest procedures. Postharvest factors that affect quality of harvested produce include the following: temperature is the most important tool to extend shelf life and maintain quality of harvested produce [75]. Delays between harvesting and cooling or processing can result in direct loss due to water loss and decay; indirect losses such as off-flavour and deterioration in nutritional quality can also occur. For instance, the temperature range and the extent of vitamin C loss depended on the type of citrus fruit. In general, the extent of loss in ascorbic acid (AA) content in response to elevated temperature was greater in vegetables than in acidic fruits such as citrus. Ascorbic acid is more stable under acidic conditions [76]. Reported that retention of vitamin C ranged from 56 to 98% for six broccoli cultivars stored at 2°C for 3 weeks.

Ethylene management: Ethylene (C_2H_4) is an odourless and colourless two-carbon natural plant hormone which is triggered at maturity in climacteric fruits. It is also known as the 'natural ageing, death or ripening hormone', it is active at small traces and its accumulation can led to fruit decay and waste during postharvest stage of harvested produce. It is regulated in various physiological processes of plant growth, germination, development, ripening, maturity and senescence. It also plays a major role in the abscission of plant organs. Several strategies of crop management, coordination of postharvest and pre-harvest factors and various techniques of plant breeding have been investigated to understand ethylene regulation pathways, biochemical and physiological processes in extending produce shelf life and improve the postharvest quality of harvested produce [77]. Some fruits are either ethylene producers or absorbers. Fruits such as apples, bananas, melons, pears and peaches are ethylene producers, while broccoli, cabbage, and cauliflower are ethylene sensitive. The respiration rate of non-climacteric crops is not influenced by the presence ethylene. Examples of non-climacteric produce are leafy vegetables, watermelon, strawberries and grapes. Non-climacteric crops will not respond to ripening with ethylene gas. Exposure of climacteric fruits to ethylene will advance the onset of an irreversible rise in respiration rate and rapid ripening. Appropriate packaging can delay the onset of climacteric and prolong shelf life of fruits by reducing ethylene production and sensitivity [28]. Ethylene production rates is influenced by type of produce, and it increases with maturity at harvest, physical injuries, disease incidence, increased temperature up to 30°C and water stress [30]. The response of various types of crops to ethylene is shown in **Table 3**.

Chemical Treatments: Calcium dips may be used to reduce physiological disorders and maintain firmness in apples and cherries. Dehydrated pineapples and guava pretreated with cysteine hydrochloride had increased vitamin C retention and reduced colour change during storage [76, 78].

Irradiation: Ionising radiation may be used for sprout inhibition, insect control or delay of ripening of certain fruits and vegetables. Irradiation of horticultural crops at relatively low doses of 75–100 krad irreversibly inhibited the sprouting of potatoes regardless of storage temperature. Losses in vitamin C were lower in potato irradiated for sprout control and subsequently stored at 15°C than in non-irradiated tubers stored at 2–4°C 78].

Class	(μl C2H4/kg-h at 20°C (68°F)	Commodities
Very low	< 0.1	Artichoke, asparagus, cauliflower, leafy vegetables, root vegetables, potato, cherry, citrus, grape, jujube, strawberry, pomegranate
Low	0.1–1.0	Cucumber, eggplant, okra, watermelon, chilli, bell pepper, pumpkin, watermelon, olive, pineapple, blueberry, raspberry
Moderate	1.0–10.0	Tomato, banana, fig, guava, honeydew melon, mango, plantain
High	10.0–100.0	Apple, passim, apricot, avocado, cantaloupe, kiwifruit, nectarine, papaya, peach, pear, plum
Very high	> 100.0	Sapota, mammee apple, passion fruit, cherimoya

Source: Adewoyin O.B.

Table 3.
Classification of crop according to ethylene production.

Respiration: The respiration rate in harvested produce will influence its metabolic activities. High temperatures can hasten the rate of respiration and CO_2 production in harvested produce. CO_2 production in stored climacteric products like tomatoes can trigger ethylene production, depending on other factors like O_2 or CO_2 levels, exposure time, and ripening stage [79].

Relative Humidity: Moisture loss from harvested produce is predominantly caused by the amount of moisture present in the air expressed as relative humidity [80]. Harvested fruits and vegetables maintain their nutritional quality, appearance, weight and flavour at very high relative humidity. On the contrary, this is adverse for grain crops such as maize, millet, wheat, rice, beans and soybean. Harvested fruits shrivel with little percentage moisture loss while grain crops require low moisture content for optimum storage conditions after harvest. The optimal values of relative humidity to maintain quality of mature green tomatoes are within the range of 85–95% (v/v) and 90–95% (v/v) for firmer ripe fruits [81].

Physical Handling: The handling of produce from the moment of detachment from the parent plant, packing from the farm gate to the market, and then the final consumption of produce are significantly associated with mechanical injuries such as bruising, scarring, scuffing, cutting or puncturing. Some of these mechanical injuries may result from careless handling, the use of inappropriate harvesting containers, inappropriate vehicles, careless loading and unloading and packaging materials. According to Miller [82], the consequences of these mechanical injuries are cumulative, leading to a total breakdown of the cell structure accompanied by unwanted metabolic activities such as increased ethylene production, accelerated respiration rates, and ripening, resulting in either reduced shelf life or poor quality [83]. It is, therefore, important to handle harvested produce with care during harvest and postharvest activities to minimise mechanical injuries to avoid losses.

Gases: The combination of different gases in a storage environment is very important in extending the shelf life of harvested produce. The optimal ambient condition required to inhibit senescence in matured harvested produce is very low supply of oxygen [84–86]. Carbon monoxide (CO) has been investigated as a gas to speed up ripening, and it is therefore necessary to balance the carbon monoxide with low oxygen to delay senescence in harvested produce [86].

5. Conclusion

Huge amount of crops are being produced annually, but most is lost at various stages of handling due to pre-harvest and postharvest factors affecting quality and shelf life of harvested produce. Every unit of produce preserved converts to added unit available for productive utilisation and food security. The use of any postharvest method or handling practices can only maintain quality. Understanding and controlling the various roles of pre-harvest factors like fertiliser application, pruning, maturity stage, cultivar selection, and irrigation can play major role in improving quality of harvested produce and prolonging their shelf life after harvest. Using best postharvest handling practices or factors such as optimum temperature, right relative humidity, right gases in storage and the best physical handling procedures to maintain the quality after harvest is also critical. Postharvest factors alone cannot maintain quality but the pre-harvest factors during production are also important. Until both factors are managed properly, quality loss will still be a major challenge in maintaining quality and prolonging shelf life of harvested produce.

Author details

Oluyinka Benedicta Adewoyin
Department of Crop Science and Horticulture, Federal University Oye, Oye, Nigeria

*Address all correspondence to: oluyinka.adewoyin@fuoye.edu.ng

References

[1] Bekele D. Role of postharvest Management for Food Security: A review. Advances in Crop Science Technology. 2021;**9**(7):3. DOI: 10.4172/2329-8863.1000475

[2] Nellemann C, MacDevette M, MandersT, EickhoutB, SvihusB, PrinsAG, et al., editors. The Environmental Food Crisis-the Environment Role in Averting Future Food Crisis. Nairobi: UNEP;

[3] Kitinoja L, Gorny JR. Postharvest technology for small-scale produce marketers: Economic opportunities, quality and food safety. In: Postharvest Horticulture. Series No. 21. University California; 1999

[4] Parfitt, J., Barthel, M. and Macnaughton, S. 2010. Food waste within food supply chains: quantification and potential for change to 2050

[5] Hodges RJ, Buzby JC, Bennett B. Postharvest losses and waste in developed and less developed countries: Opportunities to improve resource use. Journal of Agricultural Science. 2011;**2011**(149):37-45

[6] Kader AA. Postharvest Technology of Horticultural Crops. 3rd ed. Univ. Calif. Agr; 2002

[7] Hartz TK, Johnson PR, Francis DM, Miyao EM. Processing tomato yield and fruit quality improved with potassium fertigation. Horticultural Science. 2005;**40**(1):1862-1867

[8] Passam HC, Karapanos IC, Bebeli PJ, Savvas D. A review of recent research on tomato nutrition, breeding and post-harvest technology with refrence to fruit quality. The European Journal of Plant Science and Biotechnology. 2007;**1**:1-21

[9] Adams P. Nutritional control in hydroponics. In: Savvas D, Passam HC, editors. Hydroponic Production of Vegetables and Ornamentals. Athens, Greece: Embryo Publications; 2002. pp. 211-261

[10] Passam HC, Karapanos IC, Bebeli PJ, Savvas D. A review of recent research on tomato nutrition, breeding and post-harvest technology with reference to fruit quality. The European Journal of Plant Science and Biotechnology. 2007;**1**:1-21

[11] Senevirathna P, Daundasekera WAM. Effect of postharvest calcium chloride vacuum infiltration on the shelf life and quality of tomato (cv. 'Thilina'). Ceylon Journal of Science. 2010;**39**(1):35-44

[12] Parisi M, Giordano I, Pentangelo A, D'Onofrio B, Villari G. Effects of different levels of nitrogen fertilization on yield and fruit quality in processing tomato. Acta Horticulturae. 2006;**700**:129-132

[13] Heeb A, Lundegårdh B, Ericsson T, Savage GP. Nitrogen form affects yield and taste of tomatoes. Journal of the Science of Food and Agriculture. 2005;**85**(8):1405-1414

[14] Oke M, Ahn T, Schofield A, Paliyath G. Effects of phosphorus fertilizer supplementation on processing quality and functional food ingredients in tomato. Journal of Agricultural and Food Chemistry. 2005;**53**(5):1531-1538

[15] Howard LR, Smith RT, Wagner AB, Villalon B, Burns EE. Provitamin a and ascorbic acid content of fresh pepper cultivars (Capsicum annum) and processed jalapenos. Journal of Food Science. 1994;**59**:362-365

[16] Odeyemi OO. Post-harvest food problems in Nigeria: Kinds and magnitude of spoilage and agents of deterioration. In: A 3-Day Workshop on Reduction of Postharvest Food Losses in Nigeria. 20th – 22nd March 2013. Nigeria: CCE hilltop Auditorium, FUTA; 2013

[17] Picha D. Horticultural crop quality characteristics important in international trade. Acta Horticulturae. 2006;**712**:423-426. DOI: 10.17660/ActaHortic.2006.712.49

[18] Wang M, Chen L, Liang Z, He X, Liu W, Jiang B, et al. 2020. Metabolome and Transcriptome Analyses Reveal Chlorophyll and Anthocyanin Metabolism Pathway

[19] McPherson C, Farcuh M. Fruit quality: The importance of fruit textural characteristics. 2021. Available from: https://extension.umd.edu/resource/fruit-quality-importance-fruit-textural-characteristics, 3/4/2023

[20] Costa F, Cappellin L, Fontanari M, Longhi S, Guerra W. Texture dynamics during postharvest cold storage ripening in apple (malus x domestica Borkh.). Postharvest Biology and Technology. 2012;**69**:54-63

[21] Lara I, Belge B, Goulao LF. The fruit cuticle as a modulator of postharvest quality. Postharvest Biology and Technology. 2014;**87**:103-112

[22] Saei A, Tustin DS, Zamani Z, Talaie A, Hall AJ. Cropping effects on the loss of apple fruit firmness during storage: The relationship between texture retention and fruit dry matter concentration. Scientia Horticulturae. 2011;**130**:256-265

[23] Anthon GE, Barrett DM. Thermal inactivation of lipoxygenase and hydroperoxytrienoic lyase in tomatoes. Food Chemistry. 2003;**81**(2):275-279

[24] Beaulieu JC. Effect of cutting and storage on acetate and non-acetate esters in convenient, ready to eat fresh-cut melons and apples. HortScience. 2006a;**41**(1):65-73

[25] Waldron KW, Parker ML, Smith AC. Plant cell walls and food quality. Comprehensive Reviews in Food Science and Food Safety. 2003;**2**:101-119

[26] Lee SK, Kader AA. Preharvest and postharvest factors influencing vitamin C content of horticultural crops. Postharvest Biology and Technology. 2000;**20**:207-220

[27] Pantelidis GE, Vasilakakis M, Manganaris GA, Diamantidis G. Antioxidant capacity, phenol, anthocyanin and ascorbic acid contents in raspberries, blackberries, red currants, gooseberries and cornelian cherries. Food Chemistry. 2007;**102**:777-783

[28] Maestri D, Cittadini MC, Bodoira R, Martínez M. Tree nut oils: Chemical profiles, extraction, stability and quality concerns. European Journal of Lipid Science and Technology. 2020:1900450

[29] Shahzad N, Khan W, Shadab MD, Ali A, Saluja SS, Sharma S, et al. Phytosterols as a natural anticancer agent: Current status and future perspective. Biomedicine & Pharmacotherapy. 2017;**88**:786-794

[30] Ramulu P, Rao PU. Total, insoluble and soluble dietary fiber contents of Indian fruits. Journal of Food Composition and Analysis. 2003;**16**:677-685

[31] Hiza HAB, Bente L. Nutrient content of the U.S. food supply, 1909-2004: A summary report. In: Home Economics Research Report Number 57. Washington, DC: U.S. Department of Agriculture, Center for Nutrition Policy and Promotion; 2007

[32] Walsh JH, Wyse BW, Hansen RG. Pantothenic acid content of 75 processed and cooked foods. Journal of the American Dietetic Association. 1981;**78**:140

[33] García-Salinas C, Ramos-Parra PA, de la Garza RID. Ethylene treatment induces changes in folate profiles in climacteric fruit during postharvest ripening. Postharvest Biology and Technology. 2016;**118**:43-50

[34] Muley AB, Singhal RS. Extension of postharvest shelf life of strawberries (Fragariaananassa) using a coating of chitosan-whey protein isolate conjugate. Food Chemistry. 2020;**329**:127-213

[35] Lasztity R. Micro-organisms important in food microbiology. Food Quality and Standards. 2009;**3**:26-40

[36] Deckers M, Deforce D, Fraiture MA, Roosens NH. Genetically modified micro-organisms for industrial food enzyme production: An overview. Food. 2020;**9**(3):326

[37] Schouten RE. Genetic and physiological factors affecting colour and firmness In: Schouten RE, van Kooten O, Jalink H, Kappers IF, Snel JFH, Jordi W. Understanding and Measuring the Shelf-Life of Food, 69; 2004. Wageningen University, Plant Research International: The Netherlands

[38] Bamforth CW, Cook DJ. Food, Fermentation, and Micro-Organisms. John Wiley & Sons; 2019

[39] Stenseth NC, Leirs H, Skonhoft A, Davis SA, Pech RP, Andreassen HP. Mice, rats, and people: The bio-economics of agricultural rodent pests. Frontiers in Ecology and the Environment. 2003;**1**(7):367-375

[40] Mellenthin WM, Wang CY. Preharvest temperatures in relation to postharvest quality of 'd'Anjou'Pears. Journal of the American Society for Horticultural Science. 1976;**101**(3):302-305

[41] Chen XD. Drying as a means of controlling food biodeterioration. Food Biodeterioration and Preservation. 2008;**137**

[42] Saini RK, Shetty NP, Prakash M, Giridhar P. Effect of dehydration methods on retention of carotenoids, tocopherols, ascorbic acid and antioxidant activity in Moringa oleifera leaves and preparation of a RTE product. Journal of Food Science and Technology. 2014;**51**:2176-2182

[43] Durance TD. Improving canned food quality with variable retort temperature processes. Trends in Food Science & Technology. 1997;**8**(4):113-118

[44] Hussein Z, Fawole OA, Opara UO. Effects of bruising and storage duration on physiological response and quality attributes of pomegranate fruit. Scientia Horticulturae. 2020;**267**:109306

[45] Getinet H, Workneh TS, Woldetsadik K. Effect of maturity stages, variety and storage environment on sugar content of tomato stored in multiple pads evaporative cooler. African Journal of Biotechnology. 2011;**10**(80):18481-18492

[46] Tudela JA, Hernández N, Pérez-Vicente A, Gil MI. Growing season climates affect quality of fresh-cut lettuce. Postharvest Biology and Technology. 2017;**123**:60-68

[47] Wang SY, Zheng W, Galletta GJ. Cultural system affects fruit quality and antioxidant capacity in strawberries. Journal of Agriculture and Food Chemistry. 2002;**50**:6534-6542

[48] Mitchell JP, Shennan C, Grattan SR, May DM. Tomato fruit yields and quality

under water deficit and salinity. Journal of the American Society of Horticultural Science. 1991;**116**(2):215-221

[49] Hanna HY. Influence of cultivar, growing media, and cluster pruning on greenhouse tomato yield and fruit quality. Hort-Technology. 2009;**19**(2):395-399

[50] Prudent M, Causse M, Génard M, Tripodi P, Grandillo S, Bertin N. Genetic and physiological analysis of tomato fruit weight and composition: Influence of carbon availability on QTL detection. Journal of Experimental Botany. 2009;**60**(3):923-937

[51] Moneruzzaman KM, Hossain ABMS, Sani W, Saifuddin M, Alenazi M. Effect of harvesting and storage conditions on the post harvest quality of tomato. Australian Journal of Crop Science. 2009;**3**(2):113-121

[52] Kim TJ, Hyeon H, Park N, IYi TG, Lim SH, Park SY, et al. A high-throughput platform for interpretation of metabolite profile data from pepper (capsicum) fruits of 13 phenotypes associated with different fruit maturity states. Food Chemistry. 2020:127286

[53] Adewoyin OB, Babatola LA. Storability of pepper fruit as influenced by ripeness at harvest, harvesting methods and storage conditions. Acta Horticulturae. 2018;**1225**:211-216. DOI: 10.17660/ActaHortic.2018.1225.28

[54] Wang SY, Lin HS. Antioxidant activity in fruits and leaves of blackberry, raspberry, and strawberry varies with cultivar and developmental stage. Journal of Agriculture and Food Chemistry. 2000;**48**:140-146

[55] Lin Y, Huang G, Zhang Q, Wang Y, Dia VP, Meng X. Ripening affects the physicochemical properties, phytochemicals and antioxidant capacities of two blueberry cultivars. Postharvest Biology and Technology. 2020;**162**:111097

[56] Xu M, Shen C, Zheng H, Xu Y, Xue C, Zhu B, et al. Metabolomic analysis of acerola cherry (Malpighiaemarginata) fruit during ripening development via UPLC-Q-TOF and contribution to the antioxidant activity. Food Research International. 2020;**130**:108915

[57] Reymond M, Muller B, Leonardi A, Charcosset A, Tardieu F. Combining quantitative trait loci analysis and an ecophysiological model to analyze the genetic variability of the responses of maize leaf growth to temperature and water deficit. Plant Physiology. 2003;**131**:664-675. DOI: 10.1104/pp.013839

[58] Parent B, Turc O, Gibon Y, Stitt M, Tardieu F. Modelling temperature-compensated physiological rates, based on the coordination of responses to temperature of developmental processes. Journal of Experimental Botany. 2010;**61**:2057-2069. DOI: 10.1093/jxb/erq003

[59] Lobos TE, Retamales JB, Ortega-Farías S, Hanson EJ, López-Olivari R, Mora ML. Pre-harvest regulated deficit irrigation management effects on post-harvest quality and condition of V. corymbosum fruits cv. brigitta. Scientia Horticulturae. 2016;**207**:152-159

[60] Kubota C, Kroggel M, Torabi M, Dietrich KA, Kim HJ, Fonseca J, et al. Changes in selected quality attributes of greenhouse tomato fruit as affected by pre-and postharvest environmental conditions in year-round production. HortScience. 2012;**47**(12):1698-1704

[61] Kruger FJ, Claassens NJF, Kritzinger M, Claassens VE. A short

review of recent research on the impact of climatic conditions on the postharvest quality of south African export avocados. Revista Chapingo. Serie Horticultura. 1999;**58**(5):339-345

[62] Gupta UC, Gupta SC. Trace element toxicity relationships to crop production and livestock and human health: Implications for management. Communications in Soil Science and Plant Analysis. 1998;**29**(11-14):1491-1522

[63] Roy RN, Finck A, Blair GJ, Tandon HLS. Plant nutrition for food security. A guide for integrated nutrient management. FAO fertilizer and plant. Nutrition Bulletin. 2006;**16**:368

[64] Fageria NK, Baligar VC, Li YC. The role of nutrient efficient plants in improving crop yields in the twenty first century. Journal of Plant Nutrition. 2008;**31**(6):1121-1157

[65] Blossey B, Hunt-Joshi TR. Belowground herbivory by insects: Influence on plants and aboveground herbivores. Annual Review of Entomology. 2003;**48**(1):521-547

[66] Buhler DD, Netzer DA, Riemenschneider DE, Hartzler RG. Weed management in short rotation poplar and herbaceous perennial crops grown for biofuel production. Biomass and Bioenergy. 1998;**14**(4):385-394

[67] McBride MB. Toxic metal accumulation from agricultural use of sludge: Are USEPA regulations protective. Journal of Environmental Quality. 1995;**24**(1):5-18

[68] Mapanda F, Mangwayana EN, Nyamangara J, Giller KE. The effect of long-term irrigation using wastewater on heavy metal contents of soils under vegetables in Harare, Zimbabwe. Agriculture, Ecosystems & Environment. 2005;**107**(2-3):151-165

[69] Romic M, Romic D. Heavy metals distribution in agricultural topsoils in urban area. Environmental Geology. 2003;**43**:795-805

[70] Atkin OK, Loveys BR, Atkinson LJ, Pons TL. Phenotypic plasticity and growth temperature: Understanding interspecific variability. In: Meeting on Phenotypic Plasticity and the Changing Environment Held at the Society for Experimental Biology Plant Frontiers Meeting. Sheffield, UK; 2005. pp. 267-281

[71] Tiwari U, Cummins E. Factors influencing levels of phytochemicals in selected fruit and vegetables during pre-and post-harvest food processing operations. International Food Research Journal. 2013;**50**(2):497-506

[72] Sablani SS, Andrews PK, Davies NM, Walters T, Saez H, Syamaladevi RM, et al. Effect of thermal treatments on phytochemicals in conventionally and organically grown berries. Journal of the Science of Food and Agriculture. 2010;**90**:769-778

[73] Barry-Ryan C, O"Beirne, D. Ascorbic acid retention in shredded iceberg lettuce as affected by minimal processing. Journal of Food Science. 1999;**64**:498-500

[74] del Aguila JS, Sasaki FF, Heiffig LS, Ortega EMM, Jacomino AP, Kluge RA. Fresh-cut radish using different cut types and storage temperatures. Postharvest Biol. and Technol. 2006;**40**:149-154

[75] Hussein Z, Fawole OA, Opara UO. Effects of bruising and storage duration on physiological response and quality attributes of pomegranate fruit. Scientia Horticulturae. 2020;**267**:109306

[76] Albrecht JA, Schafer HW, Zottola EA. Sulphydryl and ascorbic

acid relationships in selected vegetables and fruits. Journal of Food Science. 1990;**56**(427):430

[77] Joshi MR, Srirangarajan AN, Thomas P. Effects of gamma irradiation and temperature on sugar and vitamin C changes in five Indian potato cultivars during storage. Food Chemistry. 1990;**35**:209-216

[78] Lee SK, Kader AA. Preharvest and postharvest factors influencing vitamin C content of horticultural crops. Postharvest Biology and Technology. 2000;**20**(3):207-220

[79] Kader AA. Physiological and biochemical effects of carbon monoxide added to controlled atmospheres on fruits. Acta Horticulturae. 1983;**138**:221-226

[80] Akhtar A, Abbasi NA, Hussain A. Effect of calcium chloride treatments on quality characteristics of loquat fruit during storage. Pakistan Journal of Botany. 2010;**42**(1):181-188

[81] Miller RA. Harvest and handling injury: Physiology biochemistry and detection. In: Bartz JA, Brecht JK, editors. Postharvest Physiology and Pathology of Vegetables. New York, NY, USA: Marcel Dekker; 2003. pp. 177-208

[82] Izumi H, Watada AE. Calcium treatments affect storage quality of shredded carrots. Journal of Food Science. 1994;**59**(1):106-109

[83] DeWild HPJ, Otma EC, Peppelenbos HW. Carbon dioxide action on ethylene biosynthesis of pre-climacteric and climacteric pear fruit. Journal of Experimental Botany. 2003;**54**(387):1537-1544

[84] Cantwell MI. Postharvest Handling of Specialty Vegetables. Department of Vegetable Crops: University of California, Davis, California, USA; 1997

[85] Mohammed A. Postharvest handling of horticultural crop. Technical paper on postharvest handling of pepper fruits. 2000. pp. 1-8

[86] Agar IT, Streif J, Bangerth F. Effect of high CO_2 and controlled atmosphere on the ascorbic and dehydro-ascorbic acid content of some berry fruits. Postharvest Biology and Technology. 1997;**11**:47-55

Chapter 8

Appropriate Post-Harvest Technologies for Biofortified Crops Pro Enhanced Utilization, Value Addition, and Micronutrient Retention

Olapeju Phorbee, Ganiyat Olatunde, Dotun Aderonmu, Desmond Ikerionwu, Adeola Ojo, Gift Oguzor, Lateef Sanni and Adeyinka Onabolu

Abstract

Biofortified cassava and sweet potato, targeted for vitamin A deficiency reduction in Sub-Saharan Africa, are highly perishable at post-harvest. Appropriate technologies for processing these crops should primarily be protective of their micronutrients otherwise the purpose of their biofortification is defeated. One of the value-added OFSP root products is the puree, which several techniques have been developed for its consistent quality, preservation and packaging. However, use of aseptic packaging and continuous flow microwave system of rapid sterilization have been reported most suitable, for its high temperatures (≥125°C) and short time principle. For biofortified cassava varieties, post-harvest advances have been on drying, moving from sun drying on bare floor to use of raised platform, solar and mechanical drying. Flash-drying technology is an effective and efficient drying technology that uses rapid heat transfer, which makes it suitable for biofortified cassava. With these advanced technologies, OFSP puree, wet or dried and flash-dried biofortified cassava mash can be targeted for diverse end uses in the food industry-baby foods, jam, pastries, and confectionaries. These technologies, with increased adoption through favorable policies, can enhance availability of diverse nutritious food products, utilization, consumption, and commercialization of locally produced staples, for improved food system transformation.

Keywords: biofortified crops, post-harvest, technologies, value addition, processing

1. Introduction

1.1 Prevalence of malnutrition

The challenge of malnutrition and undernutrition is long-standing globally, with slow progress in interventions despite trends of development; 1.2 billion people lack key micronutrients, 151 million children are stunted, 50.5 million children are wasted, while an estimated 2 billion and 38.3 million adults and children, respectively, are overweight or obese [1].

Similar trend with worse statistics has been reported in Sub-Saharan Africa (SSA) as stunted children under 5 years of age increased by just 23% in 24 years (1990–2014) [2]. Millions of people in SSA especially women of child-bearing age and children under 5 years from poor households are deficient in key micronutrients [3, 4]. An estimated 24% of all child deaths are due to vitamin A deficiency, out of which 48% are preschool-age children [5].

The consequences of vitamin A deficiency include a high risk of diseases such as diarrhea and measles, growth retardation, and premature death for children under 5 years of age, weakened immune system, visual impairment, and blindness [6].

1.2 Interventions on micronutrient malnutrition

Efforts to address micronutrient malnutrition using both nutrition-specific and sensitive interventions have been reported especially in SSA. On nutrition-specific interventions, there has been significant support for exclusive breastfeeding (EBF), improved Infant and Young Child feeding (IYCF) practices, micronutrient supplementation, treatment of severe malnutrition with Ready-to-Use Therapeutic Foods (RUTF), mandatory large-scale fortification of selected foods (salt, sugar, oil, and flours), and Home-Grown School Feeding in some developing countries [7]. Nutrition-sensitive agriculture, mainly biofortification, and water, sanitation, and hygiene (WASH) programs have been appreciably promoted under nutrition-sensitive interventions.

However, each of these intervention programs has its limitations, which inhibit impact at scale and sustainability. For instance, women have so many reasons for not practicing exclusive breastfeeding. Micronutrient supplementation programs are mainly funded by external donors with unguaranteed sustainability, and so unable to meet the set goals of the international health organizations. Other limitations include poor access of the poor people to markets, health-care centers, and other places where the supplements are available, as well as lack of public enlightenment on the health benefits of these nutrient supplements [8, 9]. For large-scale food fortification, being a food-based industrial approach for addressing micronutrient malnutrition, and expected to cover a wide population, has so far been largely limited in reaching most rural dwellers. The coverage of fortified foods is dependent on how developed the market infrastructure is, and most rural poor have poor access to market where the fortified foods are [7]. In Nigeria, locally processed and unfortified foods are often more readily available and affordable to the poor rural dwellers who need fortified foods more because of their poor diets. Worst still, some of the industrially produced food vehicles that are expected to be adequately fortified are often times either not fortified at all or not adequately fortified [10].

Poor knowledge of nutrient contents of many indigenous foods hinders promotion and practice of dietary diversification and nutrition-sensitive food production system. Nutrition education as a strategy is yet to reach an impactful scale on behavioral change [7].

Biofortification, therefore, came up as a sustainable agricultural-base complementary approach to address micronutrient malnutrition in developing countries, targeting the vulnerable populations.

2. Biofortification

Biofortification is the process of breeding staple crops to have higher contents of essential nutrients either through selective breeding or genetic engineering [11]. It is a cost-effective, and sustainable technique of delivering essential micronutrients to populations whose major food crops are deficient in them, and that have limited access to diverse diets and other micronutrient interventions. Biofortification is a globally recognized agricultural-based approach to addressing hidden hunger and food insecurity, especially in the SSA. Biofortification of staple crops, being within the agricultural sector, presents exceptional investment opportunities for addressing this national priority through production, processing, and marketing of diverse and more nutritious crops that can sustainably improve the nutrition status of vulnerable populations. It is cost-effective because the only major cost required is that of initial breeding and introduction; once biofortified crops are in the farmer food system, they can reach remote, rural populations that are difficult and expensive to reach with regular supplementation promotions [12–18]. According to Garg et al. [19], biofortification delivers food crops with improved key micronutrients—iron, zinc, and provitamin A that are usually lacking in the diets of the developing world. International initiatives, such as the HarvestPlus and national programs, are available to achieve these targets and so far, they have delivered crops with the potential to increase the quantity and quality of essential micronutrients in human diets, especially in staples like wheat, maize, cassava, beans, sweet potatoes, and millets [19].

Biofortification has enabled a shift in agricultural system from "increasing crop yield and productivity," which has resulted in a high rise of micronutrient-deficient food crops to "producing nutrient-rich crops in sufficient quantities," which will help address micronutrient malnutrition, especially in the developing countries [20].

2.1 Biofortification techniques

Biofortification techniques can be transgenic, conventional, and agronomic approaches, using biotechnology, crop breeding, and fertilization strategies, respectively. The three approaches have been targeted for cereals, some legumes, and vegetables (rice, wheat, maize, sorghum, common bean, potato, sweet potato, and tomato) while some crops could only be achieved by one or two of the techniques. However, more crops have been targeted by transgenic approach, but in practice, conventional breeding technique has been the highest [19].

To date, more than 400 biofortified varieties of 12 crops have been released in over 40 countries, facilitated by HarvestPlus and CIP that is exclusively on biofortified sweet potato varieties [19]. **Table 1** shows biofortified crop varieties of sweet potato available in some countries [21].

2.2 Biofortified crops

Some common staples in Africa have been successfully biofortified with provitamin A (cassava, maize, and sweet potato), beans with iron as well as sorghum and millet with iron and zinc. Biofortified cassava and sweet potato are gradually growing in awareness and adoption in Nigeria and other parts of SSA where they are staples, to address vitamin A deficiency. Biofortified crops with increased contents of essential micronutrients are delivered to consumers through the same familiar traditional practices by the key actors of the value chain, thus reaching the target populations

Countries	Orange-fleshed sweet potato	Countries	Orange-fleshed sweet potato
Uganda	Ejumula	Tanzania	Carrot C
	Kakamega		Mayai
	Vita (Naspot 90)	Nigeria	UMUSPO3-Mothers Delight
	Kabode		UMUSPO1-King J
	Naspot 120		UMUSPO4-Solo Gold
	Naspot 130	Zambia	Twatasha
Malawi	Zondeni		Kokota
	Ana Akwanire		Chiwoko
	Chipika		Zambezi
	Kadyaubwerere	Mozambique	Gaba gaba
	Kaphulira		Persistente
	Mathuthu	USA	Jewel CIP440031
Rwanda	Gihingumukungu		Resisto CIP440001
	Ndamirabana		W-151 CIP440005
	Vita		Caromex CIP440136
	Kabode		Kandee CIP440140
	Terimbere		LOS-323 CIP440185
	Cacearpedo		Cordner
Kenya	Kakamega IP441768		W-119
	K566632		

Source: Kapinga et al. [21]. Catalog of orange-fleshed sweet potato varieties for SSA.

Table 1.
Orange flesh sweet potato available/released in some African countries and USA.

of undernourished and low-income groups that have limited access to diverse diets, supplements, and fortified foods [7].

2.2.1 Biofortified versus non-biofortified crop varieties

The principle of biofortification is to nutritionally improve the existing regular staple crops without altering the traditionally known identity of the crops. No significant differences are expected, rather improvement in all the attributes of biofortified cassava, sweet potato, and white maize compared to those of their non-biofortified varieties except for orange (sweet potato) or yellow (cassava) color, which is indicative of the beta-carotene present. For biofortified millet and sorghum, which are biofortified with iron and zinc, there is no difference in the physical traits between the two breeds except that their micronutrient contents, precisely iron, and zinc are higher in the biofortified ones. However, biofortified cassava, maize, and sweet potato have been reported to have less dry matter (more moisture content) and so softer in texture than the non-biofortified varieties. Breeders are working on these so that they can match up with the regular non-biofortified varieties, to boost farmers' adoption and consumers' acceptance. Improvement on biofortified varieties of the crops is being geared towards increased dry matter and recently released varieties like "solo gold" variety of OFSP in Nigeria has higher dry matter content. Every new variety released is an advancement on the former one, based on micronutrient concentration as well as farmer and consumer-preferred quality traits especially yield, drought tolerance, and resistance to pest.

2.2.2 Cassava—overview

Cassava is a major vegetable root staple in several countries of Sub-Saharan Africa, Latin America, and Asia. Of all root crops, it is the most important crop in Africa [22]. The world production of cassava is estimated at 242 million tons, out of which 54% (130 million tons) is produced in Africa, and 52% (68 million tons) of it from west Africa alone [23]. Nigeria is the number one producer of cassava in the world with an annual production of 59 million tons in 2019, followed by the Democratic Republic of Congo, 40 million tons. Ghana is the third African producer with an annual production of approximately 22.4 million tons [24, 25]. In terms of calorie contribution, cassava is the number three source of calories in the tropics [26], with about 500 million people relying on it for at least 10% of their daily caloric intake. In West Africa, cassava is a major source of carbohydrates in human diet, a well-placed, relatively cheap staple crop in developing countries as tpulp is an important source of energy with a calorific value of 250 kcal ha^{-1} day^{-1} [26]. However, cassava, which is mainly grown for its starchy tuberous roots and is a valuable source of cheap calories for low-income earners has now gained a strategic position in world trade. Besides its direct use as food, cassava is also used as a livestock feed and a raw material in the production of starch, tapioca, and snack foods [27]. As food, it can be boiled or roasted for consumption or can be milled into flour and used in making common dishes such as *Ugali* in East Africa as well as *garri* and *fufu* in West Africa. Dried cassava chips have varied applications by end users like breweries, confectionaries, starch, and flour for food. The crop is now produced for food and income, traded in different forms targeting diverse end uses; cassava flour, dried cassava chips, and raw cassava. In Uganda, 200,000 MT of cassava flour is consumed per annum, with most of it being traded in traditional informal markets [28].

2.2.3 Biofortified cassava (yellow cassava or vitamin A cassava)

Cassava is one of the staples targeted for biofortification as it is consumed daily by populations in some SSA countries like Nigeria, Ghana, Cameroon, Sierra Leone, Uganda, and DR Congo. Biofortified (vitamin A cassava) or yellow cassava is a relatively new breed of cassava that is rich in beta-carotene for improved nutrition of the consumers. In the African continent, it is being used as a vehicle to alleviate vitamin A deficiency through its biofortification with provitamin A (beta-carotene) by HarvestPlus in collaboration with International Institute of Tropical Agriculture (IITA) [19]. Under these collaborations, six biofortified provitamin A cassava varieties have been released in Nigeria namely; TMS 01/1368—UMUCASS 36, TMS 01/1412—UMUCASS 37 and 2014; TMS 01/1371—UMUCASS 38 and NR 07/0220—UMUCASS 44, TMS 07/0593—UMUCASS 45, and TMS 07/539—UMUCASS 46) and one in the Democratic Republic of Congo (Kindisa [TMS 2001/1661]) [19, 29, 30]. The yellow cassava varieties are similar to those of white in all attributes except for the color, which is an indication of its biofortification with beta-carotene. They are also high-yielding and resistant to many pests and diseases.

2.2.4 Sweet potato—overview

Sweet potato is an important root crop globally, with an annual production of 112.8 million tons in 115 countries in 2017, and China is the leading producer, followed by Nigeria with 3.9 million metric tons per year [31], Tanzania, Indonesia, and Uganda [32]. In recent times, although SP production and consumption have significantly increased in Africa, Asia, South American continents, and Caribbean islands, it is more profusely grown in Africa. International Potato Center (CIP) [33] reported that sweet potato is the third important food crop in seven central and eastern African countries, fourth priority crop in six South African nations, and eighth in four West African countries. SP, which is known as a food security crop due to its low agriculture input requirements [34], is recently changing to a significant cash crop. Sweet potato is a versatile crop that serves the roles of food and nutrition security as well as cash crop in both raw and processed forms. SP is an important root crop that can thrive in marginal soil with wide agro-ecological adaptability. In Nigeria, it can grow in all 36 states of the country plus the federal capital territory. It has a short maturity period of 3–4 months while its roots and vines are used for both human and animal consumption [35]. Sweet potato roots contain various kinds of physiologically functional components such as polyphenolics, anthocyanins, fibers, and carotenoids.

Value addition of sweet potato roots with these functions has resulted in their commercial utilization as an ingredient in confectioneries, noodles, alcoholic drinks, and beverages [36]. All varieties of SP are good sources of vitamins C, E, and K, several B vitamins, and the key minerals of magnesium and potassium. The leaves have appreciable levels of protein, and are widely used in the dairy industry in East Africa. It is a source of livestock feed with great potential as an industrial raw material.

2.2.5 Biofortified sweet potato (orange-fleshed sweet potato)

Orange-Fleshed sweet potato is a breed of sweet potato that is additionally rich in beta-carotene, a precursor of vitamin A through biofortification using conventional breeding practices, and drawing on the rich genetic diversity of sweet potato. OFSP is a proven, effective, and sustainable source of vitamin A, significantly contributing to the fight against vitamin A deficiency (VAD) in Africa [37–39]. Just 125 g of boiled OFSP roots can meet the daily recommended intake levels of vitamin A for a child. The orange color of OFSP shows the concentration of beta-carotene present; the deeper the orange color of the root flesh, the more the beta-carotene content present. OFSP as a staple food can serve as a cheap and sustainable source of Vitamin A in developing countries, where malnutrition is a big problem, and are growing 95% of the world's sweet potato crop. OFSP also contains antioxidants that help prevent degeneration of cells, as well as natural sugars, which are slowly released into the bloodstream, thus ensuring a balanced source of energy, without spikes in blood sugar that is associated with fatigue and weight gain. It again has vital, life-promoting phytochemicals that enhance protection from peroxides [40].

HarvestPlus and International Potato Centre (CIP) have developed and released several varieties of orange sweet potato with high vitamin A across sub-Saharan Africa. In Nigeria, three OFSP varieties are available namely; UMUSPO3-Mothers Delight, UMUSPO1-King J, and UMUSPO4-Solo Gold [35]. In Uganda, six varieties have been released, namely; Ejumula, Kakamega, Vita, Kabode, Naspot 12O, and Naspot 13O); and three in Zambia (Twatasha, Kokota, and Chiwoko). Zambia

Agriculture Research Institute has successfully completed the development of 15 new varieties of vitamin A fortified sweet potatoes [41].

OFSP is widely consumed as a vegetable dish (boiled, fried or roasted) as well as in different products through processing and value addition for improved household food intake. These include pastries, beverages as well as enriching existing indigenous foods, which are described in details by Phorbee et al. [35]. OFSP products can also be a source of income as they can be commercialized at all levels for income generation, job and wealth creation for all especially women and youths.

2.2.6 Carotenoids in biofortified sweet potatoes and cassava

Carotenoids are well known for their health-promoting benefits, which include immune boosting and reduced risk of developing some non-communicable degenerating diseases like cancers, cardiovascular diseases, cataracts, and mascular degeneration [42]. Carotenoids are made up of many other components that result in provitamin A activity. These include alpha-carotene (α-carotene), beta-carotene (β-carotene), beta-cryptoxanthin, Lutein, zeaxanthin, and lycopene. Among the carotenoids, α- and β-carotenes have a high provitamin A activity [42]. Orange fleshed varieties are appreciably rich in proVitamin A [43] with some having as much as 8000 μg β-carotene per 100 g of fresh weight while some Kenyan varieties have reportedly yielded 1240–10,800 μg per 100 g of fresh weight. However, carotenoids are known to be thermal and photo sensitive as they undergo degradation when exposed to heat and light, [44] and also through some processing techniques like cooking [45]. There is, therefore, the need to process OFSP with techniques that minimize carotenoid loss, so as to achieve the purpose of its biofortification.

Yellow cassava varieties are being grown and disseminated in many West African countries especially Nigeria, for their high concentrations of beta-carotene and being used to fight vitamin A deficiency. According to Harvestplus, yellow cassava can provide up to 100% of daily recommended vitamin A intake for women of reproductive age and children when eaten regularly [46]. Since cassava is a major part of many people's diets, introducing cassava bio-fortified with vitamin A is an excellent innovation and a significant contribution towards improving the food system transformation in the SSA.

2.2.7 Biofortified maize

Maize is a versatile cash crop grown for food, feed, and industrial purposes (an important source of sugar, oil, starch, and ethanol). For example, corn starch is an important raw material in pharmaceutical, food, and textile industries. The diverse end uses of maize globally have informed the basis for breeding higher yielding varieties of maize. Further research on maize has also led to the discovery of varieties that are naturally high in beta carotene contents, which HarvestPlus uses to breed high-yielding varieties of biofortified maize. These varieties have higher contents of provitamin A, which are being used to fight vitamin A deficiency is a major output in biofortification. In Zambia, three PVA maize varieties have been commercially grown namely, GV662A, GV664A, and GV665A. Also in Nigeria, four varieties have been released out of which two are hybrid; Ife maizehyb-3, Ife maizehyb-4, and 2 open pollinated varieties-Sammaz 38 and Sammaz 39 while one OPV CSIR-CRI Honampa has also been grown in Ghana since 2013 [47]. Malawi, Zimbabwe (ZS242) and Tanzania have also released PVA maize recently [48]. In a study conducted among Zambia, an increased pupillary response was observed among children who consumed PVA maize [48]. Breeders have also assessed antioxidants like tocochromanols, oryzanol, and phenolic compounds in PVA maize, which are health-beneficial [49].

2.2.8 Biofortified sorghum and millet

The prospects of breeding for micronutrients and beta-carotene rich sorghums have been discussed by Reddy et al. [50]. ICRISAT has successfully bred and released five lines of iron biofortified varieties of sorghum in India and two in Nigeria. Three of the Indian lines are hybrids (ICSA 661 × ICSR 196, ICSA 318 × ICSR 94, ICSA 336 × IS 3760) while two non-hybrid (ICSR 14001, ICSH 14002) and those in Nigeria are 12KNICSV-22 and 12KNICSV-188 whose iron content is three times higher than typically grown sorghum. The iron biofortified varieties of sorghum are bred and targeted at boosting iron intake of the malnourished populations especially northern Nigeria where sorghum is a staple cereal with relatively high production and consumption. These new varieties involved crossing local Nigerian germplasm with improved lines from ICRISAT (Mali).

Pearl millet is reportedly the cheapest source of iron and zinc [51] and large variation has been seen in its germplasm for these micronutrients [52]. In India, ICRISAT has also released iron and zinc biofortified pearl millet variety "Dhanashakti" and a hybrid ICMH 1201 (Shakti-1201). Many other commercial varieties and hybrids containing high content of iron and zinc have also been reported [52, 53].

3. Perishability of root and tuber biofortified crops at postharvest

Postharvest losses of food crops are traced to history especially in the tropics where the temperature is relatively high. This is a big challenge in the agriculture sector as over one-third of produce is lost after harvest [54] before they reach consumers. The losses, both physical and biological are due to poor management of the produce along the value chain, which are poor packaging from field after harvest and use of inappropriate packaging materials, transportation, poor handling in marketing and display of produce for sale, exposure to heat and sunlight, which subject fruits to undue ripening, lack of good storage facilities and conditions prior to sales. Generally, the key actors of agricultural produce (producers, wholesalers, and retailers) in the SSA lack capacities and facilities to maintain highquality and safe perishable plant produce from farm to table [55]. Losses of perishable crops have implications on quality, quantity, market value, and safety of the produce. According to RAS (2015), insufficient and poorly maintained transport and market infrastructure for handling food products in both urban and rural areas have frequently resulted in high level of waste and spoilage [56].

OFSP and VAC crops, like the non-biofortified varieties are perishable after harvest. Fresh sweet potato having relatively high moisture contents are very sensitive to microbial spoilage, even at refrigerated conditions, hence they must be consumed within a few weeks after harvest or be processed into various products. Cassava can barely stay for 48 hours post-harvest before physiological deteriorations start. Also, cassava roots are bulky and therefore transportation from farm to market or other destinations within the value chain can be challenging in term of cost and stability, thus reduced quality, quantity, and market value. More so, most of the farmers are small holders who harvest manually so the roots are at risk of bruises, which stress and damage the roots. Also, during packaging and transportation, skin of the roots could remove, causing more bruises to the root and opening them up for rapid spoilage [57]. The packing sacks are also often times over filled with the crops, which can further impair the roots.

4. Postharvest technologies for enhanced shelf stability and nutrient retention

To manage postharvest losses, processing is very essential to preserve the crops as well as add value to the crops for food product diversity and improved commercial competitiveness.

4.1 Postharvest utilization and processing of OFSP

4.1.1 Utilization

The high content of provitamin A, which is important to health in OFSP has enhanced its utilization in processed forms like diced, mashed, or pureed OFSP. In school feeding program, OFSP in puree or other forms has been added to school menu in some countries like US, Nigeria, Ghana, etc. to boost the nutritional quality of those meals. CIP research has shown the importance of OFSP puree in bakery as it has been able to develop an acceptable OFSP puree-wheat flour composite bread in which 45–50% is OFSP puree. The bread is not just rich in pro-vitamin A but also cut down on quantity of sugar and oil in the formula while reducing use and dependence on wheat flour, which is often imported in some Africa countries, thus cutting down production cost significantly. The OFSP puree-based products are healthier as they retain more nutrients especially pro-vitamin A. Sweet potato purees and powders can also be used as thickening and gelling agents to impart desired textural properties, and at the same time enhance the nutritional values, antioxidant activity as well as natural color (e.g., orange and purple) of many food products. With increased urbanization, there is growing demand for convenient and healthy foods, which sweet potato purees can easily fit in as functional ingredients in processed foods. More so, the novel advancement of processing OFSP puree aseptically with continuous flow microwave heating, is an opportunity for industries to produce nutrient-rich and shelf-stable puree for institutional use in social protection programs.

Sweet potato is a major root crop utilized widely for diverse food products as well as feed and industrial products. As food, sweet potato roots (both orange and white fleshed) are mainly consumed domestically as boiled or fried. In Kenya, 90% of the sweet potato produced is used domestically as food. Use of vines as fodder and leaves as a vegetable is common in some parts of Western Kenya [58].

4.1.2 Processing of OFSP and products

Despite the importance and knowledge of the nutritional benefits of OFSP, it remains generally underutilized, possibly because the roots are perishable, which reduces their market value. Fresh sweet potato is eaten boiled, steamed, roasted, or fried in most African communities [59], so processing OFSP roots into products create more options for consumption, improve availability, and reduce losses. The fresh roots of sweet potato contain high moisture content (50–70%) and thus relatively low mechanical strength. They also have a very high respiratory rate, which generates heat that aids softening of the roots, and eventually leads to damage. The shelf-life of sweet potato roots varies from few days to months depending on the cultivar and storage conditions [60]. SP storage roots are subject to several forms of postharvest losses right from harvesting, transportation from farmers' field to market and storage. These are due to mechanical injuries, weight loss, sprouting, diseases, and pests [61]. Since sweet potato does not stay for long after harvest, there is a need for processing and value addition. Also, for diversity in use/consumption and commercialization, sweet potato is processed in various ways, using various techniques, which include cooking, fermentation, and drying. Cooking could be boiling or frying, roasting/baking of the roots while other techniques, either single or multiple, result in the development of other value-added food products for household consumption and/or income generation [35].

Processing or value addition, as a way of diversifying utilization of SP is greatly dependent on the roots' cultivars. SP cultivars' screening, targeting diverse end uses is an important activity in processing and value addition because not all SP root varieties are suitable for all end products. CIP and other related research institutions nationally and internationally test SP cultivars' suitability for diverse end uses. They develop the best food products from different cultivars of SP, promote and transfer the technologies to local farmers and processors. In some cases, they scale up the technology transfer by partnering with selected private companies (medium-large scale).

Postharvest processing of sweet potato involves primarily grading, sorting, peeling, cleaning, etc., and secondarily product making. Sweet potato processing technologies into puree and other forms are available in various parts of the world so that they can be used as functional food ingredients in many food products. The processing operations involved in these technologies and their effect on quality, storability, nutritional values and rheological properties of SP purees and powders/flours have been reviewed by Truong and Avula [62].

Available processing technologies at all levels are the key drivers of promoting SP production and consumption. All parts of SP are useful in product making; the roots can be processed into chips, crisps, flakes, flour, granules, starch, and alcoholic beverages [63]. Also, the leaves, into powders and used as functional ingredients in food products like ice cream, juices, tea drinks, and bread due to their high phenolic content and antioxidant activity [64]. In many of the SSA countries, SP has been processed into intermediate and/or finished product [63] to make both traditional and novel OFSP-based products. Some traditional foods have even been enriched with OFSP, because of its high beta-carotene content. In Uganda, an array of novel and traditional food products has been made from OFSP, namely composite flours, chapatti, mandazi, juice, bread, doughnuts, and other confectionary products. Traditional ones, which are produced by the local farmers are pit stored tubers, *Amukeke* (dry white slices), *Inginyo* (dry chips, chunks), *Amukeke* flour and *Inginyo* flour [65]. Also in Kenya, sweet potato processing is reportedly processed into different traditional products like *mandazi bhajia*, among others [58]. In Nigeria, Phorbee et al., [35] produced a recipe book on OFSP and VAC as household guide for processing the two biofortified crops into various nutritious foods. OFSP has also been used to improve carotenoid contents of some Nigerian indigenous foods. One such is the use of OFSP as a functional ingredient in making a local beverage called *Kunu* (a popular local cereal-based beverage) where it serves as both sweetener and nutrient enhancer. Others are like OFSP-enriched *Amala, fufu*, semolina, pap, etc. Although mainly at household level and micro/small to medium processing scale, OFSP has also been processed into intermediate products like chips, flour, and puree to make finished OFSP-based pastries (chin-chin, puff-puff, doughnuts, buns, etc.) and confectionaries to increase market options for sweet potato products.

The processing technologies are usually developed by the national and international institutions, and then transferred for adoption, to local NGOs, community health workers, women, and youth groups directly or through training of Trainers.

4.2 Puree overview

Purée is a cooked food, usually from vegetables, fruits, or legumes, that is ground, pressed, blended, or sieved to a creamy consistency [66]. It is usually a very smooth lump-free food made from a specific food, which the puree is named by, for example, applesauce, mashed potatoes, or tomato purée. Pureed foods are easier to digest as pureeing is like chewing, with the food partially broken down and easier for the system to absorb. Pureeing can be done manually or mechanically. Manual pureeing, depending on the food, can be achieved by simmering, boiling, until it is very soft and then, mashed with a fork or ladle, for example, potato. It can also be done by pushing the food through a strainer, or crushing it in a container until smooth and even in consistency, for example, banana. Pureeing can equally be done using a blender, food processor, or food mill, for example, tomato. However, purées generally should be cooked, either before or after grinding, to remove contaminants, reduce moisture, improve flavor and texture.

4.3 OFSP puree

Value addition through processing of agricultural products such as OFSP roots to puree is key to ensuring a stable supply of highly nutritious products to consumers. OFSP puree is an ingredient in many foods, including baby foods, casseroles, puddings, soups, pies, cakes, ice creams, breads, and other products [36, 67–69]. Sweet potato purees are also used in fruit/vegetable-based beverages and restructured products as well as commercial ones like jam, ketchup, flakes, and powders while various fermented food products have also been explored [36, 70–74].

OFSP is naturally sweet and when used in baked goods and confectionaries, there is a reduction in the sugar, fats, and oil used thus more nutritious and healthy foods for less cost. With the efforts of CIP in many developing countries, adoption of OFSP puree has empowered women and youths in developing new OFSP-based food products and created opportunities for income generation along the value chain. Farmers are becoming more confident to grow OFSP because there is market for it while puree processors are encouraged to process because there is demand for it. For instance, in Kenya alone, demand for OFSP puree is valued at more than USD 5 million annually. Creating demand for OFSP has increased the crop's market value, encouraging more farming households to grow and consume it, while making pro-vitamin A products available to consumers [75]. Similar trends are found in other African countries like Ghana, Tanzania, Rwanda, Nigeria, and Uganda although no available data on the market value.

4.3.1 Puree technology for OFSP processing

Use of puree is growing in bakeries, eateries, and restaurants globally [76] and its technology from sweet potato has been developed in industrialized countries like USA since the 1960s [77, 78] as well as in developing countries, especially Africa since the 1990s. OFSP puree processing technologies have advanced over the decades from traditional methods of manual mashing of cooked roots to highly sophisticated and automated systems.

Several techniques have been developed for puree processing in order to produce purees with consistent quality, despite variations in carbohydrate content and starch degrading enzyme activities due to cultivar differences, and postharvest handling practices [67, 79]. Process operations for pureeing of sweet potato include washing, peeling, hand-trimming, cutting, grinding, pre-cooking/finish-cooking. The cooking temperature-time must be programmed to suit the enzymatic starch conversion in order to produce puree with desired maltose levels and viscosities. For SP puree that are thermally processed, the starch is gelatinized and produces thick slurry puree with cooked feel that may not be acceptable in juicy products. This is a limitation to the heat treatment technology of SP puree but fortunately overcome with an alternative approach of grinding the raw sweet potato and treating with acid to inactivate oxidizing enzymes during juice extraction. At the same time, the ungelatinized starch and flour with high dietary fiber can be recovered as other by-products from this alternative process [80]. **Figure 1** shows the schematic representation of SP puree processing with all the unit operations highlighted for puree preserved by both freezing and use of preservative. Raw sweet potato roots are peeled either by abrasive rollers or steam flashing, followed by thorough washing, trimming, and cutting into slices, strips, chunks, or dices. The cut roots are steam cooked and then passed through a pulp finisher to make the purees. The peeled sweet potato roots are cut into desired shapes with specific sizes recommended by Walter and Schwartz [81]; For cubes, it is 2 cm; strips, 2 × 2 × 6 cm; and slices, 0.5–0.95 cm thickness or mashed using a hammer mill with rotating blades to chop and push the materials through a 1.5–2.3 mm mesh screen [81]. The cut or mashed roots are then steamed blanched at 65–75°C to activate the amylases and gelatinize the starch for hydrolysis. For the sliced, striped, and cubed roots, hammer mill is used to pulverize the blanched materials into puree. For puree that targets high maltose content, the blanched puree is pushed into a surge tank and hold at 65–75°C for more starch hydrolysis [68].

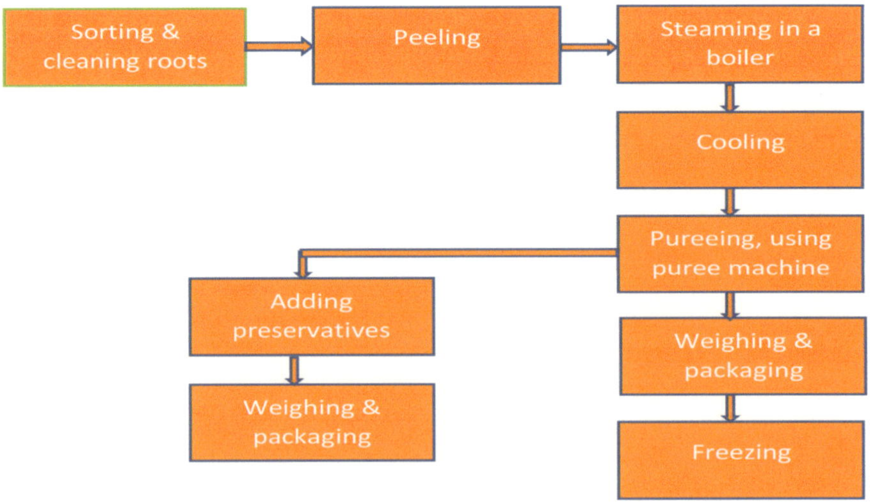

Figure 1.
Flow chart of OFSP Puree processing. Adapted from Owade et al. [95].

Hoover and Harmon developed another technique called "enzyme activation technique" for processing sweet potato puree [82, 83]. The technique, which is now commonly used in food industry uses endogenous amylolytic enzymes to hydrolyze starch in sweet potato to process the puree.

SP puree can be further processed and used in other forms for various purposes in food industry. SP flour from puree can be made by drying while extrusion technology and chemical treatment are explored for specific use of the flour. Drying of SP puree can be through high-tech drum or spray drying but in many African countries, solar and mechanical drying in cabinets is common in producing sweet potato dried chips which are then milled into flours [84, 85]. For OFSP flour from puree, choice of drying technology is important, technologies that retain the provitamin A should be the priority so as not to defeat the purpose of its biofortification. Change of flour color from orange to white after drying is an indication of carotenoid loss to drying, which should be avoided as much as possible by choosing technologies that prevent prolonged exposure to heat and sunlight. With high level of carbohydrate, B-carotene (orange-fleshed varieties), and anthocyanin (purple-fleshed varieties), SP purees and dehydrated forms can be used as functional ingredients to impart desired textural properties and phytonutrient content in processed food products [62].

Fermentation (bio-processing) of OFSP is another processing technology that can produce functional foods and beverages such as sour starch, lacto-pickle, soy-sauce, acidophilus milk, etc. through either solid-state or submerged fermentation [86]. These foods are opportunities to diversify OFSP utilization options, increase use and consumption as well as commercialization for improved health and wealth.

4.3.2 Puree technology for OFSP preservation and packaging

Purees processing technologies go hand in hand with preservation and packaging, which are achieved by various methods namely; low temperature storage (refrigeration and freezing), canning, aseptic packaging and chemical treatment of puree for prolonged shelf-life over its supply chain.

The finish-cooked puree can be packaged in can to produce a shelf-stable product or in plastic containers for a low-temperature storage (refrigeration or freezing) [79, 86, 87]. However, each of the two preservation approaches has its constraints in puree processing. For canning, in as much as it does not require special storage facilities and conditions, the finished product is prone to some sensory (flavor, texture & color) and nutrient degradation. This is because the canned SP (a low-acid food, pH 5.8–6.3) puree is subjected to high conductive heat treatment for a long time (e.g., 165 minutes at 121°C for an institutional #10 can size). On the other hand, if the sterilization is done through slow rate of heat transfer from the wall to the center of the can, there is a limit to the can size and number that can be produced, which again restricts production capacity of the industry and availability of SP puree for use as a food ingredient.

Also freezing, which is a long-known preservation technique with relatively less sensory and nutrient degradation is capital intensive in term of energy, storage space, distribution logistics. It has restricted product package sizes and above it, the puree has to be defrosted before use, which is not user-friendly. With these limitations, few food companies are into commercial production of canned and frozen puree even in developed countries and almost none in the SSA countries [84].

However, to address the limitations of canning and freezing in producing high quality shelf-stable purees, aseptic packaging and continuous flow microwave system for rapid sterilization are being used [62].

Aseptic processing uses the principle of high temperatures (≥125°C) short time (HTST) to produce a higher quality puree with comparable level of microbiological safety as that of a conventional canning system [88]. Coronel et al. developed a process for rapid sterilization and aseptic packaging of OFSP purees using a continuous flow microwave system operated at 915 MHz [88]. In this process, the SP puree is loaded into a hopper, and pumped through the system. Microwaves from a generator are delivered to sterilize the puree at 130–135°C, retain in the holding tube for 30 seconds, rapidly cool in a tubular heat exchanger, and then aseptically package in aluminum polyethylene laminated bags [89]. The process is short and produces at least 1 year shelf-stable product, packed in flexible polythene bags, with relatively less sensory and nutrient degradation. The process is protective of micronutrients so there is the possibility of retaining at least 85% of carotene and anthocyanins in the finished puree. **Figure 2** shows an overview of typical puree machine with aseptically packaged OFSP puree. CIP has also developed a shelf-stable, vacuum-packed OFSP puree that is increasing the supply of OFSP puree and making it available at all seasons.

This process is reportedly suitable for OFSP as well as purple-fleshed sweet potato puree processing [90] especially in the developing African countries. It has opened

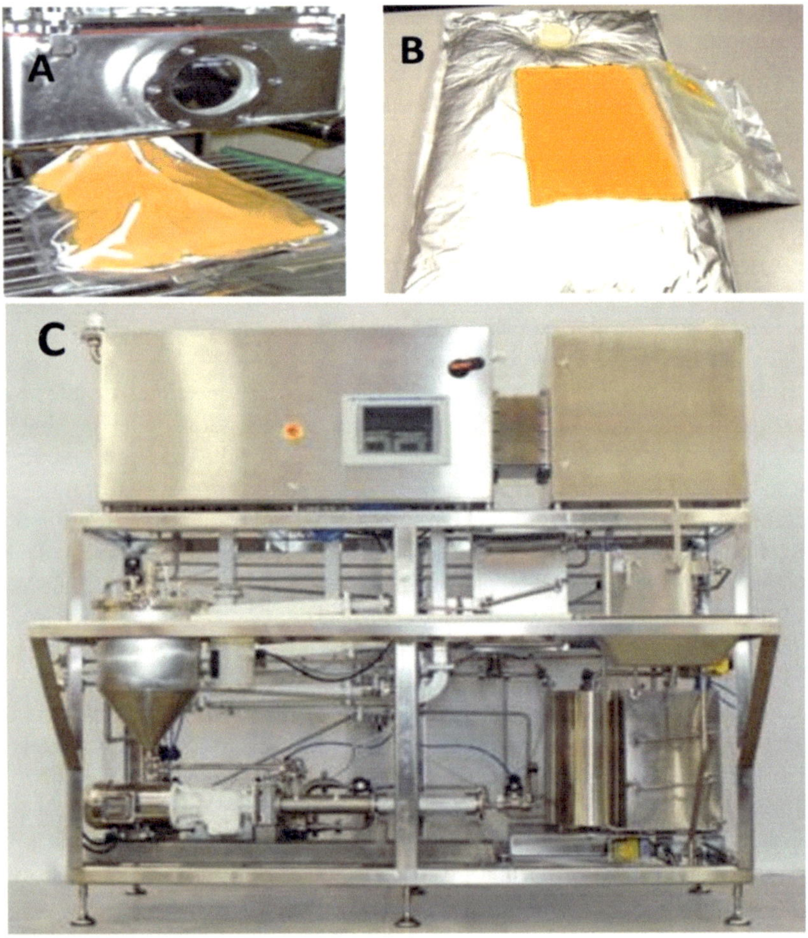

Figure 2.
A typical puree machine with the aseptically packaged OFSP puree. Source: Moyo et al. [93].

up new market opportunity for the SP industry generally and can also be applied to purees from other fruits and vegetables [91].

With the recent commercial development of the microwave-assisted processing and aseptic packaging of sweet potato purees, it is expected that more processed food products from the puree will be developed. African countries, precisely Kenya, Rwanda, Malawi, and Uganda among others, have been growing gradually in the commercialization of OFSP puree and the subsequent wheat flour substitution in bakery products. Private companies in Malawi and Kenya are now manufacturing OFSP puree and selling it to bakeries that substitute OFSP puree for up to 40% of the white wheat flour in bread and other baked goods. Recently, orange-fleshed sweet potato puree has replaced 20–50% of wheat flour in cookies, donuts, and breads by some commercial bakeries in Ghana, Kenya, Malawi, Rwanda, and Uganda [92]. Some processing companies in some African countries have been committed to OFSP processing and product development using OFSP puree-Tehila Bakery and Value Addition Center in Malawi; Organi Limited and Euro Ingredients Limited in Kenya; Sinagerard in Rwanda; Sanavita, SUGECO and Better Markets for Crop Products Limited (BMC) in Tanzania; Farmorganics Nigerian Limited in Nigeria. These have resulted in positive impacts on income generation for small-scale farmers and businesses, employment opportunities for women and youths, and improved nutritional status of target communities are some of the targeted outcomes. With these innovative processing technologies and successful piloting of the product, OFSP puree has been described as "breakthrough product" for Africa that offers the much-needed nutritious products, with consumer accepted organoleptic properties [93]. However, further work needs to be done in scaling up through more public awareness and education on its multiple health benefits for all, for consumption and commercialization.

4.3.3 OFSP puree in bread bakery

Bread has been an important, common exotic cereal product consumed by most individuals in Africa. Incorporating OFSP puree into bread would significantly increase the number of OFSP consumers and reduce VAD [94, 95]. According to Wanjuu et al., "OFSP puree can replace up to 50% of the wheat flour in bread, while reducing sugar (90%), fat (50%) and eliminating artificial colorings (egg yellow). The baked bread retains over 50% of the β-carotene, and the OFSP puree improves the texture of wheat products, making them easy to chew and digest" [96]. The composite bread has a better sensory quality (flavor, color, and soft texture), which contribute to its acceptability. Other benefits include low production costs and improved vitamin A content [97]. The OFSP puree-based bread is commercially available across sub-Saharan Africa (SSA) and is being promoted for its added nutritional benefits, which is increased β-carotene. This serves as a good medium for intake of β-carotene and to alleviate vitamin A deficiency (VAD) especially among the vulnerable populations in SSA. OFSP puree can replace some of the white, wheat flour in baked and fried products especially in the SSA where some of the countries import wheat flour at very huge costs. According to Moyo et al., African millers spend millions importing wheat, with East African countries being among the top importers. Kenya's wheat import bills are estimated at $250 million, Tanzania's at $150 million, Uganda's at $53 million, and Rwanda's at $35 million per year [93]. Substitution of wheat flour with OFSP puree up to 50% can significantly reduce dependency on imported wheat flour, enhance utilization and consumption of locally produced OFSP, create jobs for smallholder farmers, women and youths, and ultimately contribute to national economy.

OFSP puree in bread and other baked products makes puree a sustainable solution to the perishability and all-year unavailability of the crop [96]. It also encourages expanded production, utilization and consumption of OFSP-based baked products as well as meeting consumers' daily requirement of vitamin A either fully or partially. According to Wanjuu et al., (2018), bread made with OFSP puree had a longer shelf-life than the conventional white bread from 100% wheat flour, probably because of the significantly higher water activity in white bread than in the OFSP bread [96]. Using OFSP, either as flour or puree in bread making has implications on its sensory characteristics, and quality control of wheat flour-based bread [98]. OFSP pureed bread has been described by Olatunde et al. as having deeper brown colored crust, softer crumb, more uniform crumb cell, higher loaf volume (872–885 cm^3), specific volume (4.59–4.76 g/cm^3), oven spring (0.50–1.00 cm), softness (18.35–20.20 mm), crust moisture (18.05–18.17%) and consumer acceptability (7.14–7.50). In terms of consumer acceptability, bread from OFSP puree was more acceptable than that of OFSP flour [98].

4.4 VAC processing and products

Like the white varieties of cassava, yellow cassava also starts to spoil within 2 days after harvest, hence the need to process the roots into various intermediate and finished products through different processing techniques.

Yellow cassava roots can be dried in peeled, cut into chunks/sizes, or grated form for different value-added products. As much as possible, drying conditions for yellow cassava should be protective of the provitamin A (carotenoid) in it.

Similar to the white varieties, yellow cassava can be processed into traditional foods such as *gari, fufu, akpu, tapioca*, and starch; the only difference is the color of the food products, which in this case is the yellow color and it is an indication that provitamin A is present in the food. It is an added advantage over the white varieties. **Figure 3** shows various traditional products from cassava processing-*Pupuru, Garri,*

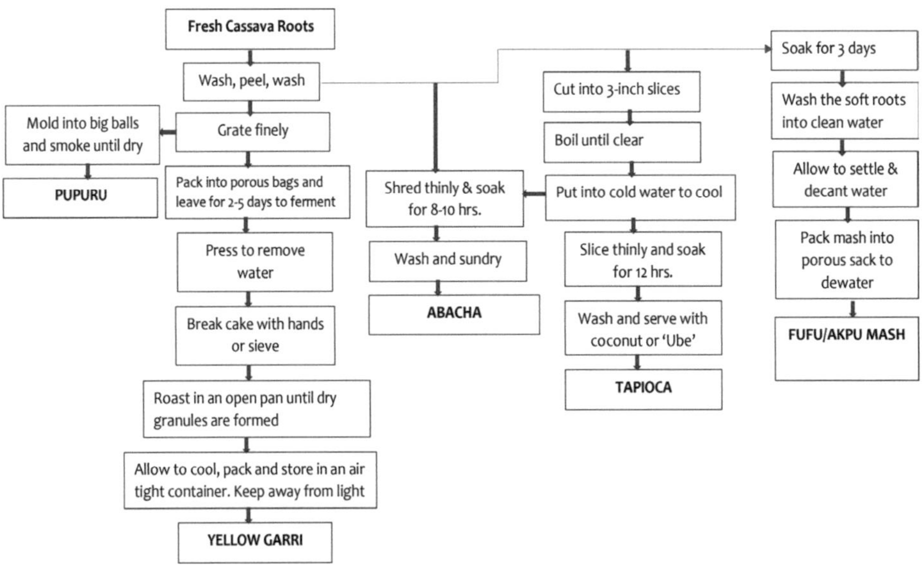

Figure 3.
Traditional VAC products. Source: Phorbee et al. [35].

Fufu/Akpu, *Abacha*, and Tapioca, which are processed through various unit operations like grating, pressing (dewatering), soaking, fermentation, roasting, etc.

Novel food products like pastries and some confectionaries have also been made with yellow cassava through high-quality cassava flour or grated and dewatered fresh roots, for both household dietary diversity and income generation/livelihood. Some Nigerian recipes have been developed for novel VAC-based foods, which use high-quality cassava flour for household consumption of carotenoid-rich foods as well as income generation at micro scale [35]. These include VAC pastries and snacks, puree, etc.

4.5 Drying technologies of cassava and VAC

Drying is a unit operation aimed at removing nearly all the free water present in a food stuff [99]. The commonly used methods of drying cassava in SSA include open sun drying on bare ground, raised platforms, road sides, roof tops, and tarpaulins as well as use of solar dryers. The selection of an appropriate drying method is necessary to ensure good quality products and prolonged shelf life.

4.5.1 Sun-drying of pressed cassava mash

Sun drying is an ancient drying technology of cassava when processing into chips, flour, or starch. Fresh cassava roots are highly perishable as they contain 65–70% moisture, and take time to sun dry. Prolonged sun drying has implications on the product quality as it is prone to spoilage and/or contamination from rain, wind, dust, insect infestation, animal attack, and microbes. Reduction of moisture is a key step in processing cassava roots into flour and must be done quickly to avoid lowering product quality. It is therefore better to remove as much water as possible by grating and pressing the cassava mash in a jack press for about 2 hours (dewatering) before drying. Sun drying is done by spreading the wet product on a clean concrete drying floor or a black plastic sheet laid on top of a concrete drying floor. The wet product needs sunlight, dry air (low humidity), and good airflow over it to dry effectively. Sun drying, therefore, depends on availability of sunlight and so it is challenging during raining season, which limits availability of cassava flour or starch all year round. Also, it is best suited for small-scale rural operations where product volumes are low (50–100 kg of dry product per day) [100]. Furthermore, sun drying exposes cassava to sunlight and so not suitable for biofortified cassava varieties.

Just like white cassava varieties, yellow cassava varieties have been reported to be relatively high in moisture content and so prone to deteriorations after harvest [101], thus necessitate post-harvest processing. However, yellow cassava being biofortified with carotenoid (pro-vitamin A), is thermal and photo sensitive so exposure to sunlight during drying can lead to significant loss of the carotenoid and therefore defeat the purpose of its biofortification.

To preserve the carotenoid content and also process yellow cassava roots into chip or flour all year round, alternative drying technologies that are protective of the biofortified vitamins in the cassava roots have been exploited over time. These include cabinet/tray dryer, drum dryer, solar dryer, and flash dryer.

4.5.2 Flash drying technology

The flash drying technology is a high-precision indoor novel drying technology that involves instant drying of a wet material by passing hot air through it, to quickly evaporate free moisture in the material, using rapid heat transfer. Flash drying is an efficient and effective drying technology as the process facilitates fast evaporation through speedy heat transfer. The technology can be applied to food, feed, and some other industrial materials. The drying process, being rapid makes it a suitable technology for nutrient-dense foods as the short exposure to heat favors retention of nutrients especially the thermally sensitive micronutrients.

4.5.2.1 Principle of flash drying

The principle of flash drying is instant evaporation of free surface moisture from a wet material mixed with hot air. Wet particle material is passed through hot air or steam where the particles are dried almost immediately and the gas or steam temperature decreases due to heat transfer [102]. The wet material, either in paste or cake is pulverized into fines and increase the surface area, to hasten the drying process. The dried material (usually in powder form) remains suspended in air and conveyed while drying [103]. Being a short-time dryer, it is very suitable for biofortified provitamin A cassava as it protects the beta carotene in the cassava unlike sun drying with a prolonged exposure to sunlight. According to Kuye et al., the output of flash drying technology is significantly improved compared to those of general drum or cabinet drying. Through it, users can achieve higher economic benefits in the short term.

4.5.3 Flash drying technology for VAC

For cassava, flash drying technology came at an opportune time, using flash dryer, which was fabricated and introduced in Nigeria by CAVA-Cassava: Adding Value in Africa project in 2016 [28]. Flash drying is achieved with flash dryer, a high-tech equipment, designed to dry pressed (dewatered) cassava mash into High Quality Cassava Flour (HQCF) within seconds, and can process 3–7 tons of HQCF per day. Flash dryers have been reported suitable for drying mash and solids of moisture content between 30 and 40% and applicable not just in food industry but also other industries like feed, chemical, and pharmaceutical [28]. A flash dryer is generally constructed with devices, which can simultaneously dry, pulverize and classify materials by particle sizes. It is a continuous drying device specially designed for drying cake, paste and muddy materials. A flash dryer must be able to resist the intense heat and rapid movement required in the drying process, so it must typically be fabricated with strong, detailed construction materials and accurate temperature control device. While other features of the dryer are specific to intended use, it is usually customized for purpose. Flash dryers, although expensive in fabrication, is low in energy consumption, high in thermal efficiency, occupies small area thus saving factory space and efficient in continuous mass production.

Specifically for drying cassava and VAC, the flash dryer by feature, typically consists of a blower, air pre-heater, feeding mechanism, (hopper, pulverizer, screw conveyor, and a rotary air lock), dryer, cyclone, and filter as shown in **Figure 4**. Wet-pressed (dewater) cassava to be dried is loaded into the feed mechanism while the air, after passing through an air filter, is heated in a hot air generator (steam can also be used). The wet cassava mash is circulated into the hot-air for thorough mixing. The pulverized cassava mash and the hot air mixture is conveyed through the dryer to the cyclone using pressure from the blower. The cassava mash gets dried through quick moisture loss and the moisture absorbed by the hot air, thus lowering the air temperature (and increasing the humidity). The mixed air and dried cassava flour

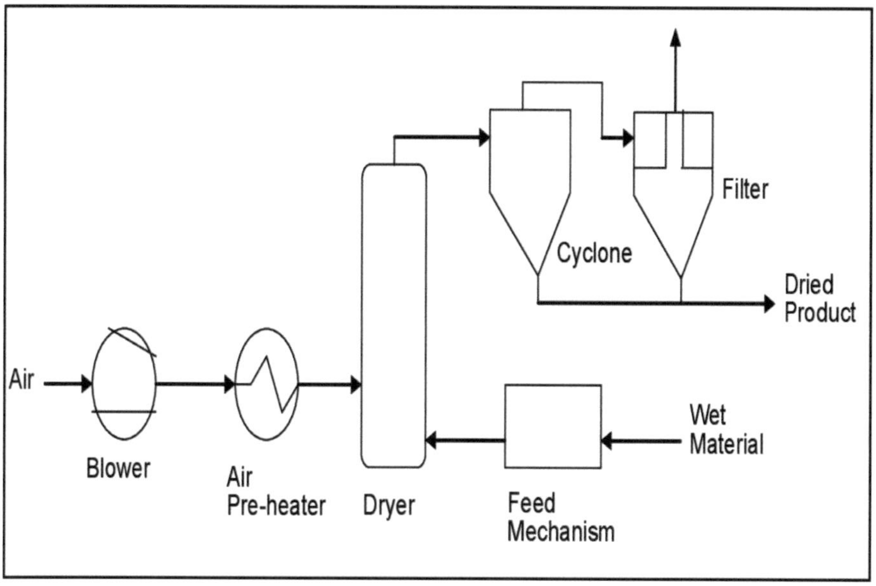

Figure 4.
Schematic diagram of a Flash Dryer: Source: Kuye et al. [103].

are separated in the cyclone, and the flour is let out from the cyclone through the discharge valves. Fine cassava flour particles that escape from the cyclone are trapped by a bag filter. Kuye et al., gave the specification of all the devices of a typical cassava flash dryer [103].

This flash dryer is not only for making high quality cassava flour, but can also be used to dry cassava starch and *fufu* cake for dried cassava starch and instant *fufu* flour, respectively. VAC can also be used to make *fufu* mash and instant *fufu* flour, which are cooked at home as a meal ("swallow") and consumed with any choice of soup in West Africa. Use of flash dryer is very suitable for VAC as it is a rapid drying process, which favors retention of the pro-vitamin A in the biofortified cassava. With flash-dried cassava flour and *fufu* from VAC, consumers' access to vitamin A is guaranteed because of high retention of the vitamin after processing. Consumption of instant *fufu* flour is growing in Nigeria and Ghana due to increased urbanization and reliance on instant convenient foods. This presents a good opportunity for the consumers to be reached with VAC *fufu,* dried with flash dryer that retains the beta carotene in the biofortified cassava and contributes to the fight against vitamin A deficiency in the consuming countries of SSA. However, increased consumers' acceptance and access to these products is still necessary.

5. Conclusion

With the advances made so far, on appropriate post-harvest technologies, processed biofortified crops have potentials to improve food and nutrition security in the sub-Saharan Africa, if fully exploited. Processing the biofortified crops into products improve availability, reduce post-harvest losses significantly, and create more options for consumption and commercialization, as the crops can be processed in various ways, using various techniques of cooking and drying. With these available technologies, the shift from non-biofortified to biofortified food consumption is a significant move towards improved food system transformation. OFSP puree with the SSA-friendly technology of aseptic packaging and continuous flow microwave system of rapid sterilization ensures a stable supply of high-quality shelf-stable nutritious product, with minimal carotenoid degradation to consumers. The technologies are indeed breakthrough within the biofortification sector in the SSA, leveraging on the versatility of the crops especially OFSP, which contains some functional components. OFSP puree is an important intermediate product with existing and potentials for diverse finished products in the food industry especially confectionaries and beverages. Similarly, biofortified cassava processing into flour, starch, and instant *fufu* flour using the flash drying technology of rapid evaporation through heat transfer, produces high-quality products that ensure delivery of pro-vitamin A to consumers. With cassava being a key staple in many SSA countries, consumer shift from white to yellow cassava will be a major revolution in household reach with vitamin A especially the rural consumers who are easily left out of the micronutrient interventions.

The shelf stability of the products from these advanced technologies also favors sustainability of consumers' access to nutritious crops thus realizing the purpose of biofortification, which is to contribute to the fight against vitamin A deficiency. In response to food insecurity and poor livelihood, the available technologies for enhanced biofortified nutritious food product development, present opportunities to significantly contribute to improved food system through enhanced utilization and consumption of locally produced crops, create jobs for smallholder farmers, women and youths, and ultimately contribute to national economy. OFSP puree makes consumer-acceptable bread, which compares very well with 100% wheat flour bread organoleptically, with increased micronutrients and functional components, reduced production cost from reduced use of wheat flour and sugar.

However, there is need for continued research on biofortified crop cultivar screening targeting more diverse end uses, with technology transferred to the end users through private-public engagement. This way, interest of the key actors of the value chain (Breeders, Farmers, Processors, and Marketers) will be stimulated towards biofortified crops. This is an awareness/adoption strategy, which should also be scaled up with favorable polices.

Acknowledgements

The authors wish to acknowledge Oluwole, Funmilola Akanbi and other staff of Food Agriculture Nutrition Network (FANN) for their support in reading through and editing the manuscript severally. Thanks to Blessing and Emmanuel for ensuring adherence to the author's guidelines.

Conflict of interest

The authors declare no conflict of interest.

Author details

Olapeju Phorbee[1,2]*, Ganiyat Olatunde[3], Dotun Aderonmu[4], Desmond Ikerionwu[4], Adeola Ojo[5,6], Gift Oguzor[7], Lateef Sanni[3] and Adeyinka Onabolu[8]

1 International Institute of Tropical Agriculture, Ibadan, Nigeria

2 Food Agriculture Nutrition Network (FANN), Abuja, Nigeria

3 Federal University of Agriculture, Abeokuta, Nigeria

4 Helen Keller International, Nigeria

5 Dorcas Eunice Foundation, Ibadan, Nigeria

6 Formerly Raw Material Research and Development Council (RMRDC), Abuja, Nigeria

7 Rivers State Primary Health Care Management Board, Port Harcourt, Nigeria

8 Federal Ministry of Agriculture and Rural Development, Abuja, Nigeria

*Address all correspondence to: ophorbee@foodagricnutritionnetwork.org; olapejuphorbee@gmail.com

References

[1] Development Initiatives. 2018 Global Nutrition Report: Shining a Light to Spur Action on Nutrition. Bristol: Development Initiatives; 2018

[2] UNICEF, WHO and World Bank Group. (2015). Levels and trends in child malnutrition: Key findings. Available from: http://www.unicef.org/media/files/JME_2015_edition_Sept_2015.pdf.

[3] Bain LE, Awah PK, Geraldine N, Kindong NP, Sigal Y, Bernard N, et al. Malnutrition in Sub-Saharan Africa: Burden, causes and prospects. The Pan African Medical Journal. 2013;**15**:120. DOI: 10.11604/pamj.2013.15.120.2535 PMID: 24255726; PMCID: PMC3830470

[4] Wirth JP, Rajabov T, Petry N, Woodruff BA, Shafique NB, Mustafa R, et al. Micronutrient deficiencies, over- and undernutrition, and their contribution to anemia in Azerbaijani preschool children and non-pregnant women of reproductive age. Nutrients. 2018;**10**(10):1483. DOI: 10.3390/nu10101483 PMID: 30314363; PMCID: PMC6213945

[5] UNICEF. Vitamin A deficiency. 2021. Available from: https://data.unicef.org/topic/nutrition/vitamin-a-deficiency/#

[6] Hodge C, Taylor C. Vitamin A deficiency. In: StatPearls [Internet]. Treasure Island, FL: StatPearls Publishing; 2022. Available from: https://www.ncbi.nlm.nih.gov/books/NBK567744/

[7] Phorbee O, Gift O, Ojo A, Olatunde G, Onabolu A. Advocacy and use of advocates as a quick win in scaling up biofortification in Nigeria: The case of Building Nutritious Food Basket (BNFB) Project. Food and Nutrition Sciences. 2022;**13**:797-810. DOI: 10.4236/fns.2022.139057

[8] Gilani GS, Nasim A. Impact of foods nutritionally enhanced through biotechnology in alleviating malnutrition in developing countries. Journal of AOAC International. 2007;**90**(5):1440-1444

[9] Perez-Massot E, Banakar R, Gomez-Galera S, Zorrilla-Lopez U, Sanahuja G, Arjo G, et al. The contribution of transgenic plants to better health through improved nutrition: Opportunities and constraints. Genes & Nutrition. 2013;**8**(1):29-41. DOI: 10.1007/s12263-012-0315-5

[10] Federal Government of Nigeria (FGN) and the International Institute of Tropical Agriculture (IITA). National Food Consumption and Micronutrient Survey 2021. Preliminary Report. Abuja and Ibadan, Nigeria: FGN and IITA; 2022. p. 288

[11] Mulongo G, Maru J, Munyua H, Kasuga R, Olapeju P, Wende M, et al. The Building Nutritious Food Baskets Project 'Insights from the Field'. Lima: International Potato Center (CIP); 2018 https://hdl.handle.net/10568/98539

[12] Bouis HE. Economics of enhanced micronutrient density in food staples. Field Crops Research. 1999;**60**:165-173. DOI: 10.1016/S0378-4290(98)00138-5

[13] Kiran A, Wakeel A, Mahmood K, Mubaraka R, Hafsa, Haefele SM. Biofortification of staple crops to alleviate human malnutrition: Contributions and potential in developing countries. Agronomy. 2022;**12**(2):452. DOI: 10.3390/agronomy12020452

[14] Virk PS, Andersson MS, Arcos J, Govindaraj M, Pfeiffer WH. Transition from targeted breeding to mainstreaming of biofortification traits in crop improvement programs. Frontiers in Plant Science. 2021;**12**:703990. DOI: 10.3389/fpls.2021.703990

[15] De Steur H, Demont M, Gellynck X, Stein AJ. The social and economic impact of biofortification through genetic modification. Current Opinion in Biotechnology. 2017;**44**:161-168. DOI: 10.1016/j.copbio.2017.01.012 PMID: 28231514

[16] Chaudhary M, Mandal A, Muduli S, Deepasree A, Abshiba. Agronomic biofortification of food crops: A sustainable way to boost nutritional security. Revisiting Plant Biostimulants. 2022:98125. DOI: 10.5772/intechopen.103750. ISBN: 978-1-80355-553-9

[17] Mkambula P, Birol E, Friesen VM, Munyua HM, Alberts D, Aytekin D, et al. Transforming food systems to deliver nutritious foods: The vital roles of fortification and biofortification [Global Alliance for Improved Nutrition (GAIN) and HarvestPlus. Discussion Paper #10]. Geneva, Switzerland, 2022. DOI: 10.36072/dp.10

[18] Hefferon KL. Can biofortified crops help attain food security? Current Molecular Biology Reports. 2016;**2**(4):180-185. DOI: 10.1007/s40610-016-0048-0

[19] Garg M, Sharma N, Sharma S, Kapoor P, Kumar A, Chunduri V, et al. Biofortified crops generated by breeding, agronomy, and transgenic approaches are improving lives of millions of people around the world. Frontiers in Nutrition. 2018;**5**:12. DOI: 10.3389/fnut.2018.00012 PMID: 29492405; PMCID: PMC5817065

[20] Khush GS, Lee S, Cho JI, Jeon JS. Biofortification of crops for reducing malnutrition. Plant Biotechnology Reports. 2012;**6**:195-202. DOI: 10.1007/s11816-012-0216-5

[21] Kapinga R, Tumwegamire S, Ndunguru J, Andrade MI, Agili S, Mwanga RO, et al. Catalogue of Orange-Fleshed Sweetpotato Varieties for Sub-Saharan Africa. Lima, Peru: International Potato Center (CIP); 2010. p. 40 https://cipotato.org/sweetpotato/sweetpotato-facts-and-figures/

[22] Olusola SR, Lordbanjou DT, Olalekan IA, Babakatcha AM, Olalekan OO. Cassava Production Enterprise in the Tropics. Tropical Plant Species and Technological Interventions for Improvement. 2023. DOI: 10.5772/intechopen.104677

[23] The method of handling cassava processing in selected villages in Edo State. 2017. https://eduproject.com.ng/accounting-education/the-method-of-handling-cassava-processing-in-selected-villages-in-edo-state/index.html

[24] FAO Stat 2019. Production indices. Available from https://www.fao.org/faostat/en/#search/Cassava%20and%20products[Accessed: 12 November 2021]

[25] TheCassava Value Chain. (2021). Sahel quarterly, nourishing Africa. Available from: https://sahelconsult.com/wp-content/uploads/2021/12/Sahel-Quarterly-_The-Cassava-Value-Chain-Volume-28-1.pdf

[26] Bayata A. Review of nutritional value of cassava for use as a staple food science. Journal of Analytical Chemistry. 2019;**7**(4):83-91. DOI: 10.11648/j.sjac.20190704.12

[27] Waisundara VY. Introductory chapter: Cassava as a staple food. In: Cassava. London, UK: InTech; 2018:1-10. DOI: 10.5772/intechopen.70324

[28] Menya J, Tumutegyereize P, Kabenge I, et al. Performance evaluation of cassava drying technologies: A case study from Uganda. MOJ Food Processing & Technology. 2020;**8**(2):46-51. DOI: 10.15406/mojfpt.2020.07.00241

[29] Amadi PE, Onwusiribe CS, Uchechukwu NU. Determinants of adoption of pro vitamin A cassava varieties by farmers in Imo State, Nigeria. Journal of Community & Communication Research. 2020;**5**(2):168-174

[30] Chavez AL, Sanchez T, Jaramillo G, Bedoya JM, Echeverry J, Bolanos EA, et al. Variation of quality traits in cassava roots evaluated in landraces and improved clones. *Euphytica*. 2005;**143**(1-2):125-133. DOI: 10.1007/s10681-005-3057-2

[31] FAOSTAT. Available from: http://faostat.fao.org. 2015

[32] FAOSTAT (2019). Food Agriculture and Organization (FAOSTAT). Available from: http://www.fao.org/faostat/en/#data/QC

[33] International Potato Center. Sweet potato in Africa. 2017. Available from https://cipotato.org/research/sweetpotato-in-africa/

[34] Ziska LH, Runion GB, Tomecek M, Prior SA, Torbet HA, Sicher R. An evaluation of cassava, sweet potato and field corn as potential carbohydrate sources for bioethanol production in Alabama and Maryland. Biomass and Bioenergy. 2009;**33**(11):1503-1508. DOI: 10.1016/j.biombioe.2009.07.014

[35] Phorbee Olapeju, Onabolu Adeyinka, Njoku Jude, Ezima Cornelia, Efuntoye Adeola, and Anoje Emeka. Orange-Fleshed Sweetpotato and Yellow Cassava: Production and Processing Guide for Household Food & Nutrition Security. Catholic Relief Services and International Potato Center. Abuja, Nigeria: 100fold Consulting; 2018. ISBN: 9789789681907

[36] Bach D, Bedin AC, Lacerda LG, Nogueira A, Demiate IM. Sweet potato (*Ipomoea batatas* L.): A versatile raw material for the food industry. Review - Food/Feed Science and Technology. Brazilian Archives of Biology and Technology. 2021;**64**:e21200568 Available from: www.scielo.br/babt

[37] Low JW, Arimond M, Osman N, Cunguara B, Zano F, Tschirley D. Ensuring the supply of and creating demand for a biofortified crop with a visible trait: Lessons learned from the introduction of orange-fleshed sweetpotato in drought-prone areas of Mozambique. Food and Nutrition Bulletin. 28 (2 supplements): The United Nations University: 2007:S258-S270. In: Qin Y, Naumovski N, Ranadheera S, D'Cunha N, editors. Nutrition-related health outcomes of sweet potato (Ipomoea batatas) consumption: A systematic review. Food Bioscience. 2022;**50**:102208. DOI: 10.1016/j.fbio.2022.102208

[38] Qin Y, Naumovski N, Ranadheera S, D'Cunha N. Nutrition-related health outcomes of sweet potato (Ipomoea batatas) consumption: A systematic review. Food Bioscience. 2022;**50**:102208. DOI: 10.1016/j.fbio.2022.102208

[39] Girard AW, Grant F, Watkinson M, Okuku HS, Wanjala R, Cole D, et al. Promotion of orange-fleshed sweet potato increased vitamin A intakes and reduced the odds of low retinol-binding protein among postpartum Kenyan women. Journal of Nutrition. 2017;**147**(5):955-963. DOI: 10.3945/jn.116.236406

[40] Abong GO, Muzhingi T, Okoth MW, Ng'ang'a F, Ochieng' PE, Mbogo DM, et al. Phytochemicals in leaves and roots of selected Kenyan Orange fleshed

sweet potato (OFSP) varieties. Hindawi International Journal of Food Science. 2020;**2020**:3567972

[41] Abong GO, Ndanyi VCM, Kaaya A, Shibairo S, Okoth MW, Lamuka PO, et al. A review of production, post-harvest handling and marketing of sweetpotatoes in Kenya and Uganda. Current Research in Nutrition and Food Science. 2016;**4**(3). DOI: 10.12944/CRNFSJ.4.3.03

[42] Eggersdorfer M, Wyss A. Carotenoids in human nutrition and health. Archives of Biochemistry and Biophysics. 2018;**652**:18-26. DOI: 10.1016/j.abb.2018.06.001

[43] Low JW, Mwanga ROM, Andrade M, Carey E, Ball AM. Tackling vitamin A deficiency with biofortified sweetpotato in sub-Saharan Africa. Global Food Security. 2017;**14**:23-30. DOI: 10.1016/j.gfs.2017.01.004

[44] Etzbach L, Pfeiffer A, Schieber A, Weber F. Effects of thermal pasteurization and ultrasound treatment on the peroxidase activity, carotenoid composition, and physicochemical properties of goldenberry (*Physalis peruviana* L.) puree. LWT. 2019;**100**:69-74. DOI: 10.1016/j.lwt.2018.10.032

[45] Schieber F. Weber. 5 – Carotenoids. In: Woodhead Publishing Series in Food Science, Technology and Nutrition, Handbook on Natural Pigments in Food and Beverages. Sawston, Cambridge: Woodhead Publishing; 2016. pp. 101-123. ISBN 9780081003718. DOI: 10.1016/B978-0-08-100371-8.00005-1

[46] HarvestPlus 2022. Vitamin A cassava. Available from: https://www.harvestplus.org/crop/vitamin-a-cassava/

[47] Gupta HS, Hossain F, Muthusamy V, Zunjare RU. Marker-assisted breeding for enrichment of provitamin a in maize. In: Qureshi A, Dar Z, Wani S, ediors. Quality Breeding in Field Crops. Cham: Springer; 2019:139-157. DOI: 10.1007/978-3-030-04609-5_6

[48] CIMMYT. Biofortification to fight "Hidden Hunger" in Zimbabwe. (2016). Available from: http://www.cimmyt.org/biofortification-to-fight-hidden-hunger-in-zimbabwe/

[49] Muzhingi T, Palacios N, Miranda A, Cabrera ML, Yeum KJ, Tang G. Genetic variation of carotenoids, vitamin E and phenolic compounds in biofortified maize. Journal of the Science of Food and Agriculture. 2016;**97**(3):793-801. DOI: 10.1002/jsfa.7798

[50] Reddy BVS, Ramesh S, Longvah T. Prospects of breeding for micronutrients and β-carotene-dense sorghums. International Sorghum Millets Newsletter. 2005;**46**:10-14

[51] Govindaraj M, Kanatti A, Rai KN, Pfeiffer WH, Shivade H. Association of grain iron and zinc content with other nutrients in pearl millet germplasm, breeding lines, and hybrids. Frontiers in Nutrition. 2022;**8**:746625. DOI: 10.3389/fnut.2021.746625

[52] Mahesh Pujar, Mahalingam Govindaraj, S. Gangaprasad, A. Kanatti, H. Shivade. (2020). Genetic variation and diversity for grain iron, zinc, protein and agronomic traits in advanced breeding lines of pearl millet [*Pennisetum glaucum* (L.) R. Br.] for biofortification breeding. Genetic Resources and Crop Evolution. 2020;**67**:2009-2022. DOI: 10.1007/s10722-020-00956-x

[53] Govindaraj M, Rai K, Cherian B, Pfeiffer W, Kanatti A, Shivade H. Breeding biofortified pearl millet varieties and hybrids to enhance millet markets for human nutrition.

Agriculture. 2019;**9**:106. DOI: 10.3390/agriculture9050106

[54] Sawicka B. Post-Harvest Losses of Agricultural Produce. In book: Filho L, et al, editor. Zero Hunger. Projects: Agro Business Social Welfare: Collection of bibliographic references and the win-win-win papakonstantinidis model. Switzerland, Springer Nature; 2020:654-669. DOI: 10.1007/978-3-319-69626-3_40-1

[55] Kumar D, Kalita P. Reducing postharvest losses during storage of grain crops to strengthen food security in developing countries. Food. 2017;**6**(1):8. DOI: 10.3390/foods6010008 PMID: 28231087; PMCID: PMC5296677

[56] Research Solutions Africa (RSA). Report of a Study on Fresh Vegetables Market in Kenya: Desk Review. 2015. In: Lengai GMW, Fulano AM, Muthomi JW. Improving access to export market for fresh vegetables through reduction of phytosanitary and pesticide residue constraints. Sustainability. 2022;**14**:8183. DOI: 10.3390/su14138183. Academic Editors: Boccia F, Bhat R

[57] Al-Dairi M, Pathare P, Al-Yahyai R, Opara U. Mechanical damage of fresh produce in postharvest transportation: Current status and future prospects. Trends in Food Science & Technology. 2022;**124**:195-207. DOI: 10.1016/j.tifs.2022.04.018. ISSN 0924-2244

[58] Were HK, Omuterema S, Ndolo P, Imbuga B, Omuse CN, Keya NCO. Sweet potato production and use in sugarcane growing areas of Western Kenya. International Journal of Disaster Management and Risk Reduction. 2013;**5**(1):121-133

[59] Mwanga ROM, Mayanja S, Swanckaert J, Nakitto M, zum Felde T, Grüneberg W, et al. Development of a food product profile for boiled and steamed sweetpotato in Uganda for effective breeding. International Journal of Food Science and Technology. 2021;**56**:1385-1398. DOI: https://doi.org/10.1111/ijfs.14792

[60] Sugri I, Maalekuu B, Kusi F, Gaveh E. Quality and shelf-life of sweet potato as influenced by storage and postharvest treatments. Trends in Horticultural Research. 2017;**7**:1-10. DOI: 10.3923/thr.2017.1.10

[61] Ray RC, Tomlins KI. Sweet Potato: Postharvest Aspects in Food, Feed and Industry. New York: Nova Science Publishers, Inc.; 2010

[62] Truong VD, Avula RY. Sweet potato purees and dehydrated powders for functional food ingredients. In: Ray RC, Tomlins KI, editors. Sweet Potato: Postharvest Aspects in Food, Feed and Industry. New York: Nova Science Publishers, Inc.; 2010. pp. 117-161

[63] Adeyemi POA, Salaam ARB. Effect of processing conditions on the quality of fried sweetpotato chips. The International Journal of Science & Technology. In: Abong GO, Ndanyi VCM, Kaaya A, Shibairo S, Okoth MW, Lamuka PO, et al, ediotrs. A review of Production, Post-harvest Handling and Marketing of Sweetpotatoes in Kenya and Uganda. Current Research in Nutrition and Food Science. 2016;**4**(3). DOI: 10.12944/CRNFSJ.4.3.03

[64] Nguyen HC, Chen CC, Lin KH, Chao PY, Lin HH, Huang MY. Bioactive compounds, antioxidants, and health benefits of sweet potato leaves. Molecules. 2021;**26**(7):1820. DOI: 10.3390/molecules26071820

[65] Kpaka C, Gugerty MK, Anderson CL. Sweet Potato Value Chain(Uganda EPAR Brief No. 217. Evans school Policy Analysis and Research. Evans School of Public Affairs, University of Washington; 2019

[66] Puree. 2022. Available from: https://en.wikipedia.org/wiki/Purée

[67] CIP. 2021. Available from: https://cipotato.org/cip-50/innovations/orange-fleshed-sweetpotato/

[68] CIP. 2020. Available from: https://cipotato.org/cip_projects/orange-fleshed-sweetpotato-puree-based-food-applications-east-africa/

[69] CGIAR: Roots, Tubers and Bananas. (2018). Making sorable orange-fleshed sweetpotato puree a commercial reality. Available from: https://www.rtb.cgiar.org/news/making-storable-orange-fleshed-sweetpotato-puree-a-commercial-reality/

[70] De Paula CD, Pastrana-Puche YI, Viloria-Benítez KM, Rubio-Arrieta JA, Simanca-Sotelo M, Álvarez-Badel B, et al. Physicochemical and sensory evaluation of sweet potato (*Ipomoea batatas* L.) restructured products produced in the Sinu Valley, Colombia. Heliyon. 2021;7(8):e07691. DOI: 10.1016/j.heliyon.2021.e07691 PMID: 34401577; PMCID: PMC8350176

[71] Vithu P, Dash SK, Rayaguru K. Post-harvest processing and utilization of sweet potato: A review. Food Reviews International. 2019;**35**(8):726-762. DOI: 10.1080/87559129.2019.1600540

[72] Weber C, Trierweiler L, Casagrande T, Trierweiler J. Economic evaluation of sweet potato distilled beverage produced by alternative route. International Journal of Development and Sustainability. 2018;7(10):2514-2527

[73] Giri NA, Sakhale BK. Sweet potato (*Ipomoea batatas* L.): A valuable tropical tuber crop: A review. The Pharma Innovation Journal. 2019;**8**(6):182-191 Available from: https://www.thepharmajournal.com/archives/2019/vol8issue6/PartD/8-5-141-696.pdf

[74] Laveriano-Santos EP, López-Yerena A, Jaime-Rodríguez C, González-Coria J, Lamuela-Raventós RM, Vallverdú-Queralt A, Romanyà J, Pérez M. Sweet potato is not simply an abundant food crop: A comprehensive review of its phytochemical constituents, biological activities, and the effects of processing. Antioxidants 2022; 11(9):1648. https://doi.org/10.3390/antiox11091648

[75] International Potato Center (CIP). (2021). Innovations: Orange-fleshed sweet potato puree for nutritious food products. Available from: https://cipotato.org/cip-50/innovations/orange-fleshed-sweetpotato/

[76] Trung, Can (2022). What is PUREE? The applications of Puree few people know Available from: https://nafoods.com/what-is-puree-the-applications-of-puree-few-people-know

[77] Kays SJ. Formulated sweetpotato products. In: Bouwkamp JC, editor. Sweetpotato Products: A Natural Resource for the Tropics. Boca Raton, FL: CRC Press Inc; 1985. pp. 205-218

[78] Moyo M et al. Orange-fleshed sweetpotato puree: A breakthrough product for the bakery sector in Africa. In: Thiele G, Friedmann M, Campos H, Polar V, Bentley JW, editors. Root, Tuber and Banana Food System Innovations. Cham: Springer; 2022 https://doi.org/10.1007/978-3-030-92022-7_5

[79] Truong VD, Avula RY, Pecota KV, Yencho GC. In: Siddiq M, Uebersax MA, editors. Sweetpotato Production, Processing, and Nutritional Quality. Handbook of Vegetables and Vegetable Processing. 2nd ed. Vol. II. Hoboken, New Jersey, USA: John Wiley & Sons Ltd; 2018

[80] Truong AN, Simunovic J, Truong VD. 2012. Process development for producing pasteurized anthocyanin-rich juice and

recovering starch from purple-fleshed sweetpotatoes. Abstract and Poster Presentation # 230-24, Annual Meeting, Institue of Food Technologists, Las Vegas, NV, June 25-28, 2012

[81] Walter WM Jr, Schwartz SJ. Controlled heat processing of jewel sweetpotatoes for purée production. *Journal of Food Quality*. 1993;**16**:71-80

[82] Hoover MW, Harmon SJ. Carbohydrate changes in sweetpotato flakes made by the enzyme activation technique. *Food Technology*. 1967;**21**:1529-1532

[83] Birungi C, Nabubuya A, Mukisa IM. Effect of sweet potato endogenous amylase activation on *in vivo* energy bioavailability and acceptability of soy-enriched orange-fleshed sweet potato complementary porridges. Food Science & Nutrition. 2018;**6**(4):1119-1127. DOI: 10.1002/fsn3.653 PMID: 29983976; PMCID: PMC6021734

[84] Sebben J, Trierweiler L, Trierweiler J. Orange-fleshed sweet potato flour obtained by drying in microwave and hot air. Journal of Food Processing and Preservation. 2016;**41**:1-8. DOI: 10.1111/jfpp.12744

[85] Abiodun AM, Enujiugha VN. The nutritional and health benefits of sweetpotatoes. International Journal of African Sustainable Development and Research. TIMBOU-AFRICA Academic Publications; 2021;**7**:53-80

[86] Aly ES, Ramesh R. Potential impacts of bio-processing of sweet potato: Review. Critical Reviews in Food Science and Nutrition. 2017;**57**:455-471. DOI: 10.1080/10408398.2014.960909

[87] Ante L, Bernarda S, Nela N, Mirela K, Vlasta P. Effect of baking and steaming on physicochemical and thermal properties of sweet potato puree preserved by freezing and freeze-drying. Croatian Journal of Food Science and Technology. 2016;**8**:90-98. DOI: 10.17508/CJFST.2016.8.2.08

[88] Coronel P, Truong VD, Simunovic J, Sandeep KP, Cartwright GD. Aseptic processing of sweetpotato purees using a continuous flow microwave system. Journal of Food Science. 2005;**70**:531-536

[89] Simunovic J, Swartzel KR, Truong VD, Cartwright G, Coronel P, Sandeep KP, Parrott D. 2014. Methods and apparatuses for thermal treatment of foods and other biomaterials, and products obtained thereby. Patent No. 8,742,305 B2.

[90] Steed LE, Truong VD, Simunovic J, Sandeep KP, Kumar P, Cartwright GD, et al. Continuous flow microwave-assisted processing and aseptic packaging of purple-fleshed sweetpotato purées. Journal of Food Science. 2008;**73**:E455-E462

[91] Kumar P, Coronel P, Truong VD, Simunovic J, Swartzel KR, Sandeep KP, et al. Overcoming issues associated with the scale-up of a continuous flow microwave system for aseptic processing of vegetable purees. Food Research International. 2008;**41**(5):454-461

[92] Bocher T, Low JW, Muoki P, Magnaghi A, Muzhingi T. From lab to life: Making storable orange-fleshed sweetpotato purée a commercial reality. The Open Agriculture. 2017;**2**(1):148-154

[93] Moyo M, Truong VD, Simunovic J, Pankuku J, Abong GO, Amagloh FK, et al. Orange-fleshed sweetpotato puree: A breakthrough product for the bakery sector in Africa. In: Thielle G, Friedmann M, Campos H, Polar V, Bentley JW, editors. Root, Tuber and Banana Food System Innovations; Value Creation for Inclusive Outcomes. Switzerland AG: Springer; 2022. pp. 145-172

[94] Awuni V, Alhassan MW, Amagloh FK. Orange-fleshed sweet potato (*Ipomoea batatas*) composite bread as a significant source of dietary vitamin A. Food Science & Nutrition. 2018;**6**(1):174-179

[95] Owade JO, Abong GO, Okoth MW. Production, utilization and nutritional benefits of orange fleshed sweet potato (OFSP) puree bread: A review. Current Research in Nutrition and Food Science. 2018;**6**(3):644-655. DOI: 10.12944/CRNFSJ.6.3.06

[96] Wanjuu C, Abong G, Mbogo D, Heck S, Low J, Muzhingi T. The physiochemical properties and shelf-life of orange-fleshed sweet potato puree composite bread. Food Science & Nutrition. 2018;**6**(6):1555-1563. DOI: 10.1002/fsn3.710 PMID: 30258598; PMCID: PMC6145253

[97] Amagloh FK. Orange-fleshed sweetpotato composite bread: The nutritional and economic potential. In: Aidoo R, Agbenorhevi JK, Wireko-Manu FD, Wangel A, editors. Roots and Tubers in Ghana: Overview and Selected Research Papers. Kumasi: KNUST Printing Press; 2019. pp. 268-283

[98] Olatunde GO, Edun AA, Shittu TA, Adeogun AI. Sensory characterization, consumer acceptability and physical properties of composite wheat bread substituted with orange-fleshed sweetpotato flour and puree. Applied Tropical Agriculture. 2020;**25**(1):97-106

[99] Prabhat Dhakal. 2022. Drying method of food preservation with types, examples. Food microbiology. Microbes notes. New Baneshwor-31, Kathmandu, Nepal; Available from: https://microbenotes.com/drying-method-food-preservation/

[100] Yustas Y. Cassava sun drying performance on various surfaces and drying bed depths. Tanzania Journal of Agricultural Sciences. 2015;**12**:31-36

[101] Onyeneke RU, Emenekwe CC, Munonye JO, Olaolu MO, Izuogu CU, Ibrahim-Olesin S, et al. Adoption of bio-fortified pro-vitamin-A cassava and health outcome of farming households in Abia and Anambra States Nigeria. Journal of Agricultural Extension. 2020;**24**(2):80-91

[102] Tran T, et al. Cost-Effective Cassava Processing: Case Study of Small-Scale Flash-Dryer Reengineering. In: Thiele G, Friedmann M, Campos H, Polar V, Bentley JW, ediors. Root, Tuber and Banana Food System Innovations. Cham: Springer; 2022:105-143. DOI: 10.1007/978-3-030-92022-7_4

[103] Kuye A, Otuu OO, Wahabi A, Bamidele A, Sanni L, Raji AO, et al. Design, fabrication and drying performance of flash dryer for high quality cassava flour. AMA, Agricultural Mechanization in Asia, Africa and Latin America. 2017;**48**:63-70

Chapter 9

Role of Traceability Systems for Food Safety within Post-Harvest Systems: Indian Context

N. Arunfred and V. Bini Marin

Abstract

The chapter deals with an understanding of the safety and traceability systems available for perishable food products in India. It represents various traceability standards for production, post-harvesting, pest control and product traceability. Moreover, global standards like GAP, GMP and HACCP are explained in the Indian context in a detailed way, right from production to post-harvesting of perishable food products. The drivers for the traceability system and the lack of infrastructural facilities that drags behind the proper implementation of traceability system in developing and vast nation like India has also been discussed. Finally, a recommendation has been made to the supply chain players in the food supply chain for the implementation of proper safety standards so that traceability and safety guidelines can be followed to meet the global standards in the Indian context. The conclusion part explores the digital advancements in India that are the driving force of the food traceable ecosystem.

Keywords: food safety, traceability, perishable food, distribution chanel, perishable food supply chain, responsiveness

1. Introduction

An organized system for tracing food production and improving food safety is the need of the hour for a country like India. The focus should be more on the need for the system in the perishable channel, board profile, food security act, management systems and guidelines to the key players. Consumers, governments and agribusiness firms always need a proper system for maintaining food quality. Every consumer expects good quality of food for the single rupee he spends [1]. It has been noticed that perishable food product meets the risk-based performance and process criteria attached to specific hazards, and also it contains hazard at levels that are harmful to human health [2].

In the channel of food distribution, traceability is a well-coordinated and well-documented movement of product and documented activities associated with the product, that is, from the producer, through a chain of intermediaries and then to the final consumer. As far as proper supply chain management is concerned, food traceability and quality are like its two eyes, as it helps to provide clear and transparent information in the chain. Food quality can be viewed from the availability of food, access to food and food absorption aspects. Availability of food deals with production and distribution of perishables. Access to food compacts with purchasing power and being able to assimilate the food consumed in order to live a healthy and long life; which is meant as food absorption. If food is not completely absorbed by the body, even if it is readily available and inexpensive, the goal is not served. Thus, the issues in food quality are the deficit or surplus of food production in relation to actual and expected consumption; which is followed by the level of staple food intake and calorie intake at the average level; then comes the effect of urbanization and food deficits on the food intake and finally the role of the public distribution system (PDS) in improving the availability and affordability of food in urban areas, respectively. Vepa et al. (2004) [3].

2. Factors affecting the quality of perishable produce

Due to the significance and need of customers for quality consideration of perishable produce, traceability issues in the food supply chain occur. This section explains the variables that affect the product's quality [4]. They are

- Bulkiness of products
- Quality variation of products
- Temperature variation
- Usage of pesticides
- Material handling practices
- Post-harvest technique used
- Impact of controlled atmosphere
- Impact of microorganism
- Physiology and Biochemical reactions
- Seasonal variation
- Availability of cold chain channel
- Size of operational holding

3. Drivers of traceability

The drivers of traceability can be broadly classified into two main types – hard driver, and soft driver traceability requirements [5]. Hard traceability criteria are those that either local or foreign marketers of perishable goods must adhere to in order to fulfill legal or contractual responsibilities under international trade agreements [6]. While soft requirements may not restrict commerce, they can have a big impact on the economics of specific supply chains. These can involve enhancing supply chain efficiency, adapting to shifting consumer demands, or addressing demands from other parties like importers or merchants. The following are the key drivers of traceability:

A. **Production standards:** These are standards that are being used as an assurance service to customers, and also as competitive points of differentiation. The outcome of such activities is the development of minimum production standards and good agricultural practice (GAP) protocols, GMP protocols, which are discussed in detail in the latter part of the chapter. Precision agriculture is another standard that involves the capture of spatial, temporal and quantitative information on production activities, such as spraying or fertilizer applications as well as crop quality and production measurements [7].

B. **Quality Assurance:** The growing need for secure track-and-trace systems for international trade is due to traceability criteria that must be met in order to transact business across international borders. Localized or smaller-scale worries about various food dangers, such as microbial, physical, and chemical dangers in the food supply, further erode consumer and merchant trust. The major consumer concern [8] in this area is the use of agrichemicals and the possible existence of high levels of pesticides in fruits and vegetables. The Government of India has regulated food industries under the Food Security Bill 2006. HACCP is also a certification standard which gives quality assurance to customers.

C. **Customers:** The Urban population is ready to pay more if the perishable product they buy is quality assured, safe and sustainable. At this level, individual supermarkets or branded marketers require traceability systems that deliver information, and products that support a particular business' market positioning. The information from the customer must be tracked by the retailer and should be sent to the producer.

D. **Chain communication:** Traceability data can be used by supply chains for a variety of purposes, one of which is to satisfy internal operational and performance improvements, giving them a commercial advantage. As openness rises, the many information exchanges between businesses made necessary by some traceability regulations may also have positive knock-on effects for the entire chain. Transparency improves vertical supply chain integration, which results in an effective food supply system [9].

E. **Feedback to producers:** Systems of traceability exist to give producers feedback. Most post-harvest systems entail a grading and sorting procedure. Each piece of food is individually measured before being sorted and graded. It is possible to give producers feedback on the overall quality, variance in quality, and yield across their farm by tying the quality information gained from sorting and grading to the individual bin or trailer and going back to a specific orchard location.

4. Safety and traceability standards

The capacity to backtrack and use recorded data to track an entity's history, use, or location based on some special identity within a business entity is known as traceability. Van Rijswijk and Frewer (2008) suggested that a useful extension to this definition is, to include the movement of products between businesses. Based on the demand for quality produce, few standards are followed for the assurance of quality to the customers [10]. This section deals with the standards that could be implemented in the food supply chain, based on four basic types. They are based on the production system, based on the post-harvest system, based on pest control and based on product (**Figure 1**).

4.1 Production system traceability: Gap standard for perishable produce

The traceability of production and post-harvest activities involves, providing information on GAP activities, linked to the production process. It also provides information on the use of fertilizers, pesticides, and water, as well as social considerations such as labour conditions [11]. Information collection will likely expand to include evidence of sustainable production systems and energy efficiency.

Consequently, food retailers and customers have forced growers to follow certain growing practices that could lessen the microbial contamination of the perishable product. Many farmers in developed and developing countries apply GAP through sustainable agricultural methods, such as integrated nutrient management, conservation agriculture and integrated pest management. According to the Food and Agriculture Organization (FAO), Good Agricultural Practices (GAP) is the application of available knowledge to address economic, social sustainability and environmental [12]. It also emphasizes the need for the development of on-farm production and post-production processes, resulting in safe and healthy food and non-food agricultural products. The important functions of GAP are as follows:

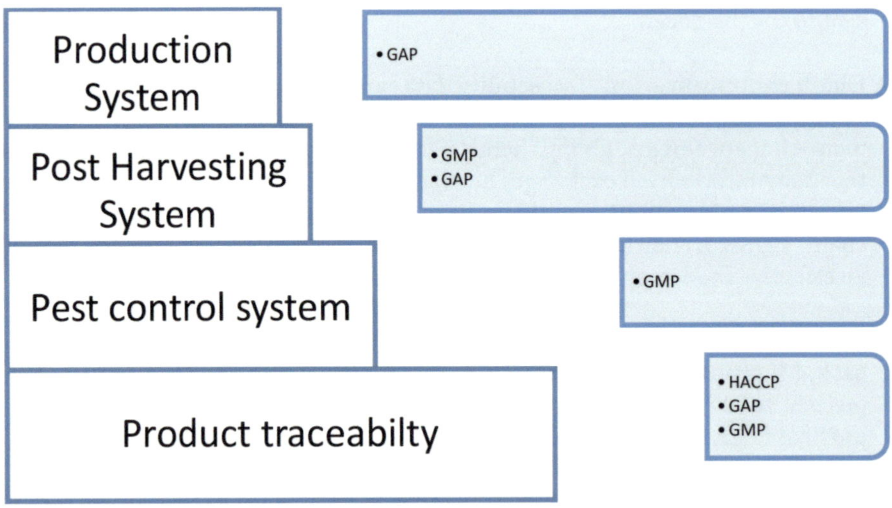

Figure 1.
Block diagram of different standards to be followed in chain process.

1. Risk evaluation

2. Guarantee to food safety at verticals

3. Preventing issues before they arise

4. Information sharing along the vertical and horizontal integration

5. Compulsory training at various operational level

6. Sanitation in equipment and in field

7. Pest Control through integrated approach

8. Monitoring and control

9. Verification by unbiased, outside audits

10. Upholding GAP Workbook for Self-Audit

The scope for Good Agricultural Practices starts from the selection of the site by a farmer, till the perishable produce is post-harvested. In the following section, a summary of the rules to be followed by farmers for quality output is listed.

a. **Pre-Planting measures:** Site selection is the first process in the pre-planting measure. Land or site for fruit and vegetable production should be selected on the basis of previous manure applications, land history and crop rotation history. After the selection of site, manure handling, livestock manure and field application are essential to cultivate the soil. Proper and in-depth composting of manure, incorporating it into the soil prior to planting, and avoiding top-dressing of plants are important steps toward reducing the risk of microbial contamination. Farmers should avoid growing root and leafy crops in the year when manure is applied to a field. The long period between application and harvest will reduce risks. All planned vegetable or fruit acreage should have manure sprayed at the end of the season, especially when the soils are warm, not waterlogged, and covered with cover crops. When applying manure at the beginning of the season, it should be distributed 2 weeks before planting, preferably on fruit and vegetable crops [13].

b. **Irrigation water quality:** Ideally, water used for irrigation or chemical spray should be free from pathogens. Drip irrigation method should be used, whenever it is possible to reduce the risk of crop contamination, as the edible parts of most crops are not wetted directly. Farmers can filter or use settling ponds, to improve the quality of fruit and vegetable crops. If side dressing is required, well-composted or well-aged (greater than 1 year) manure should be used for the application. Plant disease levels also may be reduced and water use efficiency is maximized with this method [14].

c. **Harvest:** Bins and all crop containers have to be washed and rinsed under high pressure. All crop containers should be sanitized before harvest. Bins should be properly covered, when not in use, to avoid contamination by birds and animals.

Commodity	Temperature (°F)	Relative humidity (%)
Apple	30.2–37.4	90–98
Beans, green	39.2–44.6	90–95
Cabbage, carrots, brinjal	32–35.6	95–97
Cherries	31.1–32	90–95
Cucumber	44.6–50	90–95
Grapes	30.2–33.8	85–90
Lemons, lime	39.2–59	86–88
Mango	51.8–64.4	85–90
Melon water	35.6–39.2	85–90
Orange	32–50	85–90
Potato	34.7–39.2	90–94

Source: FAO 2011.

Table 1.
Standards followed in storing major perishable crops.

Good personal hygiene is particularly important during the harvest of crops. Employee awareness, meaningful training and accessible facilities with hand wash stations encourage good hygiene [15].

d. **Post-harvest**: Loading, staging, and all food contact surfaces should be cleaned and sanitized at the end of each day. Packaging material should be stored in a clean area. Fruits and vegetables should be rapidly cooled after harvesting in order to prevent the growth of diseases and preserve quality. The temperature of the water bath used for cooling should not be more than 10°F colder than the temperature of the produce pulp. The capacity of the refrigeration chamber should not be exceeded by overload (**Table 1**).

4.2 Institutional standards and traceability - GMP for perishable produce

Good Management Practices (GMP), are guidelines, advising producers how to manage the water, nutrients, and pesticides they use, in order to minimize agriculture's impact on the state's natural resources (Raspor 2008). GMPs were developed because the agricultural activity has been linked to the contamination of watersheds with nutrients (e.g., nitrogen and phosphorus), pesticides and discharged sediments and water. The application of GMP guidelines is listed below.

- Chemical Irrigation means applying fertilizer, pesticide, or other agricultural chemicals to cropland through an irrigation system. Chemical irrigation reduces the number of passes across a field with tractors, sprayers, fertilizer applicators and machinery [16].

- Combining Tractor Operations means performing two or more cultivations, tillage, planting, or harvesting operations with a single tractor or harvester pass. Combining tractor operations reduces the number of passes or trips that a tractor.

- Integrated Pest Management means the use of a combination of techniques including organic, conventional and biological farming practices. Integrated pest management creates insect habitats that are beneficial and reduce the use of herbicides/pesticides, thereby reducing the number of passes for spraying [17]. It also reduces soil compaction and the need for additional tillage.

- Watering the soil prior to planting activities is referred to as "Planting Based on Soil Moisture." From the time of planting to crop establishment, planting based on soil moisture is successful since it is low during the planting operation.

- Precision Farming" means using GPS to precisely guide farm equipment in the field. Precision farming reduces overlap and allows operations to occur during inclement weather conditions and at night thereby generating less PM [18].

- Reduced Harvest Activity" refers to fewer mechanical harvest passes that involve cutting and removing crops from a field. Anytime work is done in a field, the soil structure may change, and some materials may be released into the air [19].

4.3 Product traceability: HACCP

Food safety is managed using the HACCP (Hazard Analysis and Critical Control Point) approach, which is recognized globally. The Codex Alimentarius Commission has approved it [20]. It is a tool that may be used to methodically identify risks unique to particular products and processes and outline ways to control such risks to assure the safety of perishable goods [21]. In order for the HACCP Plan to be implemented effectively within the industry, Good Manufacturing Practices (GMPs) procedures that effectively control general hazards to food safety must be followed. The chain must have a structural requirement stated by GMP and operational requirement stated by both GAP and GMP. The main purpose of HACCP is to prevent, reduce and eliminate hazards in food and to provide a safe product to the end customer.

4.3.1 Principles of HACCP

- Conduct a hazard analysis.

- Identify the critical control points.

- Establish critical limits.

- Establish monitoring procedures.

- Establish corrective actions.

- Establish verification procedures.

- Establish record-keeping and documentation procedures.

The operation of HACCP in the perishable food chain is described with respect to different key players of the supply chain in the next section. In India, the aforementioned Good Agricultural Practices (GAP) are still in their infancy. Due to pressure

Figure 2.
Flowchart showing process HACCP implementation.

from foreign buyers, only a small number of farmers may be engaging in it. But it needs to be clearly stated that every player in the perishable food channel is accountable for ensuring food safety from farm to fork (**Figure 2**).

5. Implementation plan for perishable food supply chain using HACCP in developing nations

In order for the HACCP plan to be implemented effectively within the establishment, it must be based on a firm foundation of Good Manufacturing Practices (GMPs) and procedures that effectively control the general hazards to food safety The step-by-step procedure to be followed for the standardized food supply chain is listed above.

5.1 Guidelines for certified growers

Based on the need for quality and safe produce, a few guidelines have been suggested to farmers based on the HACCP standard. The guidelines help farmers to integrate into the channel and help in the transparent flow of information. It also helps the farmers to integrate horizontally into the food chain. The following few steps are to be followed in order to obtain a traceable chain in the channel [22].

Step 1: Follow the GAP principle: This task involves providing a general description of all fresh produce, ingredients, processes and methods followed by the grower, starting from the pre-planting activity till harvesting. In this process, the grower's day-to-day activity must be noted or tabulated on an activity table.

Step 2: Develop a flow diagram: The next process is developing a flow diagram. The purpose of the flow diagram is to provide a clear, simple outline of the steps involved in the production of the perishable produce. The diagram must cover all steps in the growing process. It is the step-by-step procedure followed in the chain, starting from the soil selection process till harvesting, followed by a grower according to GAP guidelines (**Figure 3**).

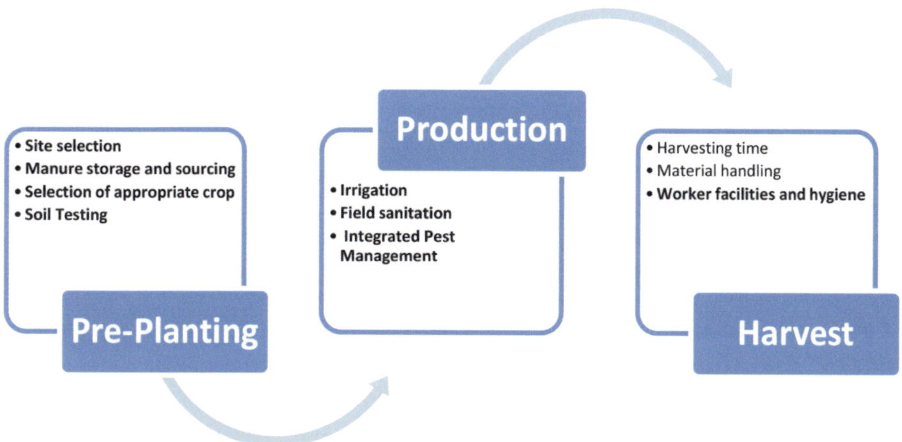

Figure 3.
Flow diagram for certified grower.

Step 3: Application of hazard principle: The first step is to identify the list of processes involved in growing, and list all the potential hazards that are not already controlled through Good Agricultural Practices, including the listing of procedures. The hazards should be classified based on three factors – biological, chemical and physical [23]. A possible source of hazards also must be tabulated. For example, while taking the situation of pest management, wrong sourcing, over usage, usage of banned chemicals, and improper dilution of pesticides are the potential hazards to be documented.

Step 4: Determine the Critical Control Points and Establish Critical Limits: An element, practice, procedure, process, or site is referred to be a critical control point (CCP) if it can be managed to prevent, control, eliminate, or decrease a hazard or to reduce the risk that it will occur [24]. One method of determining CCPs is to use a CCP Decision Tree. A Critical Limit is a limit to which a hazard must be controlled to prevent, control, eliminate or reduce the occurrence of the hazard to an acceptable level. It needs to be monitored to ensure that limits are met. For example, if over usage of pesticides is identified, then that process should be highlighted for critical analysis. The critical limit for the usage of pesticides must be found out, based on the scientific method and for the main crop to be considered.

Step 5: Establish Corrective Actions, Documentation and Record Keeping: Once the critical control point is identified and critical limits are identified, Corrective Actions are required. They are required when operations go out of Critical Limits, which ensure safety and suitability. Corrective Actions must identify and fix the problem, and investigate what can be done to prevent the problem from occurring again [25]. Finally, documentation and record keeping is the key to any standardization. Accurate record-keeping is essential to the application of a HACCP system, for auditing purposes, be it an own internal audit, verification procedures, an AQIS audit, or another external audit.

5.2 Guidelines for standardized agribusiness

The activities which are carried out after harvest till it reaches the final customer constitute the agribusiness. It is a long chain and the key processes involved are packing, distribution, transportation, and retailing. Organizing these functions is called vertical integration. To have an organized supply chain management in perishable produce, the effective implementation of standard process is essential. Following are the guidelines for the standardized agribusiness.

Step 1: Follow GMP principle: This task involves providing a general description of all fresh produce, ingredients and processes and methods followed by chain players starting from post-harvesting till the retail outlet. The process each key player does in his day-to-day activity must be noted or tabulated on an activity table. Series of product description and processes carried out are

- The source of the raw material
- The preservation method
- The packaging (e.g. vacuum packed, plastic liner in cardboard cartons, etc.)
- Transportation of the product (including the method of transportation)

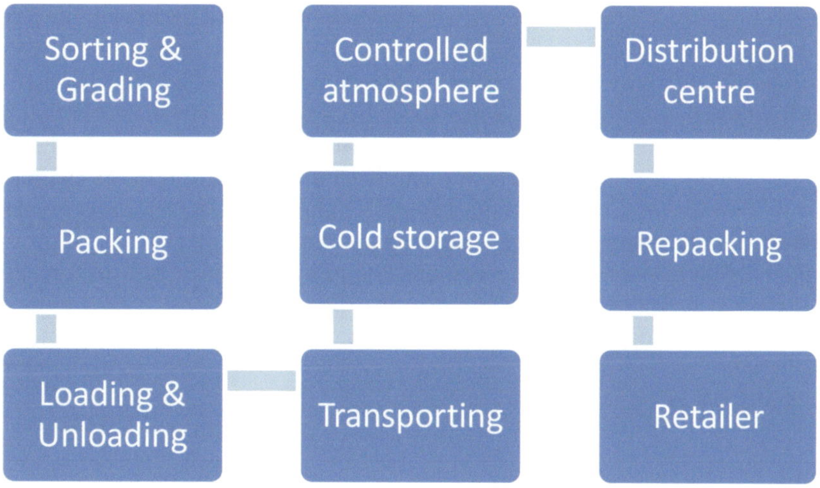

Figure 4.
Flowchart showing process in agribusiness.

- Storage conditions (e.g. frozen, refrigerated, at ambient temperature, etc.)
- The product standards applicable to the product (according to the Food Safety and Security Act 2006)

Step 2: Develop a flow diagram: The next process is developing a flow diagram. The purpose of the flow diagram is to provide a clear, simple outline of the steps involved in the operation of the perishable produce, from the farmer's farm gate to the retailer. The flow diagram must cover all steps in the post-harvest, distribution and retailing process. It is a step-by-step procedure followed in the chain starting from the soil selection process till harvesting followed by a grower according to GMP guidelines.

Step 3: Application of hazard principle: Here the list of activities involved in each functionality must be identified, and all potential hazards in each process must be listed. The hazards should be classified based on three factors – biological, chemical and physical of the product. A possible source of hazards like improper grading, unstandardized material handling and over-waxing must also be tabulated (**Figure 4**).

Step 4: Determine the Critical Control Points and Establish Critical Limits: A Critical Control Point (CCP) is arrived based on the requirement of the customer. The process which is out of control is listed.

Step 5: Establish Corrective Actions, Documentation and Record Keeping: Corrective actions are required, when operations exceed Critical Limits, which ensure safety and suitability. Corrective actions must identify and fix the problem. Accurate record-keeping makes the HACCP system effective and makes the distribution channel an efficient and effective one.

6. Conclusion

The purpose of the chapter was to understand the need for security and traceability with regard to perishable produce, to establish a proper supply chain management. Several standards which exist in the global market that are essential for maintaining the safety of perishable products have been discussed. Several definitions and principles of traceability and safety of food have been presented. An attempt has been made to draw a guideline for farmers (grower) and agribusiness that will help pursue a good traceability model. Various technologies like GPS, and RFID, are available that help to track the produce from the farm gate till it reaches the final customer, which has been briefly presented in this chapter. The traceability implementation in the chain could help to improve the quality and reduce hazards in the chain. But the costs involved in integration and implementation of technology in the diversified rural background is the challenge that exists for the government and private entities. Information system is a key driver of effective supply chain management.

Author details

N. Arunfred[1*] and V. Bini Marin[2]

1 Department of Management Studies, Nesamony Memorial Christian College, Marthandam, Tamil Nadu, India

2 Department of Business Administration, Scott Christian College, Marthandam, Tamil Nadu, India

*Address all correspondence to: arunfredeie@gmail.com

References

[1] van Wagenberg CPA, Backus GBC, van der Vorst JGAJ, Urlings BAP. A framework for the design and analysis of incentive systems for food safety control in supply chains. In: 113th EAAE Seminar, A Resilient European Food Industry and Food Chain in a Challenging World, Chania. Greece: Wageningen University and Research; 2009

[2] Alimentarius C. Code of hygiene practice for meat. Codex Alimentarius Commission. 2005;**52**:27-36

[3] Vepa S., et al. Food insecurity atlas of urban India. Food Insecurity Atlas of Urban India. 2004

[4] Watada AE, Ko NP, Minott DA. Factors affecting quality of fresh-cut horticultural products. Postharvest Biology and Technology. 1996;**9**(2): 115-125

[5] Bollen AF, Riden CP, Opara LU. Traceability in postharvest quality management. International Journal of Postharvest Technology and Innovation. 2006:93-105

[6] Bollen AF, Riden CP, Cox NR. Agricultural supply system traceability, part I: Role of packing procedures and effects of fruit mixing. Biosystems Engineering. 2007;**98**(4):391-400

[7] Zhang N, Wang M, Wang N. Precision agriculture - A worldwide overview. Computers and Electronics in Agriculture. 2002;**36**(2):113-132

[8] Opara LU. Traceability in agriculture and food supply chain: A review of basic concepts, technological implications, and future prospects. Journal of Food Agriculture and Environment. 2003;**1**:101-106

[9] Hobbs JE, Young LM. Closer vertical co-ordination in Agri-food supply chains: A conceptual framework and some preliminary evidence. Supply Chain Management: An International Journal. 2000;**5**(3):131-143

[10] van Rijswijk W, Frewer LJ. Consumer perceptions of food quality and safety and their relation to traceability. British Food Journal. 2008;**110**(10):1034-1046

[11] Sagar VR. Post-harvest management. Indian Horticulture (India). 2003;**47**

[12] Iso E. N (2005). 9000: 2005. Quality Management Systems-Fundamentals and Vocabulary (ISO 9000: 2005)

[13] Mora C, Menozzi D, Giacomini C, Cantoni C, Massari M, Morelli G. The attitude of Italian consumers towards GM foods. In: Food Culture: Tradition, Innovation and Trust—A Positive Force for Modern Agribusiness. Parma, Italy: Wageningen University and Research; 2007

[14] Bihn EA, Gravani RB, Matthews KR. Role of good agricultural practices in fruit and vegetable safety. Microbiology of Fresh Produce. 2006:21-53

[15] Swanson BE. Global Review of Good Agricultural Extension and Advisory Service Practices. Rome: FAO; 2008. p. 82

[16] Raspor P. Total food chain safety: How good practices can contribute? Trends in Food Science & Technology. 2008;**19**(8):405-412

[17] Burn AJ, Coaker TH, Jepson PC, editors. Integrated Pest Management. (No. 04; SB950, B8.). London: Academic Press; 1987

[18] Srinivasan A. Handbook of Precision Agriculture: Principles and Applications. Binghamton NY: Food Products Press; 2006

[19] Kader AA, Rolle RS. The Role of Post-Harvest Management in Assuring the Quality and Safety of Horticultural Produce. Vol. 152. FAO; 2004

[20] Codex Alimentarius Commission, & Codex Alimentarius Commission. Joint FAO/WHO Food Standards Programme. Rome: FAO/WHO; 2011. p. 122

[21] Mortimore S, Wallace C. HACCP: A Practical Approach. Vol. 1. Springer; 1998

[22] Khandke SS, Mayes T. HACCP implementation: A practical guide to the implementation of the HACCP plan. Food Control. 1998;**9**(2):103-109

[23] Gupta, A. K. (1989). Scientists' Views of Farmers' Practices in India: Barriers to Effective Interaction

[24] Hulebak KL, Schlosser W. Hazard analysis and critical control point (HACCP) history and conceptual overview. Risk Analysis. 2002;**22**(3):547-552

[25] Jevsnik M, Hlebec V, Raspor P. Meta-analysis as a tool for barriers identification during HACCP implementation to improve food safety. Acta Alimentaria. 2006;**35**(3):319-353

Chapter 10

An Overview of the Recent Developments in the Postharvest Application of Light-Emitting Diodes (LEDs) in Horticulture

Bonga Lewis Ngcobo and Isa Bertling

Abstract

The majority of losses in horticultural produce occur during postharvest storage, particularly due to poor handling. Most fruit, especially climacteric fruit, have a short postharvest life due to an increase in ethylene synthesis which signals ripening and, subsequently, senescence. Traditional practices for preserving the postharvest quality of horticultural crops are chemical-based, a practice which has lately received enormous criticism. Recently, the use of postharvest illumination with LEDs as a nonchemical and environmentally friendly technique to preserve fruit and vegetables has been reported by various authors. Unique properties of LEDs such as low radiant heat, monochromatic nature and low cost have made this lighting gain popularity in the food industry. This paper, therefore, reviews the recent development in the postharvest applications of LEDs in horticultural crops, while focusing particularly on physical characteristics, nutritional value, and overall quality alterations of fruit and vegetables. According to the recently published research, red and blue LED lights are most valuable in terms of usage, while other wavelengths such as purple and yellow are slowly gaining attention. Furthermore, LEDs have been shown to affect fruit ripening and senescence, enhance bioactive compounds and antioxidants in produce, and prevent disease occurrence; however, there are some limitations associated with the use of this novel technology.

Keywords: bioactive compounds, irradiation, LED technology, nutrition, postharvest preservation, senescence, shelf life

1. Introduction

The worldwide common challenge faced by farmers, especially in developing countries, is ensuring food security for a fast-growing world population. Recent predictions suggest that the demand for food will increase significantly as the predicted world population reaches about 9.7 billion people by 2050 [1]. On the other hand, about half (50%) of horticultural produce, mainly fruit and vegetables, is lost between harvest and consumption (postharvest) [2]. This, therefore, poses a threat and brings a serious concern to farmers to establish innovative methods and practices to increase the global food supply to provide sustainable living standards for humans by reducing the percentage of food lost in the value chain. Moreover, the consumption of fresh fruit and vegetables improves human health and well-being, because these commodities are rich sources of various vitamins, minerals, and antioxidant compounds that can prevent the occurrence of chronic diseases [3, 4]. As such, proper handling of horticultural crops is required, pre- and post-harvest, to improve product quality and yield, thereby ensuring food and nutrient security for all humans.

The most reliable techniques that are currently used to preserve fruit and vegetables are cold storage and chemical additives. These chemical additives have come lately under criticism [5, 6] as consumers are aware of the possible negative implications these compounds can have on their health; as a result, there are limitations on the use of chemical additives for the preservation of horticultural produce. Industries in the agricultural sector have, therefore, shifted to nonchemical-based approaches such as light-emitting diode (LED) technology [7, 8]. This technology was adopted after the use of lights, such as fluorescent, high-pressure sodium, and incandescent lights, came under criticism due to their large emission of radiant heat and energy inefficiency [9], whereas LEDs could provide several advantages, including durability, low emission of radiant heat, adjustable size, and cool emitting surface, resulting in an environmentally friendly technology that is also economically favorable [10]. Initially, the LEDs were only used in growth chambers and greenhouses, whereas, after some time, LED technology improved, as there was an incorporation of the new semiconductor materials and improvement of the crystal growth techniques as well

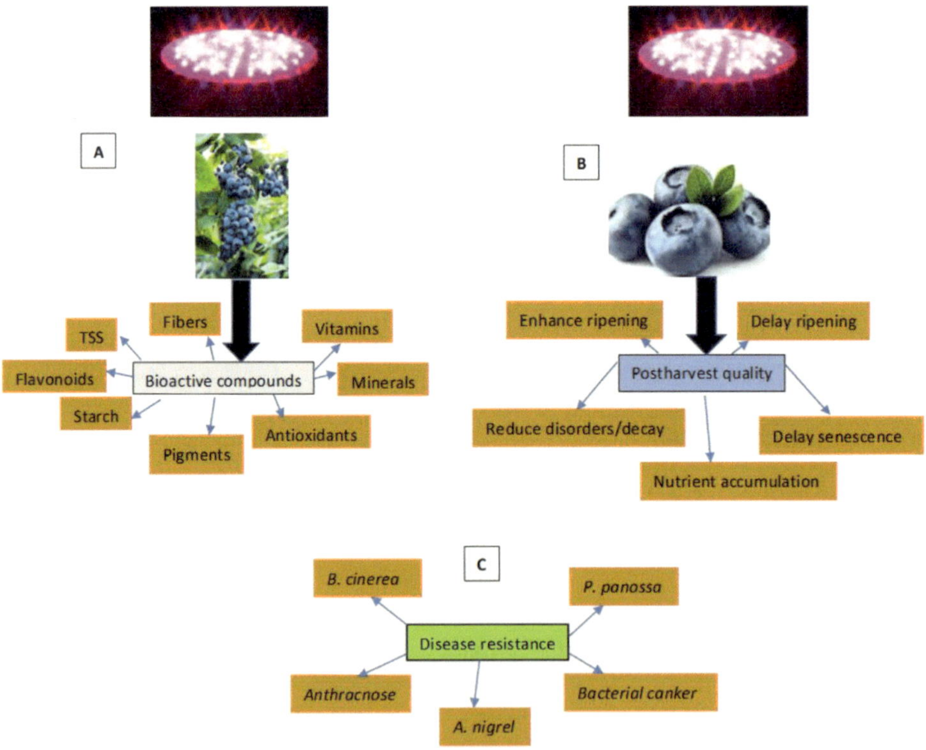

Figure 1.
Effect of LEDs on (A) the production of certain bioactive compounds, (B) postharvest quality, and (C) resistance to diseases affecting horticultural crops, adopted from] [12].

as of optics [11], resulting in LEDs being used in postharvest horticulture. Recently, research on postharvest preservation of fresh horticultural produce with the use of LEDs has gained popularity. Various studies revealed that LEDs have the potential to enhance ripening, particularly color development, as well as being able to suppress disease occurrence and improve the overall nutritional quality of fruit and vegetables exposed to various LED wavelengths [**Figure 1**] [7, 11, 13, 14]. This review, therefore, focuses on the potential of postharvest application of LED technology on horticultural crops, discussing the most significant, recent findings related to this technology. The technology and mechanism of action involved in irradiation with LEDs and the limitations of postharvest LED lighting will also be discussed.

2. Overview of LED technology used postharvest in horticultural crops

An LED consists of a semiconductor with a positive and negative junction, called p- and n-type, respectively. When an LED is connected to a power source, a flow of current starts from the positive (p-type) to the negative (n-type) junction, ultimately resulting in the flow of electrons, which causes light emission at a certain wavelength [7, 15]. The color of light emitted by the LED is determined by the band gap energy of the semiconductor material. **Figure 2** depicts the LED lighting system. Improved technology has enabled LEDs to be used in the postharvest preservation of fruit and vegetables. The unique and advantageous properties of LEDs have resulted in their use in postharvest storage of fresh fruit and vegetables [16]. The monochromatic nature, high photon efficiency, low radiant heat, durability, and prevention of thermal degradation of LEDs are favorable characteristics that make them beneficial in fruit and vegetable postharvest storage [17]. Moreover, the monochromatic nature of LEDs allows horticulturists to select specific wavelengths for the storage and preservation of horticultural produce [17]. Furthermore, LEDs operate at low direct current voltages and temperature, and their operation does not involve the use of toxic, environmentally unfriendly substances. As a result, the postharvest application of LEDs in the agricultural sector has expanded over the past years.

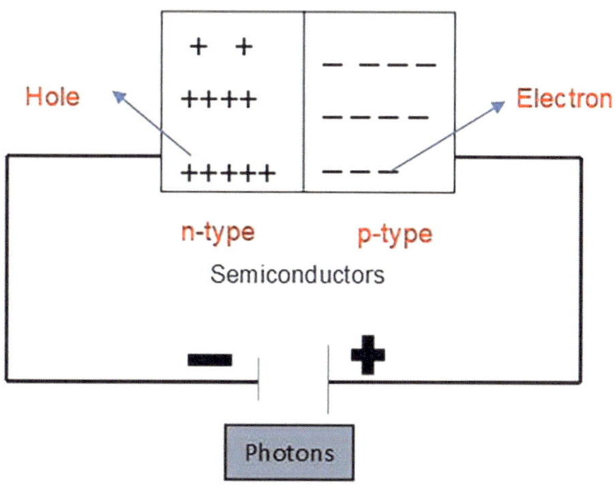

Figure 2.
Emission of photons by light-emitting diodes (LEDs), adopted from [7].

3. Mechanism of LED irradiation on horticultural crops

As mentioned above, due to their peculiar, distinctive properties, LEDs have gained popularity in the postharvest handling of horticultural crops. Even though the effect of various wavelengths, irradiation intensity, and exposure time of these LEDs on fresh fruit and vegetables has proved to be beneficial in enhancing color, bioactive compounds, antioxidants, shelf life, and overall quality [7, 11, 18–21], the mechanism(s) involved in LED irradiation technology is (are) still not clear. It is, however, known that the photosynthesis period of postharvest horticultural produce may be extended by illumination with LEDs. This can result in the long-term preservation of these commodities. The expression of genes and signaling of phytochrome may be inhibited or enhanced by irradiation with LED lights; this potentially affects fruit and vegetable senescence [22]. LEDs can reduce the breakdown of storage phytochemicals in the fruit/vegetable by delaying the emergence of ethylene and the respiratory peak [23]. Further, LED exposure can also cause a fluctuation in enzyme activity due to a change in the secondary structure of proteins. Other aspects involved in the mechanism of action of LED irradiation on fruit and vegetables require further investigations.

4. Postharvest application of LEDs in horticultural crops

The majority of horticultural crops, particularly fruit and vegetables, undergo rapid ripening postharvest due to an increase in respiration and metabolic activities, resulting in the deterioration in quality, softening, rapid water loss, tissue destruction, and senescence. This happens more so in climacteric fruit, as they continue the ripening process, even if removed from the mother plant. Postponing senescence, extending shelf life, and maintaining quality characteristics of horticultural produce are pivotal to reduce postharvest losses and ensure that appealing, high-quality produce reaches the consumer. This can be achieved by storing horticultural produce properly and exposing the produce to effective senescence-inhibiting treatments [24, 25]. The use of LED lights as a sustainable postharvest treatment offers unique opportunities to not only maintain but even improve produce characteristics.

The response of fruit and vegetables to irradiation with different spectral lights (wavelengths) varies and depends on the absorbing ability of the specific light wavelengths [10]. As such, the application of single-wavelength red and blue LEDs has been effective in enhancing bioactive compounds, phenolics, flavonoids, and other antioxidants in fresh fruits and vegetables, while maintaining their nutritional status and overall quality [7, 12] (**Table 1**). Recently, other wavelengths have started to gain interest and are producing promising results as a study by Xie et al. [30] demonstrated that purple LED light (40 μmol m^{-2} s^{-1}) enhances the concentration of ascorbic acid and carotenoids of broccoli florets, and a different study by Zhou et al. [31] revealed that irradiation with white LED light (10 μmol m^{-2} s^{-1}) is effective in maintaining postharvest quality and delaying senescence of pak choi during storage. On the other hand, mixed spectral light ratios, particularly blue:red at different ratios, have recently been adopted and proven to increase the efficiency of LED lighting [7, 12, 23, 26] (**Table 1**). Furthermore, research has shown that LED illumination can alter carotenoid accumulation and prevent fungal spoilage, which contributes

Crop investigated	LED Wavelength	Light Intensity	Effectiveness	Reference
Sweet cherries	Blue light and a ratio of white, blue, and green lights	—	Blue light enhanced the synthesis of anthocyanin and improved the quality of sweet cherries	[26]
Cherry tomatoes	Red and blue lights	Red (118 μmol m^{-2} s^{-1} at 638 nm) Blue LED light (118 μmol m^{-2} s^{-1} at 454 nm)	Both blue and red lights enhanced the health-related parameters of cherry tomatoes treated at the mature green stage	[20]
Tomatoes	Red light	—	Continuous red light irradiation accelerated ripening of green tomatoes. It also significantly increased lycopene, β-carotene, total phenolic content, and total flavonoid concentration	[27]
Tomatoes	Red light	113 μmol m^{-2} s^{-1}	Red light enhanced, color development lycopene, β-carotene, total phenolic, and total flavonoid concentration in both the outer and inner parts of tomatoes	[21]
Green chili	Red and blue lights	50 μmol m^{-2} s^{-1}	Exposure to red light accelerated color development and lycopene accumulation, whereas blue light was effective in enhancing vitamin C and total phenolics	[28]
Tomatoes	Blue light	100 μmol m^{-2} s^{-1}	30-min blue light exposure and 8-min pause enhanced lycopene, total phenolic compounds, total flavonoids, vitamin C, and soluble sugar more than other treatments	[29]
Broccoli	White, red, green, yellow, blue, and purple lights	40 μmol m^{-2} s^{-1} for each wavelength	Purple light delayed yellowing and maintained and improved the nutritional quality of broccoli during storage at 20°C.	[30]
Pak choi	White light	10 μmol m^{-2} s^{-1}	The treatment *delayed senescence* and maintained the quality of pak choi	[31]
Tomatoes	Blue and red lights	85.72 μ Einstein m^{-2} s^{-1} and 102.70 μ Einstein m^{-2} s^{-1} respectively	Blue wavelength was effective in extending the shelf life of tomatoes by delaying fruit softening and ripening	[32]
Strawberries	Blue light	—	Blue LED light prevented mold *spoilage* and preserved the physicochemical quality of strawberries	[33]

Crop investigated	LED Wavelength	Light Intensity	Effectiveness	Reference
Tomatoes	Blue light	87 W/m2	Blue light potentially maintained physicochemical quality and controlled mold growth on tomatoes during transportation and storage	[34]
Minimally processed broccoli sprouts	Red, blue, and far red	35 ± 2.5 μmol m^{-2} s^{-1}	Improved total antioxidant and decreased the microbial growth	[35]
Broccoli heads	Red light	50 μmol m^{-2} s^{-1}	The treatment delayed senescence and maintained the storage quality	[36]
Three apple cultivars: "Idared," "Fuji," and "Carjevič"	Blue light	—	Irradiation with blue light enhanced color development and nutritional quality of apples	[37]
Tomato	Red:far red (R:FR) light ratio	—	Exposure of tomato fruit to LED light with a high R:FR ratio enhanced the synthesis of lycopene	[23]

Table 1.
Postharvest effect of various LED lights on postharvest behavior of horticultural crops.

significantly to postharvest losses (**Table 1**) [11], a study by Dhakal and Baek [32] revealed that short-period exposure with blue wavelength to red tomatoes can extend the postharvest shelf life of tomatoes by delaying color development, and our recent study on cherry tomatoes also achieved the same results and documented improved phytochemical concentrations in blue-wavelength-treated tomatoes [20]. The postharvest effects of LEDs have also been tested on minimally processed food, and promising results have been achieved [35]; however, further research on such commodities is required to optimize exposure duration and type (wavelength).

5. Integration of LEDs with other treatments for postharvest quality alteration

The application of LEDs, as an environmentally sustainable and consumer-friendly approach, to preserve the quality and enhance the variety and concentration of antioxidant compounds in horticultural crops has been extensively investigated (**Table 1**). This technology has, however, some limitations depending on the wavelength used. As such, combining LEDs with other environmentally friendly treatments has recently gained attention. The aim of such combinations is to further improve the efficacy of LED treatments by generating unique properties that prevent LED limitations and contribute toward better postharvest preservation of fruit and vegetables. Hu et al. [38] investigated the combined effect of LEDs and UV light on the postharvest life of sweet oranges and revealed that different treatment combinations accelerated ripening and enhanced the nutritional quality of oranges; the study provided a potential regulation method for orange fruit quality. Hyun et al. [39] demonstrated that combining antimicrobials or photosensitizers with blue LEDs may be applied to extend the shelf life of fresh-cut apples and cherry tomatoes. LEDs can also be beneficial by reducing the occurrence of postharvest pathogens, and research conducted by [40] highlights two important mechanisms of controlling postharvest pathogens using LEDSs: one of these mechanisms is inducing the biosynthesis of specific secondary plant metabolites in fruit tissues, thereby improving the fruit/tissue resistance against pathogens, while also preventing pathogen or spore development due to the presence of photosensitizers in their cells. Zhang et al. [41] further revealed that combining blue light exposure and salicylic acid application maintained the sensory and nutritional quality of strawberries by maintaining bioactive component concentrations. Other studies also explored the effects of combining LEDs with other innovative and nonchemical-based treatments, such as heat and ethylene treatments [42–44].

6. Challenges and limitations associated with the postharvest use of LEDs

The postharvest use of LEDs has various benefits in maintaining and preserving the quality of horticultural crops (**Table 1**). However, some negative impacts are aligned with the use of this technology. Irradiation with LED lights has been reported to slightly reduce the mass of fruit and vegetables due to enhanced moisture loss [45]. The reason behind this moisture loss could be the selection of a specific, harmful wavelength and long duration or exposure of LEDs to horticultural crops. The opening of stomata can also be induced by postharvest irradiation with LED lights, and that may result in treated fruit and vegetables losing moisture. Most studies, including our recent ones, have demonstrated that postharvest application of either red or blue LED lights can improve nutritional quality and preserve the quality of horticultural crops without negatively affecting their mass [19, 20, 46]. The higher intensity, longer daily exposure, and continuous illumination have been reported to cause abiotic stress, resulting in higher mass loss; this, however, depends on the wavelength or LED light spectra used [47]. It is, therefore, very important to pay special attention when selecting the intensity, duration of exposure and spectral composition, or wavelength of LEDs to apply to a specific horticultural crop.

7. Conclusions

This review has demonstrated why LEDs are considered a novel technology in the food industry. This technology is constantly improving, and its application holds great potential in horticulture for food preservation. Importantly, LEDs are cost-effective and environmentally friendly, release minimal radiant heat, and have a monochromatic nature, which allows the selection of specific wavelengths, while excluding unwanted wavelengths that sometimes result in producing radiant heat. The application of LEDs postharvest has been shown to accelerate or delay ripening, improve color development, enhance the phytochemical concentration, improve nutritional quality, extend shelf life, and prevent fungal spoilage of various horticultural crops. It has, however, been noted that the recent research is focusing only on carotenoid-accumulating crops, such as pepper and tomato, with minimal focus on other crops. A deeper understanding of how various light spectra affect various crops and how the intensity of light and the duration of illumination affect various fruit and vegetables, especially highly perishable ones, is required. Merging of certain wavelengths still requires serious attention; it is also important to investigate the response of various crops to various ratios of wavelengths to decide on the combination that yields better results. Further studies are required to reduce moisture loss as a result of LED illumination; in this case, the application of edible coatings in different formulations after LED illumination may potentially mitigate this effect. The sensory acceptability of fruit and vegetables treated with LEDs postharvest, as well as a deeper understanding of the mechanism involved in the postharvest irradiation of LED lights, still needs to be further investigated. Lastly, research revealed that LEDs can enhance health-related compounds present in fruit and vegetables; however, further studies need to be conducted to determine optimal ratios or combinations of LEDs and investigate which one achieves this optimally without negatively affecting other quality parameters of these commodities.

Acknowledgements

None.

Conflict of interest

The authors declare no conflict of interest.

Author details

Bonga Lewis Ngcobo[1]* and Isa Bertling[2]

1 Faculty of Applied Sciences, Department of Horticulture, Durban University of Technology, Durban, South Africa

2 Horticultural Science, School of Agricultural, Earth and Environmental Sciences, University of KwaZulu-Natal, Pietermaritzburg, South Africa

*Address all correspondence to: bongalewisfuze@gmail.com and bongan@dut.ac.za

References

[1] Duro JA, Lauk C, Kastner T, Erb KH, Haberl H. Global inequalities in food consumption, cropland demand and land-use efficiency: A decomposition analysis. Global Environmental Change. 2020;**64**:102124. DOI: 10.1016/j.gloenvcha.2020.102124

[2] Lufu R, Ambaw A, Opara UL. Water loss of fresh fruit: Influencing pre-harvest, harvest and postharvest factors. Scientia Horticulturae. 2020;**272**:109519. DOI: 10.1016/j.scienta.2020.109519

[3] Slavin JL, Lloyd B. Health benefits of fruits and vegetables. Advances in Nutrition. 2012;**3**(4):506-516. DOI: 10.3945/an.112.002154

[4] More AS, Ranadheera CS, Fang Z, Warner R, Ajlouni S. Biomarkers associated with quality and safety of fresh-cut produce. Food Bioscience. 2020;**34**:100524. DOI: 10.1016/j.fbio.2019.100524

[5] Warra AA, Prasad MNV. African perspective of chemical usage in agriculture and horticulture-their impact on human health and environment. In: Agrochemicals Detection, Treatment and Remediation. Oxford, United Kingdom: Butterworth-Heinemann; 2020. pp. 401-436. DOI: 10.1016/B978-0-08-103017-2.00016-7

[6] Margenat A, Matamoros V, Díez S, Cañameras N, Comas J, Bayona JM. Occurrence and human health implications of chemical contaminants in vegetables grown in peri-urban agriculture. Environment International. 2019;**124**:49-57. DOI: 10.1016/j.envint.2018.12.013

[7] Nassarawa SS, Abdelshafy AM, Xu Y, Li L, Luo Z. Effect of light-emitting diodes (LEDs) on the quality of fruits and vegetables during postharvest period: A review. Food and Bioprocess Technology. 2021;**14**(3):388-414. DOI: 10.1007/s11947-020-02534-6

[8] Zhang X, Zhang M, Xu B, Mujumdar AS, Guo Z. Light-emitting diodes (below 700 nm): Improving the preservation of fresh foods during postharvest handling, storage, and transportation. Comprehensive Reviews in Food Science and Food Safety. 2022;**21**(1):106-126. DOI: 10.1111/1541-4337.12887

[9] Ngcobo BL, Bertling I, Clulow AD. Artificial daylength enhancement (pre-sunrise and post-sunset) with blue and red led lights affects tomato plant development, yield, and fruit nutritional quality. Horticulture, Environment, and Biotechnology. 2022:1-10. DOI: 10.1007/s13580-022-00447-1

[10] Mandal D. Light emitting diode and its application in pre and postharvest horticulture: A Review. Crop Research. 2021;**56**(6):388-401. DOI: 10.31830/2454-1761.2021.061

[11] D'Souza C, Yuk HG, Khoo GH, Zhou W. Application of light-emitting diodes in food production, postharvest preservation, and microbiological food safety. Comprehensive Reviews in Food Science and Food Safety. 2015;**14**(6):719-740. DOI: 10.1111/1541-4337.12155

[12] Hasan MM, Bashir T, Ghosh R, Lee SK, Bae H. An overview of LEDs' effects on the production of bioactive compounds and crop quality. Molecules. 2017;**22**(9):1420. DOI: 10.3390/molecules22091420

[13] Perera WPTD, Navaratne S, Wickramasinghe I. Impact of spectral

composition of light from light-emitting diodes (LEDs) on postharvest quality of vegetables: A review. Postharvest Biology and Technology. 2022;**191**:111955. DOI: 10.1016/j.postharvbio.2022.111955

[14] Poonia A, Pandey S. Application of light emitting diodes (LEDs) for food preservation, post-harvest losses and production of bioactive compounds: A review. Food Production, Processing and Nutrition. 2022;**4**(1):1-10. DOI: 10.1186/s43014-022-00086-0

[15] Ngcobo BL. Potential of pre-and postharvest illumination of cherry tomato, a climacteric fruit, to reduce the ripening period and enhance yield and quality while maintaining shelf life [Master's dissertation]. 2017. Available from: https://ukzn-dspace.ukzn.ac.za/handle/10413/16741

[16] Bantis F, Smirnakou S, Ouzounis T, Koukounaras A, Ntagkas N, Radoglou K. Current status and recent achievements in the field of horticulture with the use of light-emitting diodes (LEDs). Scientia Horticulturae. 2018;**235**:437-451. DOI: 10.1016/j.scienta.2018.02.058

[17] Yu W, Liu Y, Song L, Jacobs DF, Du X, Ying Y, et al. Effect of differential light quality on morphology, photosynthesis, and antioxidant enzyme activity in *Camptotheca acuminata* seedlings. Journal of Plant Growth Regulation. 2017;**36**(1):148-160. DOI: 10.1007/s00344-016-9625-y

[18] Loi M, Liuzzi VC, Fanelli F, De Leonardis S, Creanza TM, Ancona N, et al. Effect of different light-emitting diode (LED) irradiation on the shelf life and phytonutrient content of broccoli (*Brassica oleracea* L. var. italica). Food Chemistry;**283**:206-214 DOI: 10.1016/j.foodchem.2019.01.021

[19] Ngcobo BL, Bertling I. Effect of light combined with heat treatment on colour development and fruit quality of cherry tomato (*Solanum lycopersicum*). Acta Horticulturae. 2021;**1329**:149-156. DOI: 10.17660/ActaHortic.2021.1329.18

[20] Ngcobo BL, Bertling I, Clulow AD. Post-harvest alterations in quality and health-related parameters of cherry tomatoes at different maturity stages following irradiation with red and blue LED lights. Journal of Horticultural Science and Biotechnology. 2021;**96**(3):383-391. DOI: 10.1080/14620316.2020.1847696

[21] Panjai L, Röhlen-Schmittgen S, Ellenberger J, Noga G, Hunsche M, Fiebig A. Effect of postharvest irradiation with red light on epidermal color and carotenoid concentration in different parts of tomatoes. Journal of Food Measurement and Characterization. 2021;**15**(2):1737-1746. DOI: 10.1007/s11694-020-00770-0

[22] Zhan L, Hu J, Pang L, Li Y, Shao J. Light exposure reduced browning enzyme activity and accumulated total phenols in cauliflower heads during cool storage. Postharvest Biology and Technology. 2014;**88**:17-20. DOI: 10.1016/j.postharvbio.2013.09.006

[23] Nájera C, Guil-Guerrero JL, Enríquez LJ, Álvaro JE, Urrestarazu M. LED-enhanced dietary and organoleptic qualities in postharvest tomato fruit. Postharvest Biology and Technology. 2018;**145**:151-156. DOI: 10.1016/j.postharvbio.2018.07.008

[24] Pinela J, Ferreira IC. Nonthermal physical technologies to decontaminate and extend the shelf-life of fruits and vegetables: Trends aiming at quality and safety. Critical Reviews in Food Science and Nutrition. 2017;**57**(10):2095-2111. DOI: 10.1080/10408398.2015.1046547

[25] Sharma S, Pareek S, Sagar NA, Valero D, Serrano M. Modulatory effects of exogenously applied polyamines on postharvest physiology, antioxidant system and shelf life of fruits: A review. International Journal of Molecular Sciences. 2017;**18**(8):1789. DOI: 10.3390/ijms18081789

[26] Kokalj D, Zlatić E, Cigić B, Vidrih R. Postharvest light-emitting diode irradiation of sweet cherries (*Prunus avium* L.) promotes accumulation of anthocyanins. Postharvest Biology and Technology. 2019;**148**:192-199. DOI: 10.1016/j.postharvbio.2018.11.011

[27] Panjai L, Noga G, Hunsche M, Fiebig A. Optimal red light irradiation time to increase health-promoting compounds in tomato fruit postharvest. Scientia Horticulturae. 2019;**251**:189-196. DOI: 10.1016/j.scienta.2019.03.019

[28] Pola W, Sugaya S, Photchanachai S. Color development and phytochemical changes in mature green chili (*Capsicum annuum* L.) exposed to red and blue light-emitting diodes. Journal of Agricultural and Food Chemistry. 2019;**68**(1):59-66. DOI: 10.1021/acs.jafc.9b04918

[29] He R, Wei J, Zhang J, Tan X, Li Y, Gao M, et al. Supplemental blue light frequencies improve ripening and nutritional qualities of tomato fruits. Frontiers in Plant Science. 2022;**13**:888976. DOI: 10.3389/fpls.2022.888976

[30] Xie C, Tang J, Xiao J, Geng X, Guo L. Purple light-emitting diode (LED) lights controls chlorophyll degradation and enhances nutraceutical quality of postharvest broccoli florets. Scientia Horticulturae. 2022;**294**:110768. DOI: 10.1016/j.scienta.2021.110768

[31] Zhou F, Zuo J, Xu D, Gao L, Wang Q, Jiang A. Low intensity white light-emitting diodes (LED) application to delay senescence and maintain quality of postharvest pakchoi (*Brassica campestris* L. ssp. chinensis (L.) Makino var. communis Tsen et Lee). Scientia Horticulturae. 2020;**262**:109060. DOI: 10.1016/j.scienta.2019.109060

[32] Dhakal R, Baek KH. Short period irradiation of single blue wavelength light extends the storage period of mature green tomatoes. Postharvest Biology and Technology. 2014;**90**:73-77. DOI: 10.1016/j.postharvbio.2013.12.007

[33] Chong L, Ghate V, Zhou W, Yuk HG. Developing an LED preservation technology to minimize strawberry quality deterioration during distribution. Food Chemistry. 2022;**366**:130566. DOI: 10.1016/j.foodchem.2021.130566

[34] Chua A, Chong L, Ghate V, Yuk HG, Zhou W. Antifungal action of 405 nm light emitting diodes on tomatoes in a meso-scale system and their effect on the physicochemical properties. Postharvest Biology and Technology. 2021;**172**:111366. DOI: 10.1016/j.postharvbio.2020.111366

[35] Castillejo N, Martínez-Zamora L, Gómez PA, Pennisi G, Crepaldi A, Fernández JA, et al. Postharvest LED lighting: Effect of red, blue and far red on quality of minimally processed broccoli sprouts. Journal of the Science of Food and Agriculture. 2021;**101**(1):44-53. DOI: 10.1002/jsfa.10820

[36] Jiang A, Zuo J, Zheng Q, Guo L, Gao L, Zhao S, et al. Red LED irradiation maintains the postharvest quality of broccoli by elevating antioxidant enzyme activity and reducing the expression of senescence-related genes. Scientia Horticulturae. 2019;**251**:73-79. DOI: 10.1016/j.scienta.2019.03.016

[37] Kokalj D, Zlatić E, Cigić B, Kobav MB, Vidrih R. Postharvest

flavonol and anthocyanin accumulation in three apple cultivars in response to blue-light-emitting diode light. Scientia Horticulturae. 2019;**257**:108711. DOI: 10.1016/j.scienta.2019.108711

[38] Hu L, Yang C, Zhang L, Feng J, Xi W. Effect of light-emitting diodes and ultraviolet irradiation on the soluble sugar, organic acid, and carotenoid content of postharvest sweet oranges (*Citrus sinensis* (L.) Osbeck). Molecules. 2019;**24**(19):3440. DOI: 10.3390/molecules24193440

[39] Hyun JE, Moon SK, Lee SY. Application of blue light-emitting diode in combination with antimicrobials or photosensitizers to inactivate Escherichia coli O157: H7 on fresh-cut apples and cherry tomatoes. Food Control. 2022;**131**:108453. DOI: 10.1016/j.foodcont.2021.108453

[40] Kahramanoğlu İ, Nisar MF, Chen C, Usanmaz S, Chen J, Wan C. Light: An alternative method for physical control of postharvest rotting caused by fungi of citrus fruit. Journal of Food Quality. 2020;**2020**:8821346. DOI: 10.1155/2020/8821346

[41] Zhang Y, Li S, Deng M, Gui R, Liu Y, Chen X, et al. Blue light combined with salicylic acid treatment maintained the postharvest quality of strawberry fruit during refrigerated storage. Food Chemistry: X. 2022;**15**:100384. DOI: 10.1016/j.fochx.2022.100384

[42] Martínez-Zamora L, Castillejo N, Artés-Hernández F. Postharvest UV-B and photoperiod with blue+ red LEDs as strategies to stimulate carotenogenesis in bell peppers. Applied Sciences. 2021;**11**(9):3736. DOI: 10.3390/app11093736

[43] Ma G, Zhang L, Kato M, Yamawaki K, Kiriiwa Y, Yahata M, et al. Effect of the combination of ethylene and red LED light irradiation on carotenoid accumulation and carotenogenic gene expression in the flavedo of citrus fruit. Postharvest Biology and Technology. 2015;**99**:99-104. DOI: 10.1016/j.postharvbio.2014.08.002

[44] Azuma A, Yakushiji H, Sato A. Postharvest light irradiation and appropriate temperature treatment increase anthocyanin accumulation in grape berry skin. Postharvest Biology and Technology. 2019;**147**:89-99. DOI: 10.1016/j.postharvbio.2018.09.008

[45] D'Souza C, Yuk HG, Khoo GH, Zhou W. Light-emitting diodes in postharvest quality preservation and microbiological food safety. In: Light Emitting Diodes for Agriculture. Singapore: Springer; 2017. pp. 191-235. DOI: 10.1007/978-981-10-5807-3_9

[46] Maroga GM, Soundy P, Sivakumar D. Different postharvest responses of fresh-cut sweet peppers related to quality and antioxidant and phenylalanine ammonia lyase activities during exposure to light-emitting diode treatments. Food. 2019;**8**(9):359. DOI: 10.3390/foods8090359

[47] Pennisi G, Orsini F, Castillejo N, Gómez PA, Crepaldi A, Fernández JA, et al. Spectral composition from led lighting during storage affects nutraceuticals and safety attributes of fresh-cut red chard (*Beta vulgaris*) and rocket (*Diplotaxis tenuifolia*) leaves. Postharvest Biology and Technology. 2021;**175**:111500. DOI: 10.1016/j.postharvbio.2021.111500

Chapter 11

Recent Development in the Preharvest 1-MCP Application to Improve Postharvest Fruit Quality

Nariane Q. Vilhena, Lourdes Cervera-Chiner, Ana Moreno and Alejandra Salvador

Abstract

1-Methylcyclopropene (1-MCP), an ethylene action inhibitor, is routinely applied to fruit as a postharvest treatment prior to cold storage to extend fruit storability and posterior shelf life. Nevertheless, preharvest 1-MCP applied as a liquid spray to trees is a novel treatment for maintaining fruit quality throughout the postharvest in some crops and can be a very useful tool for improving handling operations in packing houses. This chapter aims to provide an overview of not only employing 1-MCP as a preharvest treatment in different crops, but also of its effect on the biochemical and physico-chemical parameters that influence fruit postharvest quality, storage capacity, and chilling injury development. It also intends to address the main factors related to the preharvest 1-MCP application effect, such as application time, optimum concentrations, and its combination with other preharvest treatments.

Keywords: 1-methylcyclopropene, Harvista, fruit storage, field treatment, ethylene

1. Introduction

Ethylene is the simplest natural plant ripening hormone that is involved in the regulation of many growth and development processes of horticultural crops. This hormone acts by linking itself to its action site in the cell to promote a succession of events, such as leaf abscission, fruit maturation, and the ripening process [1–3].

Fruits are generally divided into two categories: climacteric and non-climacteric [4]. It has been clearly established that all fruits produce minimal amounts of ethylene during their development. However, the ripening of climacteric fruit is characterized by an increased respiration rate and a burst of ethylene biosynthesis. Non-climacteric fruits have a different ripening pattern. They do not show a drastic change in respiration rate and the ethylene production remains at a very low level. Climacteric and non-climacteric fruits can be differentiated by not only their pattern of ethylene production during ripening but also by their response to exogenous ethylene.

The postharvest quality of horticultural crops is also influenced by ethylene, whose effect can be beneficial or detrimental depending on a number of factors like fruit type (climacteric or non-climacteric), ripening stage, and intended use [5].

The application of exogenous ethylene is often used to promote uniform ripening in climacteric fruits like bananas or apples [6, 7]. Ethylene is also applied on non-climacteric fruits to bring about certain different effects, such as accelerating citrus fruits color change (degreening) [8]. However, ethylene can negatively influence the quality of both climacteric and non-climacteric fruits by inducing the development of physiological disorders or accelerating senescence processes, especially during storage and shelf life.

Therefore, many strategies have been studied to control ethylene production or its action during the postharvest life of fruits and vegetables. Of the available methods, 1-methylcyclopropene (1-MCP) is a competitive inhibitor of ethylene perception that prevents ethylene binding and the eliciting of subsequent signal transduction and translation. As a consequence, 1-MCP applied at very low concentrations (0.5–1.0 μL L^{-1}) can delay ripening and senescence events of horticultural products that are mediated by ethylene [9–11].

Postharvest 1-MCP application has become a successful technology for controlling ripening and senescence processes to maintain fruit quality and to reduce different postharvest physiological disorders in many climacteric and non-climacteric horticultural products [9, 12]. 1-MCP is also used as an excellent tool to explore ethylene-mediated responses of plant systems, especially those involved in ripening and senescence processes.

It is known that 1-MCP lowers the highest peak value of respiration and ethylene production and delays their emergence during storage [1]. There are two known systems for regulating ethylene production in higher plants: system-1 and system-2. Auto-inhibitory system-1 functions during normal vegetative growth and is responsible for producing the basal ethylene levels detectable in all tissues in both climacteric and non-climacteric fruits. Autostimulatory system-2 is responsible for the upsurge of ethylene production that occurs during the ripening of climacteric fruits [13, 14]. The transition from system-1 to system-2 is believed to be an important step during fruit ripening [3]. For different fruits, such as apples, bananas, and tomatoes [13, 15, 16], there are reports that the expression of the genes responsible for the transition from system-1 to system-2 can be blocked by postharvest 1-MCP application.

To date, a large amount of information has been acquired about the effect of postharvest treatment with 1-MCP on many fruits and vegetables. Protocols have been developed to optimize the benefits of 1-MCP for different horticultural products because its effect depends on factors like concentration, exposure time, or application time after harvest.

Some commercial names of 1-MCP are EthylBloc®, SmartFresh®, SmartTabs®, and EthylBloc® Sachet, which contain different 1-MCP concentrations. When the product is mixed with water or buffer solution, 1-MCP gas is released to the chambers where it is applied. 1-MCP is used in combination with proper temperature and relative humidity management and can replace or be utilized in combination with controlled atmospheres.

Despite postharvest 1-MCP application being able to present lots of benefits in many fruits and vegetables, preharvest 1-MCP treatment emerges as a novel option, and one that has been tested in several crops with different effects, for example, reducing fruit drop, delaying color development and softening, ethylene production, ripening, maintaining fruit quality throughout the postharvest, and replacing the postharvest treatment in some crops [9, 17, 18].

Preharvest 1-MCP treatments can offer some advantages over conventional postharvest exposure. Besides, treatments allow a more flexible harvesting time to be obtained for some apple cultivars and might be a good option because they respond poorly or inconsistently to 1-MCP gas exposure as a postharvest treatment [19, 20].

As the preharvest 1-MCP application is considerably limited, a sprayable formulation has been developed for its utilization that facilitates its field application and is marketed as Harvista® (AgroFresh Solutions Inc., Philadelphia, PA, USA). This product allows 1-MCP to be dissolved in water and applied as a fumigation product. Today, it is the only 1-MCP formulation available for preharvest use purposes.

In some cases like bananas, the preharvest 1-MCP treatment has been tested by submerging the stem of the bunch still attached to trees in 1-MCP aqueous solution [21]. In another study conducted with figs, preharvest treatment was applied through a plastic bag with gaseous 1-MCP [22]. Despite having positive effects, these application modes would be very limited for commercial use.

Relatively few reports exist about the preharvest 1-MCP treatment effect on fruits. Most studies have focused mainly on apples and pears, since Harvista® treatment is authorized for these fruits in some countries. However, the interest generated by this treatment has led to further studies being conducted in recent years to search for a possible benefit on other fruits [23]. This chapter reports a review of the main findings of the preharvest 1-MCP application on different fruits.

2. Climacteric fruit

Increased respiration in climacteric fruit is associated with autocatalytic ethylene production, which mediates fruit ripening processes [24]. As 1-MCP inhibits ethylene perception, its field application in climacteric fruits affects fruit ripening and senescence processes, mainly by delaying harvest time and prolonging postharvest fruit quality [9]. However, effects vary according to different aspects, such as fruit cultivar, application time, and concentration.

2.1 Apple (*Malus domestica* Borkh)

1-MCP is commonly used as a postharvest treatment to prolong apple eating quality by maintaining fruit firmness, crispness, sweetness, acidity, and juiciness of cold-stored fruits. It is a proven effective treatment in many cultivars [25–27]. In recent years, and as a novel application method, numerous studies have been conducted on the effect of 1-MCP applied at preharvest on delaying maturation on trees and to maintain postharvest quality in different scenarios.

Most studies generally suggest that the preharvest 1-MCP treatment positively influences apple quality attributes. The differences observed in fruits response may vary among cultivars, mostly in relation to the ability of some cultivars to rapidly generate new ethylene receptors when fruits remain attached to trees [17, 28, 29].

Recent studies have demonstrated that preharvest 1-MCP can retard the activation of system-2 ethylene biosynthesis in apples [14]. The MdACS1 gene is necessary for system-2 activation during apple climacteric ripening. The molecular mechanisms that control the delay and suppression of the expression of MdACS1 and receptor genes after regular postharvest 1-MCP treatment are not well defined. However, the preharvest 1-MCP application is effective in suppressing its expression in the "Delicious" and "Golden Delicious" varieties, which results in the delayed activation of system-2 ethylene biosynthesis [1, 6, 14].

One effect of 1-MCP applied in the preharvest has been reported on fruit cuticular wax biosynthesis and regulation, composition, and structure for regulating ethylene biosynthesis and signaling [1]. The preharvest 1-MCP application at 150 g hm^2 lowers the contents of alcohols, acids, and esters in apple cuticular wax by reducing fruit superficial scald and decay and by maintaining fruit cuticular wax functions, such as disease resistance and water retention, after 1 cold storage month.

The preharvest 1-MCP effect on carbohydrate metabolism in apples has also been studied at harvest and during cold storage. The authors observed that the treatment at 150 g AI ha^{-1} applied 7 days before harvest inhibited starch degradation, retarded soluble sugar increase, and reduced sucrose, glucose, and fructose in "Starkrimson" apples. This was related to the ethylene regulation of related gene expressions and enzyme activities during cold storage [30].

The combined application of pre- and postharvest 1-MCP treatments has also been evaluated in apples. In this case, Harvista® was applied 10 days before harvest at 60 mg L^{-1} and, 1 day after harvest, fruits were subjected to the Smartfresh® treatment at 1 μL L^{-1}. This combination resulted in greater fruit firmness retention and longer ethylene suppression in "Golden Delicious" apples throughout cold storage [14]. The preharvest or postharvest 1-MCP treatment application led to different expression patterns of ethylene biosynthesis genes (MdACS3 and MdACS1) and receptor genes, which could result in differential effects by 1-MCP treatments.

A common preharvest treatment for apples is to apply ethephon to accelerate maturity to bring forward the harvest period and to improve color development [31, 32]. However, ethephon application leads to the activation of ethylene autocatalysis in fruit tissues, which is reflected as a drastically shortened harvest period and reduced storability. It can also lead to undesired fruit abscission before harvest and accelerated flesh firmness loss during the commercialization period [30, 33, 34]. Nevertheless in "Anna" apples treated with ethephon (50 ppm), preharvest 1-MCP treatment (1–2 mM) in the mature green stage reduced preharvest abscission and preserved fruit firmness. This treatment also mitigated the adverse influence of ethephon on flesh firmness loss during fruit cold storage at 1°C.

In order to delay fruit ripening and preharvest drop in apple, treatments with naphthaleneacetic acid (NAA), a synthetic auxin, and aminoethoxyvinylglycine (AVG), an inhibitor of ethylene biosynthesis, are applied in some production areas. Comparative studies have been conducted on the application of these treatments and the preharvest 1-MCP application [6]. In this way, sprayable 1-MCP at 396 mg L^{-1} applied 1 week before harvest to "Golden Delicious" apples had a stronger effect on delayed fruit drop than AVG or NAA [24]. Similarly with "Delicious" apples, the application of 1-MCP 15 or 7 days before harvest at a concentration of 160 or 320 mg L^{-1} delayed preharvest fruit drop more effectively than AVG or NAA used alone and had a similar effect compared to the fruit to which both AVG and NAA had been applied. In that study, the best results were obtained when 1-MCP was applied 15 days before harvest and the concentration did not affect its efficacy in reducing fruit drop [6].

Scolaro et al. [35] also compared preharvest treatments with 1-MCP or AGV on "Royal Gala" apples. They observed that the 1-MCP application had similar effects to the AVG treatment on delaying fruit ripening and also on decreasing ethylene production, starch degradation, loss of flesh firmness and acidity, epidermal yellowing, soluble solid accumulation, and red color development. The authors concluded that preharvest 1-MCP can be an alternative method to the commonly applied AVG for fruit maturation and harvest management purposes.

With "Golden Delicious" and "Law Rome" apples, 1-MCP applied 7 days or 1 day before harvest at concentrations between 75 and 155 mg L^{-1} has also been reported as an effective emergency stop treatment, similarly to NAA, and without the potential loss of firmness caused by NAA [20].

Preharvest 1-MCP application, used as a treatment to reduce the incidence of different disorders during storage, has been reported for some apple cultivars [20, 27, 36, 37]. In "Honeycrisp" apples, preharvest 1-MCP sprays reduced the incidence of both soft scald, a skin disorder characterized by brown lesions, and soggy breakdown, a flesh disorder characterized by brown and soft internal tissue [36]. In "Law Rome" apples, preharvest 1-MCP application also lowered the superficial scald incidence during prolonged cold storage (up to 120 days) [20].

Stem-end flesh browning, another disorder that develops around the shoulders of apples, is frequently manifested in some cultivars during storage. In cv. Gala, postharvest 1-MCP treatment had no effect on this disorder, while the preharvest 1-MCP application significantly reduced it, but did not prevent its development [37].

In "Fuji" apples at harvest, watercore incidence and severity, besides starch pattern indices, were lower in the fruits that underwent the preharvest 1-MCP treatment [27]. This study also evaluated the combined application of preharvest and postharvest 1-MCP treatments on fruit quality and the incidence of disorders. The incidences of flesh greasiness and watercore diminished more when the combination of both treatments was applied than by either treatment alone. Besides, preharvest and postharvest 1-MCP applications contribute to maintain fruit quality attributes during cold storage and at 20°C. The effects of preharvest 1-MCP were more consistent when the interval between spraying and harvest was 10 days compared to its application at 4 days before harvest.

The traditional postharvest 1-MCP treatment has been reported to increase the risk of certain stress-related storage disorders in apples, such as CO_2 injury, a physiological disorder that can be manifested externally and/or internally, and both injury type and susceptibility were strongly affected by apple cultivar and growing conditions [38]. A study carried out with the cv. McIntosh and cv. Empire revealed that 40 to 160 mg L^{-1} of preharvest 1-MCP applied 7 or 11 days before harvest also increased the development of external CO_2 injury during storage in a controlled atmosphere.

2.2 Pear (*Pyrus communis* L.)

Similarly to apples, in pears the effect of the sprayable preharvest 1-MCP application has been studied mainly on fruit drop, extension of the harvest window, quality maintenance during cold storage, and reduction of the incidence of different disorders [39–41].

Prevention of fruit drop with the preharvest 1-MCP application has been reported in "Santa Maria" pears, but the effect on fields was dose-dependent [40]. The most effective treatment was achieved at 150 and 200 g ha^{-1}. In this cultivar, the ripening period could be prolonged up to 4 weeks.

A similar preharvest 1-MCP treatment effect on fruit drop has been reported for "Barlett" pears [42]. In this study, the 1-MCP application was as effective as NAA in reducing premature fruit drop. 1-MCP significantly delayed ripening immediately after harvest, but this effect diminished after storing fruit at −1°C for 3.5 months. No differences in pear fruit maturity were found between the highest (100 mg L^{-1}) and the lowest (28 mg L^{-1}) applied doses. The strongest 1-MCP effect occurred when fruits were harvested soon after treatment (7 days after application). The 1-MCP preharvest application also lowered internal browning incidence during storage.

A recent study performed with the "Bartlett" and "d 'Anjou" cultivars found that the preharvest 1-MCP treatment extended the harvest window by 3–4 days without reducing the storage potential or eating quality [41]. This treatment also lowered ethylene synthesis and respiration rates, maintained fruit firmness and green color during cold storage, and retarded melting texture development in both cultivars. 1-MCP also reduced the incidence of flesh disorders by alleviating membrane lipid peroxidation, maintaining antioxidant capacity, and enhancing superoxide dismutase, catalase, and ascorbate peroxidase activity in both cultivars.

When combining the effect of the pre- and postharvest 1-MCP treatments on "Barlett" pears, the applications of 160 $\mu L\ L^{-1}$ of Harvista® and 0.15 $\mu L\ L^{-1}$ of Smartfresh® were capable of extending the melting texture life of pears up to 5 months of cold storage [43]. When the effects of Harvista® and Smarfresh® were compared in relation to the fruit firmness maintenance during the cold storage of "Abate Feel" pears, a positive result of both treatments was obtained, but the postharvest application was more effective [44]. The Harvista® application time influenced its effect on fruit firmness, since applying Harvista® 7 days before the commercial harvest time was more effective in maintaining fruit firmness after harvest than when applied before.

In "Chuhwangbae" pears, the preharvest 1-MCP application had no effect on postharvest quality attributes during cold storage and shelf life because most fruit quality attributes and specific targeted metabolites were not affected by preharvest 1-MCP application, but by storage duration [19]. Nevertheless, the sprayable preharvest 1-MCP treatment enhanced the incidence of physiological disorders compared to that of the untreated fruits.

2.3 Persimmon (*Diospyros kaki* Thunb)

The postharvest 1-MCP (Smartfresh®) application is routinely performed to allow persimmon cold storage, since it has been widely reported that it reduces flesh firmness loss as the main chilling injury symptom [45, 46]. Nevertheless, very little information is available on the preharvest 1-MCP application effect on persimmon.

In cv. Rojo Brillante, the preharvest 1-MCP application effect has been evaluated on maintaining flesh firmness in two different scenarios: 1) early in the season on the fruit treated with ethephon to advance maturity; and 2) at the end of the season on the fruit destined for cold storage, treated with gibberellic acid to delay fruit ripening [18]. The preharvest 1-MCP treatment (22 $g\ L^{-1}$) delayed fruit firmness loss induced by ethephon, extending the harvest window, and proved to be the most effective treatment when 1-MCP was applied 1 day after ethephon treatment. The preharvest 1-MCP application also maintained fruit firmness during the marketing period and applying the postharvest 1-MCP treatment was not necessary. Nevertheless, the pre- and postharvest 1-MCP combination maintained greater flesh firmness during the commercialization period than the single postharvest application.

In the fruit treated with gibberellic acid at the end of the season, the 1-MCP application performed 3 days before harvest maintained fruit firmness during cold storage to the same extent as the traditional postharvest 1-MCP application [18]. Hence in this situation, replacing the postharvest 1-MCP application with preharvest treatment can be a very useful alternative.

A positive preharvest 1-MCP treatment effect has also been observed on "Fuyu" persimmon [47]. Spraying 150 $mg\ L^{-1}$ of 1-MCP in the first commercial harvest week reduced not only premature flesh softening but also the occurrence of the translucent stain disorder at postharvest without altering fruit maturation on trees. Better results were found on fruit harvest 1 day after 1-MCP treatment, when similar efficacy of the pre- or postharvest application on both fruit firmness and the incidence of disorders was observed during storage.

2.4 Banana (*Musa spp.*)

The main postharvest banana losses are due to a short postharvest life, which is the main problem in the banana industry. The postharvest 1-MCP application has been well studied in this fruit to delay the ripening process and maintain postharvest quality [21]. Numerous studies have reported that 1-MCP applied at the 5–500 nL L^{-1} and 0.1 μL L^{-1} concentrations delays fruit ripening and skin color change during the postharvest life [48–50].

The postharvest 1-MCP treatment regulates ethylene synthesis by inhibiting the genes that regulate the aminocyclopropane-1-carboxylic acid synthase (ACS) and aminocyclopropane carboxylate oxidase (ACO) enzymes [51, 52]. The inhibition of these enzymes results in reduced fruit softening, which extends green life. Moreover, 1-MCP enhances superoxide dismutase and catalase activities and inhibits peroxidase activity, playing an important role in growth and plant development and disease resistance [11, 53]. It is noteworthy that the postharvest 1-MCP treatment effect depends on different factors such as cultivar, maturity stage, previous ethylene exposures, crop conditions, and the part of the bunch [21].

Negative 1-MCP postharvest application effects have also been reported in bananas such as irregular peel coloration, reduced volatile compound production, and delayed sugar accumulation [21, 50]. 1-MCP treatment in bananas can increase the development of chilling-related disorders, which can be trigged by the inhibition of ethylene production [10, 54]. These negative effects limit its commercial application [11].

To date, very few studies on preharvest 1-MCP applications in bananas exist. Only Manigo et al. [55] has studied the effect of preharvest 1-MCP treatment with "Cavendish" on postharvest fruit quality to identify the best and most cost-efficient application method. Three preharvest 1-MCP application methods have been evaluated: Stalk End Immersion (SEI), where the edge of bunch stalks is immersed in an aqueous solution of 1-MCP; bunch spraying (BS); and the combination of both methods (SEI-BS). In all cases, the applied dose was 400 nL L^{-1}. These treatments had a significant effect on delaying fruit ripening, retarding peel color change and fruit softening, and maintaining visual quality during storage. The fruit treated with 1-MCP by the SEI-BS method displayed lesser accumulated weight loss, and the degree of shriveling and the finger drop incidence were lower compared to the BS and SEI methods followed separately. The combined method is useful for prolonging the banana shelf life up to 19 days.

2.5 Stone fruits

Stone fruits are a diverse group, mostly of the genus Prunus, that includes peach, apricot, among others. This group is characterized by a lignified endocarp, a fleshy mesocarp and a thin exocarp or skin [56]. In "Madoka" peach (*Prunus persica* L.), the 1-MCP application in fields has been investigated with regard to fruit physiological and biochemical responses and quality attributes. Lee et al. [57] have observed not only delayed firmness loss but also an inhibition of ethylene production and respiration during the storage of the peach fruit preharvest-treated with 1-MCP. The inhibition of the expression of the genes related to sugar accumulation and cell wall softening, and of the genes responsive to ethylene receptors, has also been found. These findings suggest that the preharvest 1-MCP application can extend the shelf life of peaches by the inhibition of ethylene production and respiration.

Another assay, which compared the effect of preharvest sprayable 1-MCP (Harvista®) or postharvest fumigable 1-MCP (Smartfresh®) treatments on the quality attributes and enzymatic activities of cell wall hydrolases during the cold storage of "Hetsal Haunkeybee" peaches, has reported that fruit flesh firmness was significantly enhanced by SmartFresh®, but not by Harvista® [58]. The SmartFresh® treatment significantly reduced the enzymatic activities of α-galactosidase, β-galactosidase, β-glucosidase, β-arabinosidase, β- xylosidase, and α-mannosidase during cold storage compared to the untreated and Harvista-treated fruits.

A study on apricot (*Prunus armeniaca* L.) cv. Canino has been conducted to evaluate the effect of preharvest 1-MCP and other field treatments ($CaCl_2$ and AGV), and their combination, on fruit quality parameters during cold storage. This study showed that treatment not only improved fruit postharvest quality but also lowered the incidence of disorders throughout storage. The combination of the three compounds was the most effective treatment in maintaining fruit quality and prolonging storability up to 30 days [2].

2.6 Other fruits

With mango (*Mangifera indica* L.), the postharvest 1-MCP treatment is necessary for delaying the fast ripening that initiates before fruit harvest maturity. In this situation, preharvest treatment would be a good option for prolonging fruit storability and allow exportation [59]. In mango cv. "Carabao," the preharvest 1-MCP treatment at 10 ppm is effective in slowing external color evolution, delaying ethylene peak, and controlling both ripening and deteriorated visual quality at harvest and during storage at 13°C. However, fruit firmness does not significantly vary among treatments [60]. The fruits treated twice with 1-MCP (10 and 5 days before harvest) obtain the best results than those treated once. The authors concluded that the second application time is crucial because of the variation in the biochemical composition of fruit tissues.

Application time is also critical for the proper response of preharvest 1-MCP in Mangosteen (*Garcinia mangostana* L.). Lerslerwong et al. [61] observed that when treatment was applied in the fruit climacteric stage, it delayed the ripening process by about 1 week. This shows its potential use to retard the harvest period. However, treatment had no effect on fruit ripening when applied before the climacteric peak.

Figs (*Ficus carica* L.) are a postharvest technology challenge because of their very short shelf life and high susceptibility to diseases. The fig ripening process is classified as climacteric, with higher respiration rates and ethylene production at the beginning of the ripening phase. Yet unlike most climacteric fruits harvested before ripening onset, it does not ripen after harvest [62]. The postharvest 1-MCP treatment does not affect the ripening parameters of treated fruits (unlike other climacteric fruits), but applications to fruit on trees improve fruit storage capacity by inhibiting deterioration with minor effects on fruit growth and ripening [63, 64]. The 1-MCP application before or after harvest has also been used as a tool for studying the ethylene-related genes involved in the natural ripening process of attached fruit [22], and a possible feedback reaction has been proposed. This downstream component of ethylene signal transduction can play a role in regulating ethylene synthesis during the reaction to 1-MCP, which causes the non-climacteric behavior of fig ethylene production.

The preharvest 1-MCP influence on the development of fruit on trees and storage capacity has been studied in "Brown Turkey" figs [22]. Treatment was applied as gas in plastic bags 3 days before harvest at preclimacteric stage, which delayed fruit senescence and improved storage life up to 7 days as manifested by fruit color, firmness, internal texture, weight, size, shriveling, and decay.

Yellow pitahaya (*Selenicereus megalanthus* Haw) is a tropical fruit that undergoes physiological damage associated with cold storage, including peel browning and necrosis [65, 66]. In addition, fruit quality loss during storage has been associated with ethylene production and fruit respiration. Therefore, postharvest 1-MCP applications are done to extend fruit shelf life. Cock et al. [65] observed that the 1-MCP application 15 days before harvest in yellow pitahaya produced significant beneficial effects on chemical, physical, and sensory properties and extended fruit shelf life by 5 days with no large differences between treatment concentrations (200 or 400 μg L^{-1}). Another study conducted with yellow pitahaya under the same preharvest 1-MCP conditions showed that treatment accelerated epicarp coloration, maintained firmness, and delayed weight loss and the maturity index [66]. However, the higher applied concentration led fruit to show undesirable signs of senescence. The authors suggested that the preharvest 1-MCP application in this fruit could trigger the metabolic processes responsible for shortening fruit preservation, which has been related to both the magnitude and sensitivity to ethylene rises because fewer receptors are evaluable, and effects depend on the compound concentration, application time, and storage time.

Melon (*Cucumis melo* L.) presents a high diversity of ripening behaviors, including climacteric and non-climacteric genotypes [67, 68]. Cantaloupe melons possess typical climacteric behavior with ethylene playing a major role in the regulation of the ripening process and by affecting the ripening rate. Nevertheless, Pech et al. [3] have reported that climacteric (ethylene-dependent) and non-climacteric (ethylene-independent) regulation coexist during climacteric fruits ripening. In two cantaloupe melon cultivars (cv. Caravelle and cv. Mission), the effect of the preharvest 1-MCP application was evaluated in relation to fruit quality, harvest synchrony, and maturity [69]. Treatment was applied at a concentration between 5 and 25 g ha^{-1} and from 22 to 7 days before harvest. It presented very little or no effect on fruit quality at harvest or after cold storage. Only the cv. Mission treated with the highest concentration presented greater firmness than the other treatments after 9 storage days.

Despite most studies on preharvest and postharvest 1-MCP applications showing positive beneficial effects on maintaining fruit firmness and overall postharvest quality, in blueberry (*Vaccinium corymbosum* L.) the 1-MCP treatment applied at both the pre- and postharvests had negative effects on fruit firmness at harvest and during storage. Previous studies have demonstrated that the postharvest 1-MCP application to "Rabbiteye" blueberry led to increased ethylene production, which caused fruit softening [70]. On to the preharvest treatment, Blaker and Olmstead [71] observed in cv. 'Star" and "Sweetcrisp" that the 1-MCP application 5 days prior to harvest decreased fruit firmness, while the fruit treated 9 days before harvest did not differ from the control. However, the authors are still unclear why the preharvest treatment could result in firmness loss.

3. Non-climacteric fruits

As opposed to climacteric fruits, non-climacteric fruits are characterized by ripening transitions that do not strictly depend on a significant increase in ethylene production and an associated rise in the respiration rate [72, 73]. Although the ripening process of climacteric fruits is well documented, it is not very accurate for non-climacteric fruits. Recent studies using metabolomics, proteomics, and transcriptomics have significantly increased knowledge about molecular processes during non-climacteric fruits ripening [72]. These studies have demonstrated the involvement of different hormones, such as abscisic acid (ABA), auxin, gibberellins, among others. However, the complex mechanisms underlying the regulation and crosstalk between these hormones during fruit development and ripening require further research.

In non-climacteric fruits, applying 1-MCP at the postharvest for ripening control does not make sense because these fruits do not have a response to ethylene for maturation. However, 1-MCP has been applied to some non-climacteric fruits for other purposes to, for instance, provide insights into the occurrence of ethylene-dependent and ethylene-independent events during ripening, including changes in genes expression [10]. A positive effect of the postharvest 1-MCP application on the inhibition of senescence processes has been reported in some non-climacteric fruits, such as reducing rachis browning in grapes and delaying leaf senescence in "Shatangju" mandarins marketed with attached leaves [74, 75].

The postharvest 1-MCP treatment has also been reported to inhibit the development of some physiological disorders, such as scald development in pomegranate, pericarp browning in litchi, water soaking in watermelon, internal browning in loquat, and internal flesh browning of pineapple [76, 77]. Inhibition of degreening and color change are observed in several citrus fruits, strawberry, and olive when 1-MCP has been applied [76].

Very few studies report the preharvest effects of 1-MCP applications on non-climacteric fruits. In citrus fruits (*Citrus spp.*), the preharvest 1-MCP treatment has been reported to result in reduced undesirable tree defoliation when ethephon is applied to diminish fruit removal force [22, 78]. In "Washington" navel oranges, a higher yield per tree and increased fruit elongation have been found when a combination of preharvest 1-MCP and gibberellic acid (GA3) was applied [79]. Besides, fruit drop considerably reduced when fruits were treated with 1-MCP alone or combined with either NAA or GA3. The combination of 1-MCP with either NAA or GA3 also enhanced the maturity index compared to untreated fruits.

In "Bing" sweet cherry (*Prunus avium* L.), reduced flesh firmness during postharvest life occurred when ethephon was applied to stimulate fruit abscission during mechanical harvest. The 1-MCP treatment performed 3 days after the ethephon application counteracted ethephon-induced flesh firmness loss without inhibiting fruit removal force reduction [80].

4. Summary of the effect of preharvest 1-MCP application on different fruits

See **Table 1**

Fruits	Effect of preharvest 1-MCP application	References
Apple	Reduce fruit drop Delay color development and flesh softening Retard starch degradation Delay ethylene production and ripening Maintain soluble solids and titratable acidity Reduce disorders incidence during cold storage	[1, 6, 14, 17, 20, 24–38]
Pear	Reduce fruit drop Extend the harvest window Reduce ethylene production Maintain firmness and color during cold storage	[19, 39–44]
Persimmon	Extend the harvest window Maintain firmness and color during cold storage Reduce disorders incidence during cold storage	[18, 47]
Banana	Delay ripening, color change, softening, weight loss, and finger drop during storage	[55]
Peach	Retard flesh softening Reduce ethylene production and ripening Extend shelf life	[57, 58]
Apricot	Maintain quality and prolong storability Reduce disorders incidence during cold storage	[2]
Mango	Delay ripening and peel color change Reduce ethylene production	[60]
Mangosteen	Delay ripening and extend the harvest window	[61]
Fig	Delay fruit senescence and improved storability	[22]
Pitahaya	Extend shelf life	[65, 66]
Melon	Maintain firmness during storage	[69]
Blueberry	Decrease firmness	[71]
Citrus	Reduce fruit and leaves abscission	[22, 78, 79]
Cherry	Reduce flesh firmness during postharvest life	[80]

Table 1.
Summary of different fruits in which 1-MCP has been applied as a preharvest treatment.

5. Future research trends and new perspectives

The preharvest 1-MCP treatment has shown remarkable benefits for different fruit crops, which clearly come over for apples. Further research is needed to allow its application to a larger number of climacteric fruits, mainly those for which the post-harvest 1-MCP treatment has a positive effect on delaying ripening and maintaining fruit quality. In these cases, replacing the postharvest treatment with the preharvest 1-MCP application would be a very useful tool for improving handling operations in packing houses.

In non-climacteric fruits, further studies are necessary to elucidate the role of the preharvest 1-MCP treatment in maintaining or improving different fruits quality aspects.

Besides, continued research is required on metabolic paths, gene expression, and enzyme activities on the preharvest effects of 1-MCP on fruit metabolism.

Acknowledgements

This research was funded by the Ministerio de Ciencia e Innovación of Spain, through Project PID2021-127975OR-C21 and the IVIA-GVA project No. 52201 and by the European Union through the European Regional Development Fund (ERDF). Nariane Q. Vilhena thanks the Spanish Ministry of Science and Innovation for Grant FPI-INIA (PRE2018-085833). Lourdes Cervera Chiner thanks the "Margarita Salas" grant from the Spanish Ministry of Universities funded by the European Union-Next Generation EU.

Conflict of interest

The authors declare no conflict of interest.

Author details

Nariane Q. Vilhena[1], Lourdes Cervera-Chiner[1,2], Ana Moreno[1] and Alejandra Salvador[1*]

1 Postharvest Department, Valencian Institute for Agricultural Research, Moncada, Valencia, Spain

2 Universitat Politècnica de València, Valencia, Spain

*Address all correspondence to: salvador_ale@gva.es

References

[1] Li F, Zhang X, Wang J, Jiang Y, Zhang X, Li X. Preharvest application of 1-methylcyclopropene and Ethephon altered cuticular wax biosynthesis and fruit quality of apples at harvest and during cold storage. Horticultural. Plant Journal. 2022;**8**(2):143-152. [Internet] DOI: 10.1016/j.hpj.2021.11.008

[2] Nagy N. Effect of Preharvest applications of calcium, anti- ethylene compounds and their combinations on "Canino" apricot fruit quality and storability. Zagazig Journal of Agricultural Research. 2018;**45**(5):1609-1631

[3] Pech JC, Purgatto E, Bouzayen M, Latché A. Ethylene and fruit ripening. The Plant Hormone Ethylene. 2012;**44**:275-304

[4] Paul V, Pandey R, Srivastava GC. The fading distinctions between classical patterns of ripening in climacteric and non-climacteric fruit and the ubiquity of ethylene-an overview. Journal of Food Science and Technology. 2012;**49**(1):1-21

[5] Brizzolara S, Manganaris GA, Fotopoulos V, Watkins CB, Tonutti P. Primary metabolism in fresh fruits during storage. Frontiers in Plant Science. 2020;**11**(February):1-16

[6] Yuan R, Li J. Effect of sprayable 1-MCP, AVG, and NAA on ethylene biosynthesis, preharvest fruit drop, fruit maturity, and quality of "delicious" apples. HortScience. 2008;**43**(5):1454-1460

[7] Mohapatra D, Mishra S, Sutar N. Banana post harvest practices: Current status and future prospects-a review. Agricultural Research Communication Centre. 2010;**31**(1):56-62

[8] Sdiri S, Navarro P, Monterde A, Benabda J, Salvador A. New degreening treatments to improve the quality of citrus fruit combining different periods with and without ethylene exposure. Postharvest Biology and Technology. 2012;**63**(1):25-32. [Internet]. DOI: 10.1016/j.postharvbio.2011.08.005

[9] Valero D, Guillén F, Valverde JM, Castillo S, Serrano M. Recent developments of 1-methylcyclopropene (1-MCP) treatments on fruit quality attributes. Eco-Friendly Technology for Postharvest Produce Quality. 2016;**2004**:185-201

[10] Watkins CB. Pre- and postharvest inhibition of ethylene production and action by 1-MCP and the quality of apples and other horticultural products. Acta Horticulturae. 2016;**1120**:1-10

[11] Baswal AK, Ramezanian A. 1-methylcyclopropene potentials in maintaining the postharvest quality of fruits, vegetables, and ornamentals: A review. Journal of Food Processing and Preservation. 2021;**45**(1):1-10

[12] Watkins CB. The use of 1-methylcyclopropene (1-MCP) on fruits and vegetables. Biotechnology Advances. 2006;**24**(4):389-409

[13] Tatsuki M, Endo A, Ohkawa H. Influence of time from harvest to 1-MCP treatment on apple fruit quality and expression of genes for ethylene biosynthesis enzymes and ethylene receptors. Postharvest Biology and Technology. 2007;**43**(1):28-35

[14] Varanasi V, Shin S, Johnson F, Mattheis JP, Zhu Y. Differential suppression of ethylene biosynthesis and receptor genes in "Golden delicious"

apple by Preharvest and postharvest 1-MCP treatments. Journal of Plant Growth Regulation. 2013;**32**(3):585-595

[15] Barry CS, Llop-Tous MI, Grierson D. The regulation of 1-aminocyclopropane-1-carboxylic acid synthase gene expression during the transition from system-1 to system-2 ethylene synthesis in tomato. Plant Physiology. 2000;**123**(3):979-986

[16] Tan D, Li T, Wang A. Apple 1-Aminocyclopropane-1-carboxylic acid synthase genes, MdACS1 and MdACS3a, are expressed in different Systems of Ethylene Biosynthesis. Plant Molecular Biology Reporter. 2013;**31**(1):204-209

[17] Elfving DC, Drake SR, Reed AN, Visser DB. Preharvest applications of sprayable 1-methylcyclopropene in the orchard for management of apple harvest and postharvest condition. HortScience. 2007;**42**(5):1192-1199

[18] Vilhena NQ, Gil R, Vendrell M, Salvador A. Effect of Preharvest 1-MCP treatment on the flesh firmness of 'Rojo Brillante' persimmon. Horticulturae. 2022;**8**(5):1-7

[19] Lwin HP, Lee J. Preharvest 1-methylcyclopropene treatment effects on fruit quality attributes and targeted metabolites in 'Wonhwang' pears stored at room temperature after cold storage. Scientia Horticulturae. 2021;**289**:53-61

[20] McArtney SJ, Obermiller JD, Schupp JR, Parker ML, Edgington TB. Preharvest 1-methylcyclopropene delays fruit maturity and reduces softening and superficial scald of apples during long-term storage. HortScience. 2008;**43**(2):366-371

[21] Wasala WMCB, Benaragama CK, Kumara GDK, Sarananda KH, Dissanayake CAK. Application of 1-Methylcyclopropene (1-MCP) for delaying the ripening of Banana: A review. Asian research. Journal of Agriculture. March 2021;**14**:44-56

[22] Freiman ZE, Rodov V, Yablovitz Z, Horev B, Flaishman MA. Preharvest application of 1-methylcyclopropene inhibits ripening and improves keeping quality of "Brown Turkey" figs (Ficus carica L.). Scientia Horticulturae. 2012;**138**:266-272. [Internet]. DOI: 10.1016/j.scienta.2012.01.007

[23] Agrofresh Inc. HARVISTA™ 1.3 SC Label. 2018. Availeble from: legacy.picol.cahnrs.wsu.edu

[24] Zhang J, Ma Y, Dong C, Terry LA, Watkins CB, Yu Z, et al. Meta-analysis of the effects of 1-methylcyclopropene (1-MCP) treatment on climacteric fruit ripening. Horticulture Research. 2020;7(1):1-16. [Internet]. DOI: 10.1038/s41438-020-00405-x

[25] Tomala K, Grzeda M, Guzek D, Głabska D, Gutkowska K. Analysis of possibility to apply preharvest 1-methylcyclopropene (1-MCP) treatment to delay harvesting of red Jonaprince apples. Sustainability (Switzerland). 2020;**12**(11):1-10

[26] Tomala K, Grzęda M, Guzek D, Głąbska D, Gutkowska K. The effects of preharvest 1-methylcyclopropene (1-MCP) treatment on the fruit quality parameters of cold-stored 'szampion' cultivar apples. Agriculture (Switzerland). 2020;**10**(3):1-13

[27] Lee J, Kang IK, Nock JF, Watkins CB. Effects of preharvest and postharvest applications of 1-methylcyclopropene on fruit quality and physiological disorders of "Fuji" apples during storage at warm and cold temperatures. HortScience. 2019;**54**(8):1375-1383

[28] McArtney SJ, Obermiller JD, Hoyt T, Parker ML. "Law Rome" and "Golden delicious" apples differ in their response to preharvest and postharvest 1-methylcyclopropene treatment combinations. HortScience. 2009;**44**(6):1632-1636

[29] Doerflinger FC, Nock JF, Miller WB, Watkins CB. Preharvest aminoethoxyvinylglycine (AVG) and 1-methylcyclopropene (1-MCP) effects on ethylene and starch concentrations of 'empire' and 'McIntosh' apples. Scientia Horticulturae. 2019;**244**(September 2018):134-140

[30] Sun Y, Shi Z, Jiang Y, Zhang X, Li X, Li F. Effects of preharvest regulation of ethylene on carbohydrate metabolism of apple (Malus domestica Borkh cv. Starkrimson) fruit at harvest and during storage. Scientia Horticulturae. 2021;**276**:September 2020, 109748. [Internet]. DOI: 10.1016/j.scienta.2020.109748

[31] Brackmann A, Thewes FR, dos Santos LF, Machado EP, Ludwig V, da Artmann L, et al. Effect of growth regulators application on the quality maintenance of 'Brookfield' apples. Bragantia. 2015;**74**(4):453-456

[32] Soethe C, Steffens CA, Hawerroth FJ, do Amarante CVT, Heinzen AS. Maturation of 'Baigent' apples protected by antihail nets and sprayed with aminoethoxyvinylglycine and ethephon. Pesquisa Agropecuaria Brasileira. 2021;**56**

[33] Ferrara G, Mazzeo A, Matarrese AMS, Pacucci C, Trani A, Fidelibus MW, et al. Ethephon as a potential abscission agent for table grapes: Effects on pre-harvest abscission, fruit quality, and residue. Frontiers in Plant Science. 2016;**7**(May 2016):1-7

[34] Farag KM, Nagy NMN, Haikal AMRM, Derhab SS. Mitigation of ethephon and protone influence while improving "anna" apples coloration, fruit quality and storability by preharvest application of sprayable 1-mcp. Journal of Agriculture and Environment Sciences. 2015;**14**(August):1-23

[35] AMT S, Argenta LC, DO ACVT, Petri JL, Hawerroth FJ. Controle Da Maturação Pré-Colheita De Maçãs 'Royal Gala' Pela Inibição Da Ação Ou Síntese Do Etileno. Revista Brasileira de Fruticultura. 2015;**37**(1):38-47

[36] DeEll JR, Ehsani-Moghaddam B. Preharvest 1-methylcyclopropene treatment reduces soft scald in "Honeycrisp" apples during storage. HortScience. 2010;**45**(3):414-417

[37] Doerflinger FC, Sutanto G, Nock JF, Shoffe Y Al, Zhang Y, Watkins CB. Stem-end flesh browning of 'Gala'apples is decreased by preharvest 1-MCP (Harvista) and conditioning treatments. Fruit Q. 2017;**25**(3):9-14

[38] DeEll JR, Ehsani-Moghaddam B. Effects of preharvest and postharvest 1-methylcyclopropene treatment on external CO2 injury in apples during storage. Acta Horticulturae. 2012;**945**:317-324

[39] Kc AN, Rasmussen AL, DeShields JB. 1-Methylcyclopropene Preharvest application and its effect on storability of "Comice" and "bosc" pears. HortScience. 2022;**57**(1):24-31

[40] Sakaldas M, Gundogdu MA, Gur E. The effects of preharvest 1-methylcyclopropene (Harvista) treatments on harvest maturity of 'Santa Maria' pear cultivar. Acta Horticulturae. 2019;**1242**:287-294

[41] Li Y, Zhang S, Dong Y. Improving storability, physiological disorders, and

antioxidant properties of 'Bartlett' and 'd'Anjou' pears (Pyrus communis L.) by pre-harvest 1-methylcyclopropene spraying. International Journal of Food Science and Technology. 2021;**56**(1):115-125

[42] Villalobos-Acuña MG, Biasi WV, Flores S, Mitcham EJ, Elkins RB, Willits NH. Preharvest application of 1-methylcyclopropene influences fruit drop and storage potential of "Bartlett" pears. HortScience. 2010;**45**(4):610-616

[43] Li M, Zhi H, Dong Y. The influence of pre- and postharvest 1-MCP application and oxygen regimes on textural properties, cell wall metabolism, and physiological disorders of late-harvest 'Bartlett' pears. Postharvest Biology and Technology. 2021;**173**: 111429-111439. (May 2020)

[44] Defilippi BG, Manríquez D, Robledo P. Use of 1-methylcyclopropene (1-MCP) as a strategy to improve post-harvest life of "abate Fetel" pears. Acta Horticulturae. 2011;**909**:739-744

[45] Salvador A, Arnal L, Monterde A, Cuquerella J. Reduction of chilling injury symptoms in persimmon fruit cv. "Rojo Brillante" by 1-MCP. Postharvest Biology and Technology. 2004;**33**(3):285-291

[46] Tessmer MA, Appezzato-da-Glória B, Kluge RA. Evaluation of storage temperatures to astringency 'Giombo' persimmon: Storage at 1 °C combined with 1-MCP is recommended to alleviate chilling injury. Scientia Horticulturae. 2019;**257**(July):108675. [Internet]. DOI: 10.1016/j.scienta.2019.108675

[47] Vieira MJ, Argenta LC, do Amarante CVT, Steffens CA, de Souza EL. Conservação de caqui "Fuyu" com o tratamento em pré-colheita e pós-colheita com 1-metilciclopropeno. Pesquisa Agropecuaria Brasileira. 2016;**51**(3):197-206

[48] Blankenship SM, Dole JM. 1-Methylcyclopropene: A review. Postharvest Biology and Technology. 2003;**28**(1):1-25

[49] Sisler EC, Serek M, Dupille E, Goren R. Inhibition of ethylene responses by 1-Methylcyclopropene and 3-Methylcyclopropene. Plant Growth Regulation. 1999;**27**(2):105-111

[50] Harris DR, Seberry JA, Wills RBH, Spohr LJ. Effect of fruit maturity on efficiency of 1-methylcyclopropene to delay the ripening of bananas. Postharvest Biology and Technology. 2000;**20**(3):303-308

[51] Zhang MJ, Jiang YM, Jiang WB, Liu XJ. Regulation of ethylene synthesis of harvested banana fruit by 1-methylcyclopropene. Food Technology and Biotechnology. 2006;**44**(1):111-115

[52] Kranthi Kumar T, Bhagavan BVK, Mamatha K, Rani AS, Subbaramamma P. Response of 1-MCP on physiological and physical characteristics in Banana (Musa paradisiaca L.) cv. Tella Chakkerakeli during storage. International Journal of Current Microbiology and Applied Sciences. 2018;**7**(12):3135-3144

[53] Xu F, Liu Y, Xu J, Fu L. Influence of 1-methylcyclopropene (1-MCP) combined with salicylic acid (SA) treatment on the postharvest physiology and quality of bananas. Journal of Food Processing and Preservation. 2019;**43**(3):1-7

[54] Jiang W, Zhang M, He J, Zhou L. Regulation of 1-MCP-treated Banana fruit quality by exogenous ethylene and temperature. Food Science and Technology International. 2004;**10**(1):15-20

[55] Manigo BI, Paul J, Matuginas L. Department of Agriculture RFO XI high value crops development program Davao City effects of pre-harvest application of 1-Methylcyclopropene (1-MCP) on the postharvest quality of "Cavendish" Banana (Musa cavendishii). Southeastern Philippines Journal of Research and Development. 2020;**135**:135-152

[56] Brady CJ. Stone fruit. Biochemistry of Fruit Ripening. Dordrecht: Springer; 1993. pp. 379-404

[57] Lee DB, Lee GJ, Kim SH, Kang IK, Choi SJ, Yun HK. Changes in fruit characteristics and expression of ripening-related genes in 'madoka' peaches treated with 1-methylcyclopropene and aminoethoxyvinylglycine. Horticultural Science and Technology. 2020;**38**(1):32-43

[58] Yoo J, Ryu S, Park C, Win NM, Choi HW, Lee J, et al. Effects of preharvest and postharvest 1-methylcyclopropene treatments on fruit quality attributes and cell wall metabolism during short-term cold storage in 'hetsalhaunkeybee' peaches. Horticultural Science and Technology. 2019;**37**(5):609-617

[59] Gaikwad SS, Sakhale BK, Chavan RF. Effect of 1– MCP concentration, exposure time and storage temperature on post-harvest quality of mango fruit cv. Alphanso. Food Research. 2020;**4**(3):746-752

[60] Castillo-israel KAT, Serrano EP, Esguerra EB, Merca FE. Dual preharvest application of 1-methylcyclopropene (1-MCP) aqueous spray controls mango (Ma ngif er a ind ic a L. cv. 'Carabao') ripening Katherine. Philippine Science Letters. 2003;7(2):324-330

[61] Lerslerwong L, Rugkong A, Imsabai W, Ketsa S. The harvest period of mangosteen fruit can be extended by chemical control of ripening-a proof of concept study. Scientia Horticulturae. 2013;**157**:13-18. [Internet]. DOI: 10.1016/j.scienta.2013.03.027

[62] Dagan B, Stover E, Flaishman MA. The fig: Botany , horticulture. Horticulutral Reviews. 2008;**34**:113-180

[63] Sozzi GO, Abraján-Villaseñor MA, Trinchero GD, Fraschina AA. Postharvest response of "Brown Turkey" figs (Ficus carica L.) to the inhibition of ethylene perception. Journal of the Science of Food and Agriculture. 2005;**85**(14):2503-2508

[64] Owino WO, Manabe Y, Mathooko FM, Kubo Y, Inaba A. Regulatory mechanisms of ethylene biosynthesis in response to various stimuli during maturation and ripening in fig fruit (Ficus carica L.). Plant Physiology and Biochemistry. 2006;**44**(5-6):335-342

[65] Cock LS, Valenzuela LST, Aponte AA. Physical, chemical and sensory changes of refrigerated yellow pitahaya treated preharvest with 1-MCP. DYNA (Colombia). 2013;**80**(178):11-20

[66] Serna CL, Tores VLS, Ayala AAA. Effect of pre- and postharvest aplication of 1-methylcyclopropene on the maturation of yellow pitahaya (Selenicerus megalanthus haw). Vitae. 2012;**19**(1):49-59

[67] Flores F, El Yahyaoui F, De Billerbeck G, Romojaro F, Latché A, Bouzayen M, et al. Role of ethylene in the biosynthetic pathway of aliphatic ester aroma volatiles in Charentais cantaloupe melons. Journal of Experimental Botany. 2002;**53**(367):201-206

[68] Beaulieu JC. Within-season volatile and quality differences in stored fresh-cut cantaloupe cultivars. Journal of Agricultural and Food Chemistry. 2005;**53**(22):8679-8687

[69] Leskovar D, Agehara S, Goreta S. (4) 1-MCP Preharvest spray application to synchronize harvest and improve fruit quality of cantaloupe. HortScience. 2019;**41**(4):1015A-11015A

[70] MacLean DD, Nesmith DS. Rabbiteye blueberry postharvest fruit quality and stimulation of ethylene production by 1-methylcyclopropene. HortScience. 2011;**46**(9):1278-1281

[71] Blaker KM, Olmstead JW. Effects of preharvest applications of 1-Methylcyclopropene on fruit firmness in southern highbush blueberry. Acta Horticulturae. 2014;**1017**:71-76

[72] Giovannoni J. Maturation and ripening. Annual Review of Plant Biology. 2001;**52**:725-749

[73] Cherian S, Figueroa CR, Nair H. "Movers and shakers" in the regulation of fruit ripening: A cross-dissection of climacteric versus non-climacteric fruit. Journal of Experimental Botany. 2014;**65**(17):4705-4722

[74] Li Q, Wu F, Li T, Su X, Jiang G, Qu H, et al. 1-Methylcyclopropene extends the shelf-life of "Shatangju" mandarin (citrus reticulate Blanco) fruit with attached leaves. Postharvest Biology and Technology. 2012;**67**:92-95. [Internet]. DOI: 10.1016/j.postharvbio.2012.01.001

[75] Li L, Kaplunov T, Zutahy Y, Daus A, Porat R, Lichter A. The effects of 1-methylcyclopropane and ethylene on postharvest rachis browning in table grapes. Postharvest Biology and Technology. 2015;**107**:16-22

[76] Li L, Lichter A, Chalupowicz D, Gamrasni D, Goldberg T, Nerya O, et al. Effects of the ethylene-action inhibitor 1-methylcyclopropene on postharvest quality of non-climacteric fruit crops. Postharvest Biology and Technology. 2016;**111**:322-329

[77] Selvarajah S, Bauchot AD, John P. Internal browning in cold-stored pineapples is suppressed by a postharvest application of 1-methylcyclopropene. Postharvest Biology and Technology. 2001;**23**(2):167-170

[78] Burns JK. 1-Methylcyclopropene applications in preharvest systems: Focus on citrus. HortScience. 2008;**43**(1):112-114

[79] Farag KM, Nagy NMN. Effective reduction of postset and preharvest abscissions and increasing the yield of "Washington" navel orange fruits by 1-MCP, GA3 and NAA. Journal of Applied Sciences Research. October. 2012:5132-5141

[80] Elfving DC, Auvil TD, Castillo F, Drake SR, Künzel H, Kupferman EM, et al. Effects of preharvest applications of ethephon and 1-MCP to 'Bing' sweet cherry on fruit removal force and fruit quality. Journal of the American Pomological Society. 2009;**63**(3):84-100

Section 3

Special Considerations for Selected Fruits

Chapter 12

Pomegranate: Postharvest Fungal Diseases and Control

Annamaria Mincuzzi and Antonio Ippolito

Abstract

Due to well-known nutraceutical properties, pomegranate (*Punica granatum* L.) cultivation is recently increasing in various areas of the world including Italy. Fungal diseases are the major causes of postharvest yield and economic losses. Most of the fungi infect pomegranates in the field during the blooming stage remaining latent until fruit ripening, others infect fruit during harvest and postharvest handling through rind injuries. Main postharvest fungal diseases of pomegranates are gray and blue molds caused by *Botrytis* spp. and *Penicillium* spp., respectively, black heart and black spot due to *Alternaria* spp., anthracnose related to species ascribable to *Colletotrichum* genus, and Coniella rot, due to *Coniella granati*. Few fungicides are allowed for pre- and postharvest treatments, making it extremely difficult to control fungal infections. In this scenario, especially in organic fruit production, alternative control means may be a desirable solution to reduce pomegranate losses during the production chain. This chapter focuses on the most important postharvest diseases of pomegranates and possible strategies and means to reduce spoilage.

Keywords: balausta, *Aspergillus*, *Talaromyces*, chitosan, seaweed extract, *Pilidiella*, minor crop, pomegranate loss, pomegranate market, fludioxonil

1. Introduction

Pomegranate (*Punica granatum* L.) belongs to the order of Myrtales, but the family type is uncertain between Punicaceae and Lythraceae [1]. Regarding the binomial name, the genus *Punica* reflects the feminized Roman name of Chartage, instead the species epithet *granatum* referred to the seeded arils; the common English name "pomegranate" describe its anatomical features of "grainy apple" [2, 3]. The Transcaucasian-Caspian region [4] is the center of diffusion of pomegranate shrubs, which early spread in the Mediterranean basin and then gradually reach the New World. Being a heliophilous plant, pomegranates tolerate temperature until 45–48°C with hot wind, even though temperatures below −18°C are harmful [1]. Pomegranate cultivation is promoted by the cool winter and hot summer; these features match with Mediterranean climate, although these shrubs are fully spread in tropical and sub-tropical areas too [2].

1.1 An outline of botanical features

Some botanical features of pomegranate plants are relevant in plant pathology. First and foremost, flower biology. Being an andromonoecy shrub, it brings both hermaphroditic and functionally male flowers to the same plant [5]. Male flowers bloom earlier than bisexual ones and have a *vexillum* function attracting insects, and a selective advantage improving the genetic variability related to cross-pollination instead of self-pollination [5]. Fertile hermaphrodite flowers have well-developed *gynoecium* and *androecium*, urceolate vase-shaped calyx, and prominent U-shaped ovary; often they show big ovary and long style, the *stigma* emerges at the same height as the anthers or above them. On the other hand, male flowers are smaller, since the *androecium* is well-developed, but the *gynoecium* (including ovary and ovules) is rudimental: the style is shorter than anthers; the calyx is bell-shaped and the ovary is V-shaped. Furthermore, several intermediate flower forms exist that may be more or less close to the chief sexual forms showing different degrees of anatomical functionality [1, 4, 5]. Usually, hermaphroditic flowers (and occasionally a few intermediate forms) are fertilized [1, 5]. This is a delicate phase because latent fungal pathogens infect fruit during this stage passing through the open connection with the ovary [6]. According to the cultivar, between 5 and 8 months after fertilization and fruit set, pomegranate fruit, named *balausta*, ripen [4]. The most significant morphological feature of this berry-like and fleshy fruit is the persistence of the residues of the thick floral calyx [1] creating optimal ecological niche for fungal settlement and growth. In addition, the persistence of necrotic stamens acts as secondary source chiefly for wound fungal pathogens [6]. The leathery and woody rind and the richness in polyphenolic compounds guarantee physical and chemical protection from pathogenic fungi to a certain degree [1]. Internally, the carpel, made of light yellow and spongy tissue, as the mesocarp, acts as preferential way for latent fungi diffusion. Internally the *balausta* is reparteed in asymmetrical chambers, called "*locules,*" by membranes made of papery tissue [1, 2, 4]. Arils, attached to the carpel and wrapped by membranes, are the edible portion of pomegranates, constitute 40–60% of fruit weight, and are the most susceptible part of the fruit. Over time, pathogen diffusion happens chamber by chamber. Pomegranate is a non-climacteric fruit, so it must be harvested at the optimum maturity stage [4].

1.2 An outline of horticultural features and disorder prevention

Few horticultural characteristics are important to control disorders, improve disease resistance, and enhance healthy fruit production which is the key point for extended storage. Pomegranates could grow in different soils but prefer fertile and humus-rich ones with a medium-deep density and well-drained avoiding water stagnation [1]. During the dry season, especially in arid and semi-arid environments, pomegranate plants need a regular daily irrigation to prevent water stress; also considering reduction of groundwater resources, drip irrigation is one of the favorite techniques [7]. Furthermore, within 0.5% of soil mass, pomegranate is a saline-tolerant plant: sodium, chlorine, and potassium are accumulated in root tissues, preventing diffusion of toxic salts [1, 4]. Also, fertilization is important to prevent both pomegranate disorders, like cracking, and related diseases due to wound pathogens; nitrogen and calcium are the most involved chemical elements. Soil application of watery nitrogen enhances vegetative growth, fruit size, and yield; in addition, foliar application of potassium chloride, potassium sulfate, and microelements improves both yield and growth of pomegranate [4, 8]. Among foliar treatments, calcium application is the most significant; it induces physiological resistance and stabilization of the cell membrane, preventing low-quality production and supporting postharvest storage [9]; in addition, if applied at the blooming stage and 1 month later, it prevents fruit cracking [10]. Indeed, nitrogen and calcium are key chemical elements for disease prevention; also pruning practices may influence disorder and disease development. In some parts of the world, multiple trunk method is the traditional growing practice for pomegranates, from three to five main trunks are developed and branches are open-vase pruned. This favors the branch maintenance year-after-year and allows the replacement of diseased one [4]. The single stem method provides a single trunk of up to 30 cm in height, after which divided into three or four main branches; this way ensures better vase-shape maintenance and adequate penetration of light that is fundamental for pomegranate veraison [4]. However, sunrays surplus could cause sunburn damaging fruit, indeed the perfect light balance is noticeable [1].

1.3 An outline of pomegranate trade

In the last decades, pomegranate cultivation increased worldwide to face consumer requests; pomegranates display nutraceutical properties due to the high content of active secondary metabolites (i.e., alkaloids, terpenoids, and polyphenols). These compounds are well-known for both antimicrobial and antioxidant properties, and therapeutic ones that feed this trade [11]. Consequently, the request for fresh fruit and related processed products is rapidly growing reaching a global production of about 8.1 million tons [12]. In the world, the chief producers are India, China, and Iran which produced 70% of pomegranates [12], although official data are not available. Particularly, Indian production represents 41% of the global trade having 288,000 ha of pomegranate orchards and producing 3,256,000 tons of fruit gathered in the Maharashtra state and in the Solapur district [13]. In India, the most spread cultivars are Bhagwa and Arakta available throughout the year; these pomegranates support export to European, Middle East, and Asian countries, especially during the production gap (December, January, and March) [14]. Even though China is the second worldwide producer, 70% of produced soft-seeded 'Tunisia" pomegranates are headed to the national markets [14]. A similar scenario distinguishes the Iran market, which mainly aspires to internal commerce of "Malas Yazdi" pomegranates due to export troubles [14]. Regarding Europe, the main producers are Spain and Italy, detailing Spanish production aims at export, instead Italian is not enough to satisfy internal requests [12]; indeed, Italy imports 4% of the global production of pomegranates. Being high-quality, flavorful, and royalty-free cultivars [15], chiefly cultivated plants belong to the Israeli "Akko," the American "Wonderful" and the Spanish "Mollar de Elche," this last is featured by sweetness and herbaceous seeds. These cultivars, such as their wide-spread clones, sequentially ripening in September and October, so are available almost till January [4, 16]. Most updated Italian data referred to 2021 when pomegranate orchards are 1420 ha and have a total production of 192,485 [17]. As displayed in **Table 1**, comparing 2011 and 2019 production years, the present production is about 800- and 3600-fold increased, respectively. Almost the entire production happens in southern Italy, especially in Apulia and Sicily regions, where the Mediterranean climate is favorable. Pomegranate cultivation is promoted by the cool winter and hot summer, although these shrubs are spread in tropical and sub-tropical areas too [2]. Being a minor crop, worldwide analysis of

Measured parameters	2011	2016	2021
Total surface (ha)	62	622	1420
Productive surface (ha)	60	402	1249
Total production (q)	5131	46,343	192,485
Harvested production (q)	4968	45,717	186,972

Table 1.
Pomegranate Italian production. Pomegranate surface (ha) and fruit production (q) are compared in different years.

pomegranate market is lacking due to both the absence of consistent and updated data and grouping within a single Harmonized Code1 (HS code) [18].

1.4 An outline of yield and economic losses

Yield losses caused by fungal postharvest rots may significantly reduce pomegranate production [19] partially justifying differences between produced and harvested amount of fruit (**Table 1**). Yield losses are reflected in economic ones. Particularly, fruit losses are distributed among the field, wholesale, and retail sites referring to related transportation too; similar data are obtained for Indian [20, 21] and South African markets [22, 23]. In the field importance of injuries, cracking, and fungal infections is highlighted rather than secondary infections and physical damages (dehydration, overripening, etc.) mainly involved during the other phases [20, 21]. In India, 10% of yield losses occur in the field and wholesale and 15% are in retail [20, 21], whereas, in South Africa, the percentages involved are 18, 23, and 21%, respectively [22, 23].

2. Postharvest pomegranate diseases

A careful eye in presenting postharvest fungal diseases of pomegranates is needed, according to the mode of infection that gives rise to "latent" infections and "wound" ones. In the field, latent pathogens infect the future fruit during the blooming stage and remain latent until environmental and physiological optimal conditions let them to develop; usually, this happens during postharvest stages. Chief diseases ascribable to this group are gray mold, black heart and black rot, soft rot, and anthracnose, which represent around 65% of total infections [6]. The remaining percentage belongs to wound pathogens that penetrate pomegranate rind and infect fruit following traumatic events caused by bad handling (i.e., no close-cropped peduncle), pests (i.e., borers), weathering (i.e., hail and rain), and abiotic damages (i.e., cracking) occurring from the field till the retail. Blue mold and aspergillosis are the main diseases related to these events [6]. Both groups of pathogens may be responsible also for secondary infections, generally related to nesting or infected stamens. Treatments to control primary and secondary infections, as well as good agronomical practices (i.e., debris and mummy removal) are needed to reduce yield and economic losses [19]. To this aim, a reliable identification of the chief genera and species of fungi is needed to face them with specific/effective substances. Pathogen/disease characterization is primarily based on fruit symptom evaluation, followed by pathogen isolation and observation of their specific macroscopic/microscopic structures on a broad spectrum medium, such as potato dextrose agar (PDA). Finally, molecular analyses are applied to univocally confirm their identity.

2.1 Latent infections

2.1.1 Alternariosis

Alternaria genus is the etiological agent of black spot and black heart diseases, even though the latter is more hazardous to human health. Indeed, *Alternaria* spp. produce more than 30 mutagenic or genotoxic mycotoxins, among which the most relevant are alternariol, alternariol monomethyl ether, altenuene, altertoxins, tentoxin, and tenuazonic acid [24]. Distinguishable symptoms feature both diseases. Black spot symptoms are necrotic areas in pomegranate rind, which appear as small circular black blotches, reddish-black in the middle, and surrounded by yellowish-green halos. Internally fruit are healthy and edible (**Figure 1A**). On the other hand, pomegranates affected by black heart are apparently asymptomatic, but the inner part is rotted. Often, black-heart alternariosis is related to darker rind color and/or lighter weight due to dehydration and aril disintegration; occasionally fruit may appear asymmetric and irregularly shaped. Internally, brown and soft-rotted arils become grayish-black in color (**Figure 1B,C**). Infection symptoms spread from the calyx area (crown) to the entire fruit along the carpel via; this diffusion pathway is shared by all latent fungal pathogens. Alternariosis is one of the most widespread pomegranate diseases, and these have been reported all over the world: Italy, Israel,

Figure 1.
External symptoms of (A) black spot and (B) black heart caused by Alternaria *spp. (C) Close-up on heart rot showing mycelium growth. Colonies on PDA of* Alternaria alternata*: (D) dark and (E) whitish strains. (F) Colony of* Alternaria arborescens *on PDA [6].*

Greece, California, Spain, and India. Species involved in Alternaria diseases belong to *A. alternata* and *A. arborescens* species-complex, their distribution is about overlapped; in addition, just in India is recorded *A. burnsii* [13, 25–29]. Although, within this genus conidial morphology is easily recognizable, the sporulation pattern is different among species and/or morphotypes, but intraspecific variability of colony morphology is high (**Figure 1C-E**). Colonies varied both in color (whitish, brown, deep green, and/or black) and texture (flat, fluffy, and/or wooly); sometimes *Alternaria* colonies grown on artificial media display typical patterns including different colors and textures. Within *A. alternata* complex, *alternata* is the most spread morphotype that has dark brown conidia (20 ± 10 μm in diameter), oval-ellipsoidal shaped, with 4 ± 1 transverse septa; conidia are arranged in branched chains. Colonies belonging to morphotype *tenuissima* are greenish with white margins; their conidia are elongated with a long-tapered beak. *A. alternata* morphotype *limoniasperae* is less common but observed as pomegranate pathogen too; colonies are flat, light brown, and granulated with undulating edges, while conidia are long, ellipsoidal, and display 1–3 transverse septa. The second *Alternaria* species recorded on pomegranate fruit is *A. arborescens* [26] whose colonies are greenish-gray or brown in color, characterized by a slow growth rate; conidia shape ranges from oval to ellipsoidal, with 1–4 transverse and 1–2 longitudinal septa. Being erect and straight, branched, bended, and geniculate, *A. burnsii* conidiophores are typical whereas conidia are ununiformly slight but featured by a short beak [30]. Due to morphological heterogeneity, molecular approach is useful to identify species and morphotypes, particularly it may be advantageous using OPA 1–3 or OPA 10–2 barcoding regions [31] and/or a multilocus approach as described by Aloi et al. [26].

2.1.2 Gray mold

Gray mold is among the main postharvest diseases of pomegranates all over the world [32–34]. As proved by Testempasis and colleagues [35], gray mold is caused by species belonging to *B. cinerea* complex as *B. cinerea s.s.*, *B. pseudocinerea,* and *B. cinerea* group *S*., being the former two the most spread etiological agents. The infection starts in the crown area showing small tan-colored spots, which rapidly spread to the whole fruit (**Figure 2A-C**). Developing, spots became darker and softer until causing rind collapse; finally, gray fluffy mycelium and black sclerotia grow (**Figure 2D**). Internally, softening and browning of the arils and the development of gray mycelium feature this infection. Aiming to control postharvest losses, nesting among contiguous fruit is important because this significantly enhances pathogen diffusion. Early, colonies are whitish (**Figure 2E, F**), then become brownish-gray, and finally covered by sclerotia circle-arranged. Lemon-like conidia measure 7.7 ± 2.4 × 6.8 ± 2.5 μm on average, while sclerotia are 2.9 ± 1.5 × 2.1 ± 0.6 mm. Morphological identification within the species complex is not possible, therefore a duplex PCR assay to evaluate indels in the *mrr1* gene is the optimal solution; suggested primer pairs are BcinN-in-F/BcinN-in-R and Mrr1-spez-F/Mrr1-spez-R [35].

2.1.3 Coniella rot

Considering symptoms, especially at early stage, Coniella fruit rot resembles gray mold. Hence, specific features must be considered. Coniella rot is caused by *Coniella granati* which is synonymized with *Pilidiella granati*; this host-specific fungus is well-known since the end of the nineteenth century when it was isolated

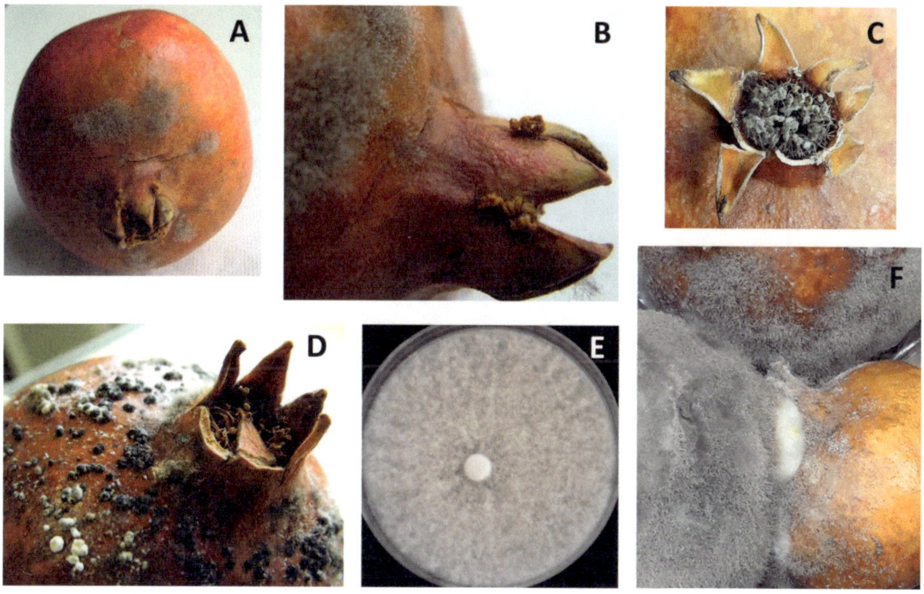

Figure 2.
Gray mold caused by Botrytis cinerea*. (A) Early stage of rot. (B) Close-up image of sporulated gray mycelium, (C) infected stamens, and (D) black sclerotia. (E)* B. cinerea *young colony on PDA. (F) Nesting secondary infection among fruit [6].*

in Italy [36]. Nowadays, it is worldwide spread from Spain and Greece till Tunisia and South Africa causing both shoot blight and canker disease and the above-mentioned crown rot [37–40]. Crown rot symptoms consist of small circular rind spots, spread around the calyx area; these rapidly broaden and make lesions softer and darker till reaching brown color (**Figure 3A**). At maturity, a thin whitish mycelium with spherical pycnidia, ranging between dark brown and black in color, covers the lesions. Internally fruit decay is soft and brown and involves arils too (**Figure 3B,C**). Usually, due to softness, decayed part of each fruit splits. PDA colony ranges between white and creamy in color, texture is leathery and covered by abundant dark pycnidia (110 ± 30 μm in diameter, **Figure 3D**) with thin membranous walls. In addition, hyphae are septate and conidia (13.75 ± 3.750 × 3.5 ± 1.5 μm) are unicellular and hyaline, ellipsoid to fusiform in shape, straight or slightly curved. Even though morphological identification is relevant, molecular confirmation based on Internal Transcribed Spacer ITS5/ITS4 is useful [6].

2.1.4 Anthracnose

Anthracnose is increasingly widespread, especially in tropical and subtropical areas where rainfall and damp wind could enhance its dissemination when temperatures increase [41]; although climate change is favoring its diffusion in the Mediterranean region, it is retained as a minor disease. Species belonging to *Colletotrichum* species-complex are the etiological agents; usually, pomegranate fruit symptoms are superimposable among different species displaying typical anthracnose (**Figure 4A**). Both species-complex and species are differently distributed in the world, indeed in the southeastern United States *C. theobromicola*, *C. siamense*,

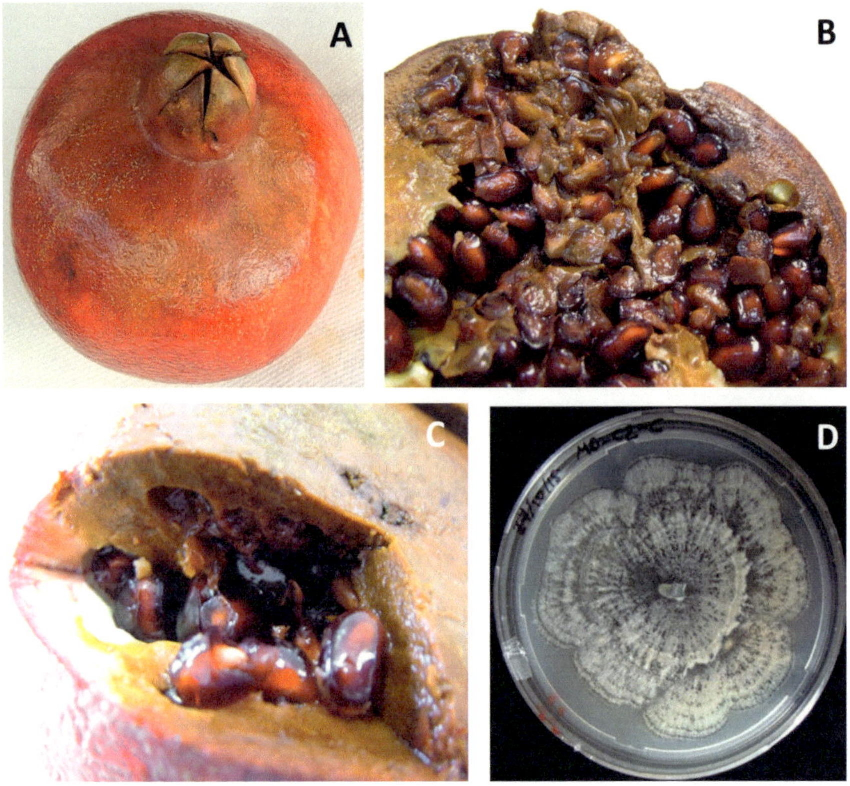

Figure 3.
Coniella granati rot. (A) External symptoms. (B) View of internal decay. (C) Close-up of the soft rot. (D) Colony on PDA plate [6].

C. gloeosporioides, *C. nymphaeae*, *C. fiorinae*, and *C. simmondsii* are common post-harvest pathogens of pomegranates [42]. Similarly, in Brazil, *C. theobromicola*, *C. siamense*, *C. gloeosporioides*, and *C. tropicale* are recorded [43, 44], instead in the old countries most widespread species belong to *C. acutatum* complex and *C. gloeosporioides* complex. For instance, *C. gloeosporioides* is recorded in Turkey and in Albania [44] and *C. acutatum* is registered in Italy [6]. Generally, anthracnose causes the typical soft sunken lesions, which merge as they grow; progressively, white mycelium develops onto and lesions become circular, concentric, and brown with darker spots (**Figure 4B**). On PDA, strains of *C. acutatum* appear fluffy, early white with a reverse ranging from salmon till grayish color, then peachy-pink with pinkish-salmon conidial masses (**Figure 4C**). On the other hand, mycelium of *C. gloeosporioides* displays a texture ranging from faintly aerial to dense cottony; color of colonies ranges from white to pale-olivaceous reaching gray. Acervuli are abundant, small dark-based with sprinkled setae. Meanly, conidia of *C. acutatum s.s.* and *C. gloeosporioides s.s.* measure $11.3 \pm 2.8 \times 4.2 \pm 1.1$ μm and $12 \pm 2.9 \times 4.9 \pm 2.1$ μm, respectively, although, generally, the size is host specific (**Figure 4C**). Both chief species are one-celled, but the first has elliptical-fusiform conidia, instead the last shows oval to oblong, end-pointed conidia. Being species-complex, morphological features are not enough to identify each species within the complex, therefore is needed a molecular multilocus approach according to species-complex they belong [45].

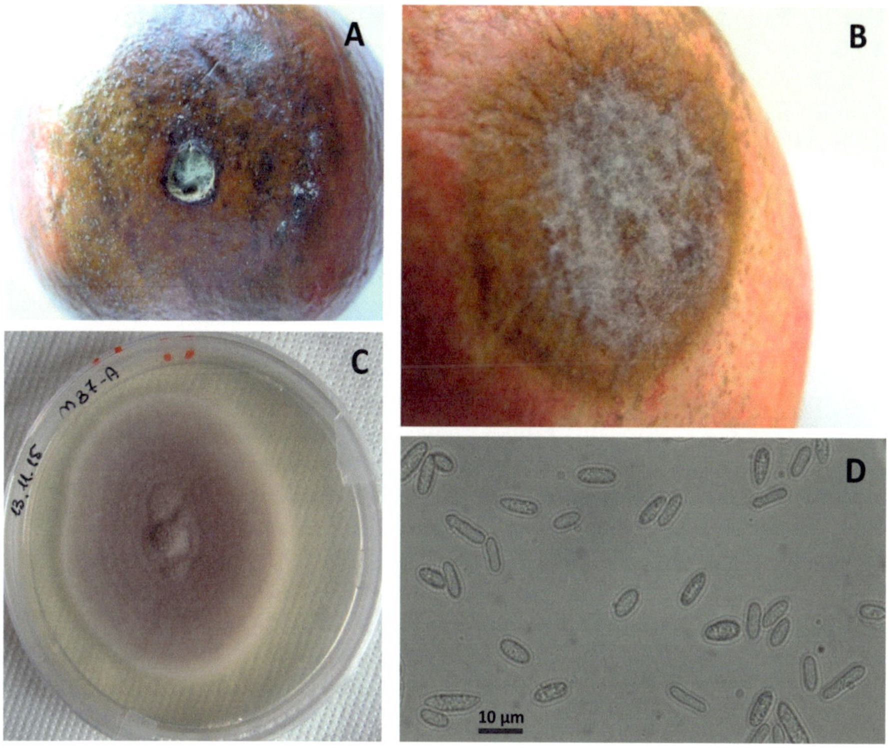

Figure 4.
Anthracnose caused by Colletotrichum acutatum s.s. *Artificial (A) and natural (B) infections. (C)* Colletotrichum acutatum s.s. *colony grown on PDA plate and (D) conidia [6].*

2.2 Wound infections

2.2.1 Blue-green mold

The main fungal group involved in pomegranate wound infections belong to *Penicillium sensu lato (s.l.)* that includes *Penicillium s.s.* and *Talaromyces* genera; within this last genus, just *T. albobivertillius* species is recorded, which results spread both in Italy and China [46, 47]. In *Penicillium s.s.* group, the most abundant species are *P. glabrum*, *P. adametzioides*, and *P. brevicompactum.* Minor species belonging to this genus are *P. jhonkrugii*, *P. pagulum*, and *P. citrinum*. Among these, being not pathogenic for pomegranate fruit, *P. brevicompactum* and *P. jhonkrugii* are not relevant; however, producing cytotoxic compounds named brevianamide A and mycophenolic acid [48], *P. brevicompactum* is putatively hazardous for human health. *P. glabrum* has been reported with high incidence also in Greece, Spain, Uzbekistan, and Slovak Republic [37, 49–52], while *P. adametzioides* has been isolated in Israel [53] too. Although in Italy is believed a minor species, *P. citrinum* (synonymized with *P. implicatum*) is spread in Slovak Republic [51] and Pakistan [53]; being widespread and a citrinin producer [54], its presence is relevant; indeed, citrinin is a nephrotoxic mycotoxin [55, 56]. According to most up-date literature, other *Penicillium* species involved as postharvest pathogens of pomegranate fruit are *P. expansum*, *P. sclerotiorum*, and *P. minioluteum*, the last being synonymized with *Talaromyces minioluteus* and re-arranged in *T. minioluteus*-complex, that includes eight species [57].

Usually, lesions due to these fungi appear as brownish and circular necrotic areas, which became darker, deeper, and occasionally irregularly shaped until reach and infect arils. Finally, blue-green sporification grow inside or onto infected pomegranates. Infected stamens can serve as source of secondary inoculum increasing fungal diffusion (**Figure 5A-D**). Cultured colonies of *Penicillium s.l.* display different textures, from powdery to crustose or velvety; in general, they are blue-green colored and rounded by concentric whitish margins of different thicknesses and/or roughness. The reverse side of each colony has different colors according to pigments and/or metabolites produced, although, often, shades range between white-yellow or red-brownish. Due to species-specificity of produced metabolites, the metabolomic profile is useful for species identification. Regarding micromorphology, features vary based on species, although typical brush-like conidiophores have spherical and unicellular conidia and are arranged as unbranching chains on the top of the phialides. Conidia diameter, which meanly ranges between 2.5 and 5 μm, and wall ornamentation should be useful for species identification. However, morphological identification of *Penicillium s.l.* species is complex; indeed, molecular confirmation by PCR is needed. Bt2a/Bt2b primer pair is used to identify *Penicillium s.l.* species amplifying a portion of the β-tubulin gene [48].

2.2.2 Aspergillosis

Often, aspergillosis of pomegranates is caused by species of *Aspergillus* belonging to the section *nigri*, also called "black aspergilli." Almost 27 species belong to this section, displaying high inter- and intra-specificity genomic variability; their taxonomy is really

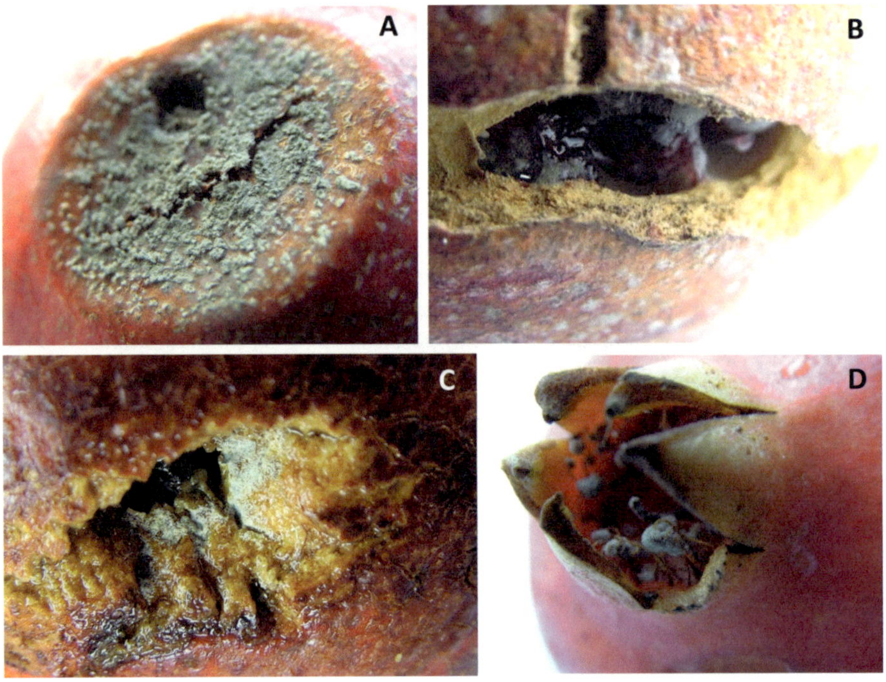

Figure 5.
Penicillium s.l. decay. (A) Necrotic lesion covered with blue-greenish sporification. (B) Cracking, (C) wound, and (D) infected stamens [6].

complicated and not completely solved [58, 59]. Species may be misidentified, and it is not easy evaluating the incidence of each species within the section [59]; furthermore, species identification is relevant especially from a sanitary point of view since some species are potential mycotoxin producers [59]. In general, black aspergilli are well-known as wound pathogens but they could cause internal decay too [6, 60]. *A. niger,* the most spread species within this genus, is recorded in Turkey, Greece, India, Pakistan, and China [61–64], but no evidence of its presence in Italy, where black aspergilli isolated from pomegranate belong to the species *A. tubingensis*, *A. welwitschiae*, *A. uvarum*, and *A. japonicus* [6], being the first identified also in China [65]. *A. niger* and its sister species *A. welwitschiae* are the most hazardous species since they are putative producers of ochratoxin A (OTA) and fumonisins. According to IARC classification, these last are nephrotoxic and hepatotoxic with potential carcinogenic effects on rat and mice; instead, OTA has nephrotoxic and teratogenic, carcinogenic, and immunotoxic properties in rats and possibly in humans [59]. Unfortunately, regulations still do not report limits for mycotoxin amount in fresh pomegranates and their derivatives. Regarding distinctive symptoms, early stage of infection is featured by rind concentric discoloration, from yellow to red-brownish colored (**Figure 6A**-**C**). Internally, pomegranates show a soft brownish-black rot of the arils that may be covered by black powdery sporulation (**Figure 6D**). Black aspergilli cultures appear cottony or velvet-like in texture (**Figure 6E**). Hyphae are septate and hyaline, and mycelium is whitish, but spore development leads to a black appearance; in general, yellow and white shades characterize colony reverse. Characteristic conidiophores are aspergillum-shaped; metulae maintain the phialides and related vesicles. Conidia (2–5 μm in diameter) are arranged in radial chains that give it the characteristic shape. As in the case of *Penicillium s.l.*,

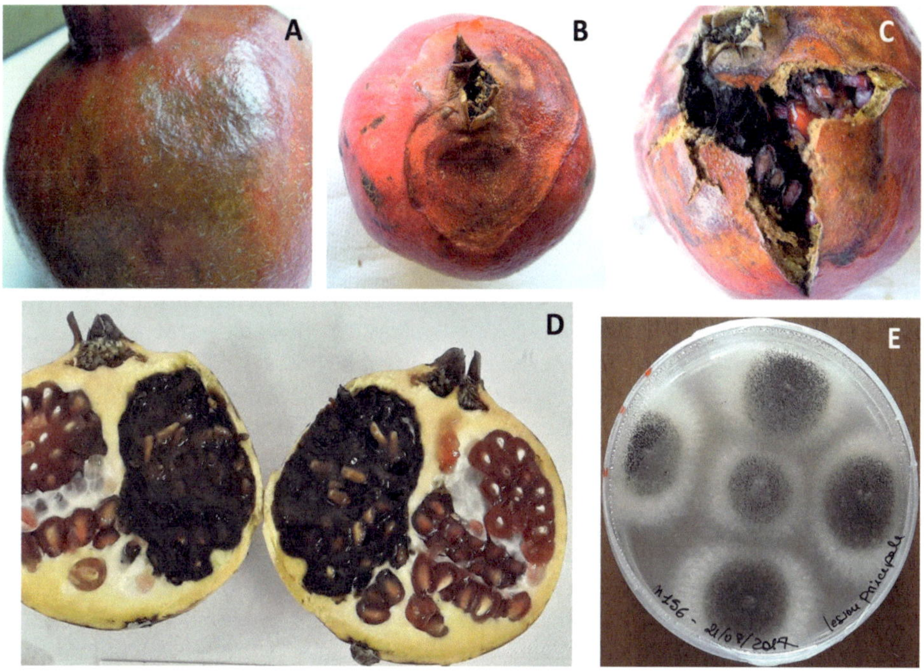

Figure 6.
Aspergillus sect. Nigri rot. (A) Early and (B) advanced stages of the disease. (C) Rind cracking of diseased fruit. (D) View of internal decay; significant is the soft texture of the tissues. (E) Front of a PDA plate of Aspergillus *spp. [6].*

species identification according to macro- and micromorphology is very difficult, requesting molecular confirmation by PCR amplification of a calmodulin portion with CMD5/CMD6 primer pair [48].

2.3 Other fungal diseases

Among fungi involved in pomegranate diseases, there is *Cytospora* spp., which can cause wood canker, branch dieback, and postharvest fruit rot. Although in Spain has been reported *C. annulata* [32], the most widespread species is *C. punicae*. This latter has been reported in various pomegranate cultivation areas such as United States, South Africa, Greece, Cyprus, and Italy [6, 66–69]. Postharvest fruit decay is identifiable by circular soft lesions of the rind, creamy-brownish colored, and centrally darker; related subcutaneous area shows a yellowish and corky appearance. In culture, colonies are whitish, then become olive green; at maturity, colonies are dark brown and covered by globose pycnidia (375 ± 125 μm). Allantoid, aseptate, and hyaline conidia meanly measure 5 ± 1 × 1.5 ± 0.5 μm. Finally, are occasionally identified fungal pathogens belonging to *Fusarium* genus in Egypt and Tunisia [70–72]; being a mycotoxigenic genus that potentially produces deoxynivalenol, 3-acetyl deoxynivalenol, 15-acetyl deoxynivalenol, nivalenol, fusarenon X, T-2 toxin, HT-2 toxin, neosolaniol, diacetoxyscirpenol; zearalenone, fumonisin B1, fumonisin B2, and fusaric acid [73], fusariosis presence needs to be monitored. Also, *Cladosporium* spp. is described as the etiological agent of fruit rots both in Spain and in China [32, 74].

3. Disease control

Obtaining healthy pomegranates is the early step to ensure an extended postharvest life, so fruit sorting during the production chain is relevant. Cold-storage temperature significantly influences the incidence of postharvest diseases and related storage life. Although optimal cold-storage temperature slightly changes among cultivars, in general, 6.5 ± 1°C is fine for most of them in presence of 90–95% relative humidity (RH). Weight loss and decay incidence are increased by higher temperatures, instead chilling injuries are enhanced by lower temperatures; similarly, higher RH favors fungal growth and lower one causes weight loss [75]. In addition, arranging pomegranates in microperforated plastic bags can reduce fruit dehydration allowing transpiration and minimizing condensation, indeed it is possible to further extend the storage life of fruit. These bags create a modified atmosphere [76].

Latent and wound pathogens display different modes of infection, so it is important to take action otherwise; particularly, to control postharvest latent pathogens is needed preharvest application of fungicides during the blooming stage; instead to reduce infections caused by wound pathogens during postharvest processing is useful to act on harvested pomegranates. However, good agronomical practices based on pruning, mummy and debris removal, adequate irrigation, and fertilization are key steps in fungal disease prevention. Nevertheless, in general, fruit are just postharvest treated [77], also because often are no chemical fungicides labeled for preharvest application, like happen in Florida and Italy [6, 77] except for temporary emergency registration. In agreement with the United Nations Priorities, disease prevention is basic to reduce food waste from 30 to 15% and discarded fruit by 20%; finally, UNO requests the reduction of postharvest pesticides by 20% by 2030. This "One Health" approach is useful to reduce fruit-waste and economic losses, defend human health by reducing exposure to chemical fungicides, and decrease fungal resistance to chemical molecules. Regardless of cultivar susceptibility and conventional/alternative nature of the treatment, fungicide application timing is significant to nip-in-the-bud infections.

Tested chemical fungicides aim to stroke main pomegranate postharvest fungal pathogens, such as *B. cinerea*, *A. alternata*, *C. granati,* and *C. gloeosporioides*. In addition, being a minor crop, no fungicides are fully registered in most of the producer countries, such as Italy, where regulations change year by year. As an example, in this country in 2019 and in 2020, to reduce gray mold incidence, fludioxonil postharvest application was allowed, and in 2021 preharvest employment of the beneficial microorganism *Bacillus amyloliquefaciens* sbs. *plantarum,* strain D747, was temporarily approved in addition to sulfur and copper. Particularly, since 2021 employment of copper, well-known for its broad-spectrum anti-bacterial properties, has been banned due to toxicity to the whole ecosystem (humans, animals, plants, and environment). Therefore, the already allowed microorganism, together with boscalid, fludioxonil, commercial formulates based on *Trichoderma asperellum* or *Trichoderma atroviridae* strain T11, and essential oils (geraniol, thymol, and eugenol) have been temporarily approved for 2022. Finally, for treatments on leaves, *Coniothyrium minitans* is temporarily allowed as well as dazomet, metam-potassium, and metam-sodium for treatment in soil.

Fludioxonil belongs to phenylpyrroles that originated from pyrrolnitrin antibiotic, which is produced by different species within the *Pseudomonas* genus [78]. It is a non-systemic fungicide, exploitable both pre- and postharvest, and broad-spectrum; it is effective against *B. cinerea* and other fungal pathogens, such as *Alternaria* spp. and *Aspergillus* spp. [75, 79]. Although its mode of action is not fully understood, fludioxonil inhibits spore germination, germ tube elongation, and mycelium growth of *B. cinerea*. Its mode of action involves the hyperactivation of the high osmolarity glycerol (HOG) signaling pathway through group III hybrid histidine kinases (HK). Being part of the multistep phosphorelay systems (MSP), HKs of group III are needed for the variability of cellular signal transduction in eukaryotic organisms. These signaling systems allow interaction and response between microorganisms and environments through cellular homeostasis regulation, but HOG is responsible for fungicide action also. In the HisKA domain, the histidine H736 is the putative signaling switch [80]. As argued by Xavier and colleagues [77] treating fungicide effectiveness, almost two treatments during the blooming stage are enough to significantly reduce gray mold incidence; in addition, he suggests to alternatively use different active molecules to avoid fungal resistance.

Concerning boscalid, it strikes different stages of fungal development from germ tube elongation and spore germination till appressoria formation or mycelial growth; in addition, during absorption through leaf surface, it is translaminarily and acropetally transported and distributed. Boscalid is a carboxamide fungicide that blocks fungal respiration inhibiting succinate dehydrogenase (SDH) activity. It acts by binding itself to the ubiquinone reduction site within complex II of the mitochondrial electron transport chain. Its site-specific mode of action makes boscalid a broad-spectrum fungicide, but this implies high risk to develop resistant fungal strains also [81]. Generally, these resistant strains are related to point mutations that reach to the substitution of histidine to arginine (H272R) or tyrosine (H272Y) within SDH.

The US Environment Protection Agency classified fludioxonil as reduced-risk compound, but some researchers described its potentially hazardous effects due to mitochondrial oxidative damages that are reflected in organ-specific responses and diseases acting as comorbid factors. Indeed, European Union (EU) set a maximum residue limit (MRL) of 3 ppm within 7 days after pomegranate treatment [82, 83]. However, fludioxonil residues half from the 7th till the 30th day after application [75] and, as observed by Usanmaz et al., fludioxonil concentration decreases below 0.25 ppm after 7 days from the application [79]. Similar considerations concern boscalid that is considered a new-generation fungicide; as other members belonging to SDH inhibitors, it is considered safe. Although, due to its brief lifetime (it is tuned in 2003), EU set 2 mg/kg as MRL value for pomegranate [84].

Hence, in this scenario, the balance of advantages and disadvantages should be considered when choosing the chemical fungicides employment. Therefore, alternative control means may be the safe solution to control postharvest losses caused by fungal pathogens, reducing putative mycotoxin contamination, and avoiding health hazardous compounds like chemical fungicide residues. As stated above, among approved and commercially available microorganisms, there is *B. amyloliquefaciens*. Its efficacy is related to competition for space and nutrients, induction of resistance, and antibiosis. Lipopeptides, such as iturines, fengycin, and surfactine, are produced by *B. amyloliquefaciens* to control gray mold [85]. During the blooming stage, in the field application of *B. amyloliquefaciens* significantly allows controlling postharvest losses caused by *Botrytis* spp.; treated pomegranates enhance the activity of enzymes involved in defense mechanisms, like polyphenol oxidase, peroxidase, phenylalanine ammonia lyase, superoxide dismutase, and catalase (86). By these defense mechanisms, substances (i.e., quinones, reactive oxygen species) and lignin, which are active against fungi, are produced [86]. Furthermore, being elicitors of the induced systemic resistance, chitinase and β-1,3-glucanase are enhanced and boosted by iturin A and surfactine [87]. Repeated treatments during the blooming stage should improve *B. amyloliquefaciens* efficacy due to gradual physiological reduction of defense mechanisms during fruit ripening [88, 89].

Among EU-approved basic substances there is chitosan hydrochloride. Basic substances are "active substances, not predominantly used as plant protection products but which may be of value for plant protection and for which the economic interest of applying for approval may be limited" [https://www.efsa.europa.eu/en/supporting/pub/en-1900]. Chitosan is a D-glucosamine linear polymer available in several commercial formulates differing in composition. It acts in three different ways: it elicits host defense mechanisms, it discloses antimicrobial activity, and it has coating properties [90]. Its effectiveness changes based on the origin (it can be obtained by exoskeleton of crustaceans or insects or fungal mycelium), deacetylation degree (from 60 to 100%), molecular weight (between 3800 and 20,000 Da), production process (chemical or biological synthesis), environmental features (like pH and temperature), and not least sensitivity of fungal species [90–92]. Its water solubility, bio-adhesive properties, and related user-friendliness depend on just mentioned parameters and hydrochloride. Antifungal properties are also highly influenced by its composition [92], although its effectiveness in controlling postharvest rots of pomegranates is proven. Therefore, chitosan hydrochloride display both direct and undirect antifungal effects as described by Munhuweyi et al. [78]: halving of mycelial growth of *C. granati*, *Botrytis* sp., and *Penicillium* sp. is validated by *in vitro* trial; *in vivo* assay displays a reduction by 18–66% of rot incidence caused by the same fungi. Main postharvest fungal pathogens of pomegranates, such as *Botrytis* spp., *Penicillium* spp., and *C. granati*, are controlled by using chitosan concentration between 0.5 and 2.2 g/L, with an efficacy comparable to fludioxonil chemical control [78]. Chitosan could be applied as fruit coating also. It changes respiration and transpiration rate, so delaying ripening, reducing decay incidence, and maintaining qualitative parameters till 6 months [93]. However, its employment is more popular regarding fresh arils. Often chitosan effectiveness is improved by combination with essential oils for both whole fruit and arils [78, 94]. On the other hand, geraniol, thymol, and eugenol have recently been approved, indeed no reliable data are available, although preliminary *in vitro* trials regarding their efficacy is in progress [95].

Chitosan is also usable combined with sulfur nanoparticles, this element is historically known for its antifungal properties: Greek and Romans used it as drug and disinfectant [96, 97]. Its broad-spectrum efficacy contributed to sulfur success, which over time has been included in other fungicide formulations, such as nanoparticles and dots, often displaying stronger antimicrobial activity and greater environmental friendliness [96, 98, 99]. Sulfur success is related to wide efficacy, high sensitivity, and selectivity to fungi, in addition, fungal resistance due to its employment is limited. Based on low toxicity to humans and animals and rapid dissipation, elemental sulfur is safe enough also, but hydrogen sulfide and sulfur dioxide are hazardous to humans and animals [97]. Common sulfur mode of action is not clearly understood; the most popular theory suggests involvement of sulfur permeability of fungal cell walls due to ergosterol content. Hence, sulfur passes into the cell cytoplasm and damages both cytochrome *b* and *c* and the proteic and non-proteic sulfhydryl groups implicated in the mitochondrial electron transport chain [97].

4. Conclusions

Although in recent years pomegranate is worldwide spreading due to high market value and nutraceutical properties, it is still considered a minor crop, so conventional and alternative fungicides registered for this crop are scarce. This entails important postharvest yield and economic losses mainly related to fungal diseases. Albeit losses are evident in the post-harvest phase, most of the infections start in the field during the blooming stage, so preharvest treatments are needed to control them. Remaining infections are caused by wound pathogens that attack fruit through injuries; these infections being reparteed along the whole processing chain imply a particular care of fruit from harvest to the retail. The effects of good agronomical practices, preharvest treatments during the blooming stage, fruit sorting along the production chain, optimal storage conditions, and good hygiene conditions decrease the incidence of postharvest pomegranate rots and extend commercial pomegranate availability. Alternative control means deeply supporting the reduction of postharvest pomegranate disease incidence defending human and animal health by both fungicide residues and fungal mycotoxins and ensuring a One Health approach to saving food waste production.

Acknowledgements

This work was conducted within PRIMA StopMedWaste "Innovative Sustainable technologies to extend the shelf-life of Perishable Mediterranean fresh fruit, vegetables, and aromatic plants and to reduce WASTE," which is funded by the Partnership for Research and Innovation in the Mediterranean Area (PRIMA), Project ID: 1556, a program supported by the European Union, and Euphresco Basics "Basic substances as an environmentally friendly alternative to synthetic pesticides for plant protection" projects. Authors aknowledge "Masseria Fruttirossi Srl" (Castellaneta, Taranto, Italy), for their collaboration in the research activities.

Appendices and nomenclature

EU	European Union
HK	hybrid histidine kinases
HOG	high osmolarity glycerol
HS code	Harmonized Code
IARC	International Agency for Research on Cancer
MRL	maximum residue limit
MSP	multistep phosphorelay systems
OTA	ochratoxin A
PDA	potato dextrose agar
RH	relative humidity
UNO	United Nations Organization

Author details

Annamaria Mincuzzi* and Antonio Ippolito
Department of Soil, Plant, and Food Sciences, University of Bari Aldo Moro, Bari, Italy

*Address all correspondence to: annamaria.mincuzzi@uniba.it

References

[1] Teixeira da Silva JA, Rana TS, Narzary D, Verma N, Meshram DT, Ranade SA. Pomegranate biology and biotechnology: A review. Scientia Horticulturae. 2013;**160**:85-107

[2] Stover E, Mercure EW. The pomegranate: A new look at the fruit of paradise. HortScience. 2007;**42**(5):1088-1092

[3] Chandra R, Babu KD, Jadhav VT, Jaime A, Silva TD. Origin, history and domestication of pomegranate. Fruit, Vegetable and Cereal Science and Biotechnology. 2010;**4**(2):1-6

[4] Holland D, Hatib K, Bar-Ya'akov I. Pomegranate: Botany, horticulture, breeding. In: Janick J, editor. curatore. Horticultural Reviews [Internet]. Hoboken, NJ, USA: John Wiley & Sons, Inc.; 2009. pp. 127-191

[5] Wetzstein HY, Ravid N, Wilkins E, Martinelli AP. A morphological and histological characterization of bisexual and male flower types in pomegranate. Journal of the American Society for Horticultural Science. 2011;**136**(2):83-92

[6] Mincuzzi A, Sanzani SM, Palou L, Ragni M, Ippolito A. Postharvest rot of pomegranate fruit in Southern Italy: Characterization of the Main Pathogens. Journal of Fungi. 2022;**8**(5):475

[7] Parvizi H, Sepaskhah AR, Ahmadi SH. Effect of drip irrigation and fertilizer regimes on fruit yields and water productivity of a pomegranate (*Punica granatum* (L.) cv. Rabab) orchard. Agricultural Water Management. 2014;**146**:45-56

[8] Dhillon WS, Gill PPS, Singh NP. Effect of nitrogen, phosphorus and potassium fertilization on growth, yield and quality of pomegranate 'Kandhari'. Acta Horticulturae. 2011;**890**(890):327-332

[9] Korkmaz N, Aşkın MA. Effects of calcium and boron foliar application on pomegranate (*Punica granatum* L.) fruit quality, yield, and seasonal changes of leaf mineral nutrition. Acta Horticulturae. 2015;**1089**:413-422

[10] Davarpanah S, Tehranifar A, Abadía J, Val J, Davarynejad G, Aran M, et al. Foliar calcium fertilization reduces fruit cracking in pomegranate (*Punica granatum* cv. Ardestani). Scientia Horticulturae. 2018;**230**:86-91

[11] Wink M. Current understanding of modes of action of multicomponent bioactive phytochemicals: Potential for nutraceuticals and antimicrobials. Annual Review of Food Science and Technology. 2022;**13**(1):337-359

[12] Pienaar L. The Economic Contribution of South Africa's Pomegranate Industry [Internet]. Stellenbosch, South Africa: Elsenburg: WCDOA; 2021. Available from: https://www.sapomegranate.co.za/wp-content/uploads/2021/05/PomegranteReport_2021F.pdf [Accessed: 2022-09-20]

[13] Manjunatha N, Sharma J, Pokhare SS, Agarrwal R, Patil PG, Sirsat JD, et al. Characterization of *Alternaria* and *Colletotrichum* Species Associated with Pomegranate (*Punica granatum* L.) in Maharashtra State of India. Journal of Fungi. 2022;**8**(10):1040

[14] Kahramanoglu İ. Trends in pomegranate sector: Production, postharvest handling and marketing. International Journal of Agriculture

Forestry and Life Sciences. 2019;**3**(2):239-246

[15] Özgüven AI, Gültekİn U, Gözlekçİ S, Yİlmaz I, Yİlmaz C, Küçük E, et al. A review of the economics and the marketing of the pomegranate industry in Turkey. Acta Horticulturae. 2015;**1089**:221-228

[16] Stiletto A, Trestini S. Factors behind consumers' choices for healthy fruits: a review of pomegranate and its food derivatives. Agricultural Economics. 2021;**9**(1):31

[17] Istat.it. [Internet]. Available from: https://www.istat.it/ [Accessed: 2022-11-13]

[18] Pienaar L, Vanessa B-J. The economic contribution of South Africa's pomegranate industry. Ağrı. 2021;**18**(4):57-64

[19] Mincuzzi A. personal communication

[20] Murthy DS, Gajanana TM, Sudha M, Dakshinamoorthy V. Marketing and post-harvest losses in fruits: Its implications on availability and economy. Indian Journal of Agricultural Economics. 2009;**64**(2):259-275

[21] Shanabhoga MB, Suresha SV, Varadaraju GM, Dechamma S. Marketing of pomegranate by the growers using public and private extension service. Research Journal of Agricultural Sciences. 2017;**8**:705-707

[22] Opara IK, Fawole OA, Kelly C, Opara UL. Quantification of on-farm pomegranate fruit postharvest losses and waste, and implications on sustainability indicators: South African case study. Sustainability. 2021;**13**(9):5168

[23] Opara IK, Fawole OA, Opara UL. Postharvest losses of pomegranate fruit at the packhouse and implications for sustainability indicators. Sustainability. 2021;**13**(9):5187

[24] Myresiotis CK, Testempasis S, Vryzas Z, Karaoglanidis GS, Papadopoulou-MourkidouE.Determination of mycotoxins in pomegranate fruits and juices using a QuEChERS-based method. Food Chemistry. 2015;**182**:81-88

[25] Gat T, Liarzi O, Skovorodnikova Y, Ezra D. Characterization of alternaria alternata causing black spot disease of pomegranate in Israel using a molecular marker. Plant Disease. 2012;**96**(10):1513-1518

[26] Aloi F, Riolo M, Sanzani SM, Mincuzzi A, Ippolito A, Siciliano I, et al. Characterization of *Alternaria* species associated with heart rot of pomegranate fruit. Journal of Fungi. 2021;**7**(3):172

[27] Tziros GT, Lagopodi AL, Tzavella-Klonari K. *Alternaria alternata* fruit rot of pomegranate (*Punica granatum*) in Greece. Plant Pathology. 2008;**57**(2):379

[28] Luo Y, Hou L, Förster H, Pryor B, Adaskaveg JE. Identification of *Alternaria* species causing heart rot of pomegranates in California. Plant Disease. 2017;**101**(3):421-427

[29] Berbegal M, López-Cortés I, Salazar D, Gramaje D, Pérez-Sierra A, Garcia-Jimenez J, et al. First report of *Alternaria* black spot of pomegranate caused by Alternaria alternata in Spain. Plant Disease. 2014;**98**(5):689-689

[30] Singh NK, Patel DB, Chaudhari SR, Morad BG, Rabari SM, Chattopadhyay A, et al. Characterization of the plant pathogenic isolates of *Alternaria burnsii*. National Academy of Agricultural Science. 2016;**34**(6):1461-1465

[31] Garganese F, Schena L, Siciliano I, Prigigallo MI, Spadaro D, De Grassi A,

et al. Characterization of citrus-associated *Alternaria* species in Mediterranean areas. PLoS One. 2016;**11**(9):e0163255

[32] Palou L, Taberner V, Guardado A, Del Río MÁ, Montesinos-Herrero C. Incidence and etiology of postharvest fungal diseases of pomegranate (*Punica granatum* cv Mollar de Elche) in Spain. Phytopathologia Mediterranea. 2013;**52**(3):478-489

[33] Bardas GA, Tzelepis GD, Lotos L, Karaoglanidis GS. First report of *Botrytis cinerea* causing gray mold of pomegranate (*Punica granatum*) in Greece. Plant Disease. 2009;**93**(12):1346-1346

[34] Tedford EC, Adaskaveg JE, Ott AJ. Impact of Scholar (a new postharvest fungicide) on the California pomegranate industry. Phytopathology News. 2005;**39**(2):19

[35] Testempasis S, Puckett RD, Michailides TJ, Karaoglanidis GS. Genetic structure and fungicide resistance profile of *Botrytis* spp. populations causing postharvest gray mold of pomegranate fruit in Greece and California. Postharvest Biology and Technology. 2020;**170**:111319

[36] Alvarez LV, Groenewald JZ, Crous PW. Revising the Schizoparmaceae: *Coniella* and its synonyms *Pilidiella* and *Schizoparme*. Studies in Mycology. 2016;**85**(1):1-34

[37] Palou L, Guardado A, Montesinos-Herrero C. First report of *Penicillium* spp. and *Pilidiella granati* causing postharvest fruit rot of pomegranate in Spain. New Disease Reports. 2010;**22**(21):2044-0588

[38] Thomidis T. Pathogenicity and characterization of *Pilidiella granati* causing pomegranate diseases in Greece. European Journal of Plant Pathology. 2015;**141**(1):45-50

[39] Jabnoun-Khiareddine H, Ibrahim N, Aydi Ben Abdallah R, Mars M, Daami-Remadi M. Response of Tunisian pomegranate (*Punica granatum* L.) cultivars and several plant hosts to *Coniella granati* (Saccardo). The Horticulture Journal. 2018;**5**:2376-0354

[40] Lennox CL, Mostert L, Venter E, Laubscher W, Meitz-Hopkins JC. First report of *Coniella granati* fruit rot and dieback on pomegranate in the western cape of South Africa. Plant Disease. 2018;**102**(4):821

[41] Joshi MS, Sawant DM, Gaikwad AP. Isolate variations in *Colletotrichum gloeosporioides* infecting pomegranate. The Journal of Plant Protection Sciences. 2014;**6**(1):21-26

[42] Silva-Cabral JRA, Batista LRL, de Costa JO, de Ferro MM, Silva SJC, Lima de GA, et al. First report of *Colletotrichum tropicale* causing anthracnose on pomegranate in Brazil. Plant Disease. 2019;**103**(3):583-583

[43] Silva-Cabral JRA, Batista LRL, de Oliveira Costa JF, da Rocha Amorim EP, de Andrade Lima GS, Assunção IP. Three species of *Colletotrichum* are associated with anthracnose of pomegranate in Brazil. Comunicata Scientiae. 2020;**11**:e3126-e3126

[44] Cara M, Mincuzzi A, Merkuri J, Vrapi H, Cara O, Ippolito A, et al. *Colletotrichum gloeosporioides sensu stricto* as causal agent of anthracnose on pomegranate fruit in Albania. Crop protection. 2020;**137**:105291

[45] dos Santos Vieira WA, Bezerra PA, da Silva AC, Veloso JS, Câmara MPS, Doyle VP. Optimal markers for the identification of Colletotrichum species. Molecular Phylogenetics and Evolution. 2020;**143**:106694

[46] Huang S, Li T, Yang T, Zheng X, Yang D, Fu G, et al. First report of pulp rot in externally asymptomatic pomegranate fruit caused by *Talaromyces albobiverticillius* in Henan Province, China. Plant Disease. 2022;**106**(7):1989

[47] Mincuzzi A, Ippolito A, Montemurro C, Sanzani SM. Characterization of *Penicillium s.s.* and *Aspergillus* sect. *nigri* causing postharvest rots of pomegranate fruit in Southern Italy. International Journal of Food Microbiology. 2020;**314**:108389

[48] Rand TG, Giles S, Flemming J, Miller JD, Puniani E. Inflammatory and cytotoxic responses in mouse lungs exposed to purified toxins from building isolated *Penicillium brevicompactum* Dierckx and *P. chrysogenum* Thom. Toxicological Sciences. 2005;**87**(1):213-222

[49] Bardas GA, Tzelepis GD, Lotos L, Karaoglanidis GS. First report of *Penicillium glabrum* causing fruit rot of pomegranate (*Punica granatum*) in Greece. Plant Disease. 2009;**93**:1347

[50] Hammerschmidt L, Wray V, Lin W, Kamilova E, Proksch P, Aly AH. New styrylpyrones from the fungal endophyte *Penicillium glabrum* isolated from *Punica granatum*. Phytochemistry Letters. 2012;**5**(3):600-603

[51] Labuda R, Hudec K, Piecková E, Mezey J, Bohovič R, Mátéová S, et al. *Penicillium implicatum* causes a destructive rot of pomegranate fruits. Mycopathologia. 2004;**157**(2):217-223

[52] Ezra D, Kirshner B, Hershcovich M, Shtienberg D, Kosto I. Heart rot of pomegranate: Disease etiology and the events leading to development of symptoms. Plant Disease. 2015;**99**(4):496-501

[53] Khokhar I, Bajwa R, Nasim G. New report of *Penicillium implicatum* causing a postharvest rot of pomegranate fruit in Pakistan. Australasian Plant Disease Notes. 2013;**8**(1):39-41

[54] Houbraken J, Frisvad JC, Samson RA. Taxonomy of *Penicillium* section *citrina*. Studies in mycology. 2011;**70**:53-138

[55] Pfohl-Leszkowicz A, Manderville RA. Ochratoxin A: An overview on toxicity and carcinogenicity in animals and humans. Molecular nutrition & food research. 2007;**51**(1):61-99

[56] Pascual-Ahuir A, Vanacloig-Pedros E, Proft M. Toxicity mechanisms of the food contaminant citrinin: Application of a quantitative yeast model. Nutrients. 2014;**6**(5):2077-2087

[57] Pyrri I, Visagie CM, Soccio P, Houbraken J. Re-evaluation of the taxonomy of *Talaromyces minioluteus*. Journal of Fungi. 2021;7(11):993

[58] Vesth TC, Nybo JL, Theobald S, Frisvad JC, Larsen TO, Nielsen KF, et al. Investigation of inter- and intraspecies variation through genome sequencing of *Aspergillus* section *Nigri*. Nature Genetics. 2018;**50**(12):1688-1695

[59] Perrone G, Gallo A. *Aspergillus* species and their associated mycotoxins. In: Moretti A, Susca A, editors. Mycotoxigenic Fungi [Internet]. New York, NY: Springer; 2017. pp. 33-49. (Methods in Molecular Biology; Vol. 1542 Available: http://link.springer.com/10.1007/978-1-4939-6707-0_3

[60] Yehia HM. Heart rot caused by *Aspergillus niger* through splitting in leathery skin of pomegranate fruit. African Journal of Microbiology Research. 2013;7:834-837

[61] Pala H, Tatli A, Yilmaz C, Özgüven AI. Important diseases of pomegranate fruit and control possibilities

in Turkey. In: I International Symposium on Pomegranate and Minor Mediterranean Fruits 818. 2006. pp. 285-290

[62] Thomidis T. Fruit rots of pomegranate (cv. wonderful) in Greece. Australasian Plant Pathology. 2014;**43**(5):583-588

[63] Munhuweyi K, Lennox CL, Meitz-Hopkins JC, Caleb OJ, Opara UL. Major diseases of pomegranate (*Punica granatum* L.), their causes and management—A review. Scientia Horticulturae. 2016;**211**:126-139

[64] Jatoi GH, Muhammad S, Metlo WA, Al-Ani LKT, Haseenullah Abro MA, Gadhi MA, et al. Efficacy of different essential oils, fungicides and biocontrol agents against *Aspergillus niger* the causal agent of fruit rot in Pomegranate International. Journal of Biosciences. 2020;**16**(3):51-65

[65] Guo MJ, Wang QT, Cheng YH, Hou CL. Identification of *Aspergillus tubingensis* causing pomegranate fruit rot in China. Australasian Plant Pathology. 2021;**50**(2):233-240

[66] Peduto Hand F, Choudhury RA, Gubler WD. First report of *Cytospora punicae* causing wood canker and branch dieback of pomegranate (*Punica granatum*) in the United States. Plant Disease. 2014;**98**(6):853-853

[67] Venter E, Lennox CL, Meitz-Hopkins JC. First report of *Cytospora punicae* causing post-harvest fruit rot on pomegranate in South Africa. Plant Disease. 2017;**101**(4):631-631

[68] Palavouzis SC, Tzamos S, Paplomatas E, Thomidis T. First report of *Cytospora punicae* isolated from pomegranate plants with symptom of collar rot in northern Greece. Journal of Plant Pathology. 2015;**97**(1):216

[69] Samouel S, Kanetis L. First report of *Cytospora punicae* causing trunk canker of pomegranate (*Punica granatum*) in Cyprus. Plant Disease. 2016;**100**(1):222-222

[70] Ammar MI, El-Naggar MA. Screening and characterization of fungi and their associated mycotoxins in some fruit crops. International Journal of Advanced Research. 2014;**2**(4):1216-1227

[71] Benfradj N, Mannai S, Fathallah A, Boughalleb-M'Hamdi N. First report of *Fusarium oxysporum* and *Globisporangium ultimum* as causal agents of pomegranate fruit rot in Tunisia. New Disease Reports. 2022;**45**(1):e12055

[72] Shi W, Tan Y, Wang S, Gardiner D, De Saeger S, Liao Y, et al. Mycotoxigenic potentials of *Fusarium* species in various culture matrices revealed by mycotoxin profiling. Toxins. 2016;**9**(1):6

[73] Zhou Y, Yang Y, Guo J, Bai J, Hu X. Black spot disease of pomegranate caused by *Cladosporium cladosporioides* in China. IOP Conference Series: Earth and Environmental Science. 2018;**170**:022131

[74] Kahramanoğlu İ, Aktaş M, Gündüz Ş. Effects of fludioxonil, propolis and black seed oil application on the postharvest quality of "Wonderful" pomegranate. PLoS One. 2018;**13**(5):e0198411

[75] Lufu R, Ambaw A, Opara UL. The influence of internal packaging (Liners) on moisture dynamics and physical and physiological quality of pomegranate fruit during cold storage. Food. 2021;**10**(6):1388

[76] Xavier KV, Kc AN, Vallad GE. Fungicide application timing essential for the management of leaf spot and fruit rot on pomegranate (*Punica granatum* L.) in Florida. Plant Disease. 2020;**104**(6):1629-1637

[77] Wang W, Fang Y, Imran M, Hu Z, Zhang S, Huang Z, et al. Characterization of the field fludioxonil resistance and its molecular basis in *Botrytis cinerea* from Shanghai Province in China. Microorganisms. 2021;**9**(2):266

[78] Munhuweyi K, Lennox CL, Meitz-Hopkins JC, Caleb OJ, Sigge GO, Opara UL. Investigating the effects of crab shell chitosan on fungal mycelial growth and postharvest quality attributes of pomegranate whole fruit and arils. Scientia Horticulturae. 2017;**220**:78-89

[79] Usanmaz S, Alas T. Degradation of postharvest applied fludioxonil in pomegranate fruit cv. Wonderful and efficacy on the gray mould. Journal of Stored Products and Postharvest Research. 2019;**10**(2):20-23

[80] Bersching K, Jacob S. The molecular mechanism of fludioxonil action is different to osmotic stress sensing. Journal of Fungi. 2021;**7**(5):393

[81] Hu S, Zhang J, Zhang Y, He S, Zhu F. Baseline sensitivity and toxic actions of boscalid against *Sclerotinia sclerotiorum*. Crop Protection. 2018;**110**:83-90

[82] EU Pesticides Database - European Commission [Internet]. Available from: https://ec.europa.eu› screen › mrls [Accessed: 2022-11-13]

[83] Brandhorst TT, Klein BS. Uncertainty surrounding the mechanism and safety of the post-harvest fungicide fludioxonil. Food and Chemical Toxicology. 2019;**123**:561-565

[84] European Food Safety Authority (EFSA). National summary reports on pesticide residue analysis performed in 2018. EFS3 [Internet]. 2020;**17**(4):1-228. Available from: https://data.europa.eu/doi/10.2903/sp.efsa.2020.EN-1814

[85] Calvo H, Mendiara I, Arias E, Blanco D, Venturini ME. The role of iturin A from *B. amyloliquefaciens* BUZ-14 in the inhibition of the most common postharvest fruit rots. Food Microbiology. 2019;**82**:62-69

[86] Elanchezhiyan K, Keerthana U, Nagendran K, Prabhukarthikeyan SR, Prabakar K, Raguchander T, et al. Multifaceted benefits of *Bacillus amyloliquefaciens* strain FBZ24 in the management of wilt disease in tomato caused by *Fusarium oxysporum* f. sp. *lycopersici*. Physiological and Molecular Plant Pathology. 2018;**103**:92-101

[87] Yamamoto S, Shiraishi S, Suzuki S. Are cyclic lipopeptides produced by *Bacillus amyloliquefaciens* S13-3 responsible for the plant defence response in strawberry against *Colletotrichum gloeosporioides*? Letters in Applied Microbiology. 2015;**60**(4):379-386

[88] Purrington CB. Costs of resistance. Current opinion in plant biology. 2000;**3**(4):305-308

[89] Bekir J, Bouajila J, Mars M. The effect of cultivar and ripening on antioxidant system and PAL activity of pomegranate (*Punica Granatum* L.) grown in Tunisia. Journal of Food Biochemistry. 2015;**39**(5):575-584

[90] Romanazzi G, Moumni M. Chitosan and other edible coatings to extend shelf life, manage postharvest decay, and reduce loss and waste of fresh fruits and vegetables. Current Opinion in Biotechnology. 2022;**78**:102834

[91] Velásquez CL, Avelizapa LR. A Review on the physicochemical and biological aspects of the chitosan antifungal activity in agricultural applications. Journal of Research Updates in Polymer Science. 2020;**9**:70-79

[92] Shahrajabian MH, Chaski C, Polyzos N, Tzortzakis N,

Petropoulos SA. Sustainable agriculture systems in vegetable production using chitin and chitosan as plant biostimulants. Biomolecules. 2021;**11**(6):819

[93] Candir E, Ozdemir AE, Aksoy MC. Effects of chitosan coating and modified atmosphere packaging on postharvest quality and bioactive compounds of pomegranate fruit cv. 'Hicaznar'. Scientia Horticulturae. 2018;**235**:235-243

[94] Singla M, Pareek S, Kumar N, Sagar NA, Fawole OA. Chitosan-cinnamon oil coating maintains quality and extends shelf life of ready-to-use pomegranate arils under low-temperature storage. Journal of Food Quality. 2022;**2022**:1-18

[95] Ranjbar A, Ramezanian A, Shekarforoush S, Niakousari M, Eshghi S. Antifungal activity of thymol against the main fungi causing pomegranate fruit rot by suppressing the activity of cell wall degrading enzymes. LWT - Food Science and Technology. 2022;**161**:113303

[96] Kim YH, Kim GH, Yoon KS, Shankar S, Rhim JW. Comparative antibacterial and antifungal activities of sulfur nanoparticles capped with chitosan. Microbial Pathogenesis. 2020;**144**:104178

[97] Griffith CM, Woodrow JE, Seiber JN. Environmental behavior and analysis of agricultural sulfur: Environmental behavior and analysis of agricultural sulfur. Pest Management Science. 2015;**71**(11):1486-1496

[98] Shankar S, Jaiswal L, Rhim JW. New insight into sulfur nanoparticles: Synthesis and applications. Critical Reviews in Environmental Science and Technology. 2021;**51**(20):2329-2356

[99] Priyadarshi R, Riahi Z, Rhim JW, Kim JT. Comparison of physicochemical and functional properties of elemental sulfur, sulfur nanoparticles, and sulfur quantum dots for sustainable biological applications. Materials Today Sustainability. 2022;**20**:100243

Chapter 13

Prunus spp. Fruit Quality and Postharvest: Today's Challenges and Future Perspectives

Sara Ricardo-Rodrigues, Marta Laranjo, Miguel Elias and Ana Cristina Agulheiro-Santos

Abstract

Prunus is a genus of trees and shrubs that date to the Eocene. Some species are known for their health benefits and for their exceptional role in international trade. Several *Prunus* species are widely cultivated all over the world, such as sweet cherry (*Prunus avium* L.), sour cherry (*Prunus cerasus* L.), plums (*Prunus salicina* L.), prunes (*Prunus domestica* L.), peaches (*Prunus persica* L.) or almonds (*Prunus amygdalus*, syn. *Prunus dulcis*). In this work, we review the most important quality parameters and sensory attributes for the abovementioned main *Prunus* species. Moreover, we focus on the postharvest challenges that are posed today to producers and retailers, as well as on consumer preferences. Finally, we discuss some new commercialization perspectives considering that the final aim agronomic activity is to produce fruits of good nutritional and sensory quality, with the least environmental impact possible and in a sustainable manner, according to the Sustainable Development Goals (SDGs) of 2030 Agenda of the United Nations.

Keywords: stone fruits, quality, shelf-life, consumers, trade

1. Introduction

Peaches (*Prunus persica* L.), Japanese plums (*Prunus salicina* Lindl), prunes (*Prunus domestica* L.), apricots (*Prunus armeniaca* L.), sweet (*Prunus avium* L.) and sour (*Prunus cerasus* L.) cherry fruits, and even some well-known varieties such as flat peach (*Prunus persica* var. platycarpa, also called var. compressa) and nectarines (*Prunus persica* var. nucipersica or var. nectarina Batsch), are fleshy fruits from the *Prunus* genus, whose culture is widespread, and these fruits are broadly consumed and appreciated all over the world. Within this genus, almonds (*Prunus amygdalus* Batsch) are an exception because they are a dry fruit, whose production and consumption have great expression.

Drupes can be of two types, namely dry or fleshy and succulent fruits. Drupes or stone fruits usually have one seed per carpel [1]. Fleshy or succulent fruits have a fleshy and thick pericarp, with three layers, an exterior protective epicarp, a fleshy

IntechOpen

and edible mesocarp and woody inner stony endocarp, which adhere to the seed [1]. On the other hand, dry fruits have an entire pericarp with enclosed seeds, which is dry stony at full maturity [1].

2. Respiratory pattern and maturity

The fruits continue metabolic processes such as respiration, transpiration, ethylene biosynthesis, carbohydrate metabolism, among others, even after being harvested and end up rotting naturally. The detachment from the mother plant speeds up all those processes. In the fruit respiration processes, oxygen is absorbed, and carbon dioxide is released using the accumulated carbohydrates (starch and sugars). The ultimate goal of this reaction is energy production. Respiration is a continuous process that takes place in fruits both in the field and after harvesting [1]. When the fruit is separated from the mother plant, it cannot replace carbohydrates and water. Therefore, respirations stops when the reserves of carbohydrates and water are depleted, followed by fruit senescence.

Considering the physiological behaviour of fruits during ripening, and from the practical point of view, fruits have been classified into two groups: climacteric and non-climacteric [2, 3].

Climacteric fruits are characterised by a marked increase in respiration and ethylene production at the beginning of ripening, while non-climacteric fruits do not exhibit such respiratory behaviour [4, 5]. It has been known for a long time that climacteric fruits have in common the presence of ethylene to regulate maturation in [1]. Moreover, the production of ethylene in climacteric fruits is autocatalytic, and the application of exogenous ethylene is able to ripe climacteric fruits [6]. On the contrary, the absence of ethylene can effectively stop their ripening. Already in 1934, Franklin Kidd [7] explained the climacteric ripening process of fruits and the existence of a climacteric peak.

Therefore, the non-climacteric fruits must be ripe at harvest. Non-climacteric fruits have a quite different ripening behaviour, without a peak of respiration or of ethylene production [5]. The behaviour and key regulators for non-climacteric fruits are poorly understood until today. Abscisic acid (ABA) has been suggested as one of the potential key regulators in ripening process of non-climacteric fruits [5]. In short, climacteric fruits can ripen after harvest, where non-climacteric fruits cannot do that.

This classification, although oversimplified, is very useful for practical reasons. Nevertheless, analyses of ripening and related changes in CO_2 and ethylene levels, at different fruit growth and maturation stages, have challenged this basic classification [4].

According to Cambridge dictionary, ripeness is the quality of being ready to be collected or eaten [8]. This is the popular concept that does not reflect the physiological knowledge of fruit ripening.

The knowledge of respiration and ripening pattern of each fruit is determinant to choose optimum harvest date, proper management strategies and storage practices, to achieve good nutritional and sensory quality, as well as decrease loss and waste [1].

Storage strategies involve temperature control, to decrease respiration rates and humidity control, to avoid transpiration, and eventually to control gases such as CO_2 and O_2 and eliminate ethylene in the atmosphere of the storage chambers [9].

Non-climacteric fruits should be kept away from any ethylene source to avoid the possible adverse effects on their ripening and/or quality, until more detailed information on the role of ethylene in these fruits is available. Moreover, the coexistence of ethylene-dependent and ethylene-independent pathways in climacteric and non-climacteric fruits has been described [4].

The mechanical or physiological damage, during harvest or storage, will cause an increase in respiration and consequently shorten shelf-life. Inappropriate temperatures (high temperatures and freezing) and lack of humidity may also cause physiological damages or disorders and an undesirable boosting of ripening [10].

Finally, it has been clearly demonstrated that many regulators of fruit ripening are common to both climacteric and non-climacteric fruits; namely, low O_2 and high CO_2 in the fruit microenvironment can delay the increase in ethylene and respiration, and consequently fruit ripening [5].

2.1 Ripening of Prunus species

Considering the wide and diversified range of fruits included in the genus *Prunus*, apricot, peach, plum, nectarine and durian are climacteric, whereas sweet and sour cherry are non-climacteric fruits. However, it is far too simplistic [4].

Plums are generally classified as climacteric fruits, but some cultivars vary markedly in there ripening behaviour. Some cultivars are known for their rapid softening, while others remain firm enough for commercial purposes, exhibiting a longer shelf-life. Actually, there are two distinct patterns of ripening, with some cultivars behave typically as climacteric fruits and others showing a suppressed-climacteric behaviour due to their reduced capacity of converting 1-aminocyclopropane-1-carboxylic acid (ACC) to ethylene [11, 12]. El-Sharkawy and colleagues [6, 13–15] studied cultivars of Japanese plums that show different patterns: 'Early Golden' behaves as a climacteric fruit, and 'Shiro' presents a suppressed climacteric pattern. These differences can be caused by a wide range of factors such as different ethylene receptors [15], in the ACC-synthase genes [13], and in the role of auxins in mediating gene expression for ethylene-responsive transcriptional factors (ERFs) [6, 14].

Peaches are clearly classified as climacteric fruits, so the ripening process is controlled by ethylene. Moreover, an increase in auxins production in the fruits leads to the induce expression of the ACC-synthase [16, 17]. Additionally, Trainotti and co-workers [18] used a genomic approach and reported that there is a cross-talk between auxins and ethylene; that is, auxin genes are regulated by ethylene and vice versa, which confirm the role of auxins in regulating the ripening of peaches was confirmed.

2.2 Development, ripening and quality of Prunus spp.

Since the work of Chalmers et al. [19], it is assumed that a double sigmoid curve characterises the development and growth of some fruits, such as *Prunus* drupes (**Figure 1**).

Stage I is an initial exponential phase, from the fecundation of ovary throughout the morphological changes that occur until the formation of the fruit. During the first half of stage I, an increase in the number of cells is observed, while in the last part of that stage there is an increase in cell size. In cherries, the cuticle reaches its full thickness. Stage II corresponds to a plateau, characterised by no cell division and by a slight tangential elongation of cells, cell wall thickening and endocarp lignification. Stage III corresponds to the second exponential growth phase, when cells start to enlarge to the final fruit size, and it ends with the complete and mature fruit. Finally, stage IV corresponds to the physiological ripening of drupes, so that fruits achieve a

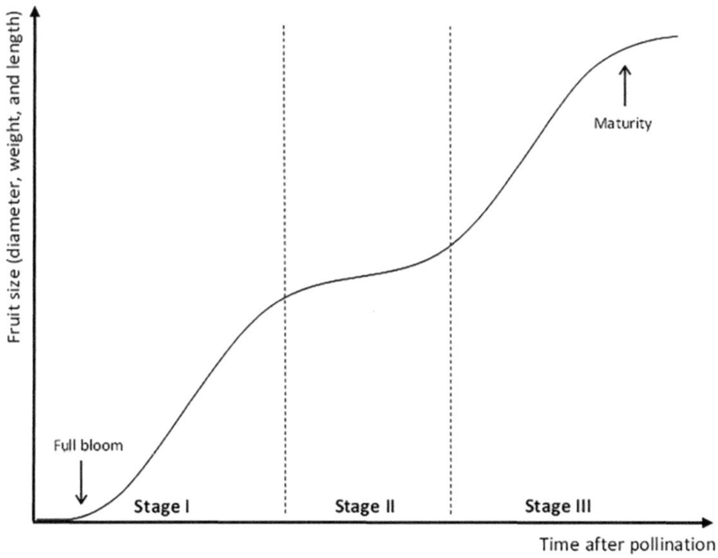

Figure 1.
Generalisation of the double-sigmoid growth curve typical of fruits of the genus Prunus.

high nutritional and organoleptic value, and are appropriate for human consumption, reflected in their economic valorisation.

During ripening, the morphological and physiological changes occur at different levels, the more evident being changes in colour, texture and biochemical composition [20]. Colour changes due to the degradation of chlorophylls, reflected in the loss of green colour, and the increase in anthocyanins and carotenoids, non-photosynthetic pigments, which confer a reddish and/or orange pigmentation to the fruits [21]. The decrease in fruit firmness that occurs throughout the ripening is a very complex process, which involves the breakdown of complex carbohydrates into sugars, and the activity of cell wall-modifying enzymes, causing reduction in intercellular adhesion, depolymerization and solubilisation of pectins, depolymerization of hemicelluloses, and loss of pectic galactose side chains [20]. All the aforementioned factors are responsible for the loss of firmness and the consequent increase in succulence of ripe fruits [22–24]. Changes in taste are very important for consumers' acceptance, and generally correspond to an increase in perceived sweetness, caused by a decrease in acidity and increase in sugar content. Moreover, in many fruits, the aroma becomes exquisite due to the release of volatile compounds, very appealing to the consumer. Carbohydrates, amino acids and fatty acids are the major fruit flavour precursors. The biosynthesis of volatile compounds is related to metabolic changes that occur during fruit ripening and have different profiles in the different ripening stages, and in different cultivars. Mihaylova and colleagues [25] studied the volatile compounds of eight peach varieties (*Prunus persica* L.) and identified 65 volatile compounds (aldehydes, esters and fatty acids), in different relative quantities depending on the variety.

The adequate ripening refers to fruits that present the best sensory and nutritional characteristics for consumers, and simultaneously allow the proper harvest management, storage and transport minimising loss and waste.

This concept of ripening, and consequently quality, although seemingly simple is, in fact, very complex and variable, depending among others, on consumer profile, agronomical practices, and the available facilities, and obviously on the fruit species.

The health-promoting properties of stone fruits also contribute to their quality and are due to the presence of vitamins, namely A, C, E, and folates, dietary fibres, and phenolic compounds, mostly flavonoids [26]. Phenolics display antimicrobial properties that are important in the preservation of fresh fruits [26]. Moreover, flavonoids may protect against chronic diseases and play a preventive role in neurological disorders [27, 28].

All fleshy fruits of the genus *Prunus* share their short shelf-life. The use of cold storage is undoubtedly mandatory. There are many variations of cold storage and many methods that can be applied to maintain quality and increase shelf-life in a sustainable way. It should be remembered that fruits are complex and dynamic biological systems, whose thermophysical properties vary with numerous factors, such as temperature, moisture content, species and even cultivar. These data are essential for the study and optimization of postharvest handling processes, such as pre-cooling and cold storage, predicting practical situations, namely cooling rate, cooling time, cooling uniformity and cooling energy utilisation, and allowing the monitoring of temperature-induced changes in fruit quality [28].

3. Some facts about the international trade of stone fruits

According to a recent report published by the United States Department of Agriculture (USDA), the production of peach and nectarine in 2022/23 will increase by 1 million tonnes, reaching 23.7 million tonnes [29]. This forecast is justified by the expected increase in production in China, the European Union and Turkey, which are the largest producers in the world.

China's peach production is expected to rise to 16.8 million due to higher yields despite area declines, because growers are switching to currently more profitable crops, such as cherries.

Russia lifted restrictions in February 2022, so exports are expected to increase. Remarkable is the investment of the China government in new smart farm tools, namely online selling. However, the political instability that is being witnessed at the moment, due to the invasion of Ukraine, leads to fears of some unexpected embargoes with unpredictable consequences on the international market. EU production is expected to improve to 3.1 million tonnes, with a significant recovery in French and Greek production. However, Spain has suffered a cold spring this year, reducing production, and this will have a negative effect. As most EU exports come from Spain, EU shipments are also expected to decrease.

Turkey is the world's third largest producer of peaches and nectarines. The investment of government programs in these commodities is remarkable. Turkey's production is expected to reach a record 940,000 tonnes, with nectarine supply increasing steadily. Exports decreased slightly due to the reduction of shipments to Russia.

According to the 2021 statistical data of the Food and Agriculture Organisation of the United Nations (FAO) [30], the world production of peaches and nectarines was of 1504682.00 ha of harvested area, with a yield of 166110.00 hg ha^{-1}, corresponding to 24994352.05 tonnes, whereas the harvested area for the European Union is 194050.00 ha, with a yield of 157908.00 hg ha^{-1}, corresponding to 3064200.00 tonnes.

Plums are one of the most important commodities in *Prunus* genus, occupying the second place and evidencing rapid worldwide growth in popularity [31]. According to FAO, the world's plum production reached 12 million tons in 2021, and the leading producer being China (5 million tons per year), followed by Romania and USA [30].

Sweet cherry (*Prunus avium* L.) is a highly valued fruit, with a large international market. In 2021, the top exporters of fresh cherries were Chile, Hong Kong, USA, Turkey and Spain [30].

Almonds are in high demand around the world and their long shelf-life makes them easy to store and transport. Increasingly health-conscious consumers are driving the rise in demand for almonds. According to the FAO 2021 statistical data, there is an harvested area of 2,283,414 ha for almonds that correspond to a yield of 17,491 hg ha^{-1} [30]. The top three producer are the United States, Spain and Australia [30]. The United States is undoubtedly largest producer and marketer of almonds, with a current production figure of 2 M million tonnes that remain stable since 2019 [32]. For the first time in more than a quarter century in California almond area has declined in 2022, due to a faster rhythm of orchard removals than new planting grow, with a drop of 1.2% relative to 2021. There appears to be a declining trend in the area under almond cultivation in California [32]. A 10% decrease is expected for the 2022/2023 season, around 1.5 million tonnes, without the shell. Even so, the USA is by far the largest producer with 79% of the world production, followed far behind by Australia, with 7% and Spain with 4% of the world production. Iran, Morocco, Syria and Turkey are traditional producers of almond. The large increase of areas cultivated with almonds in Turkey and Portugal is remarkable. The consumption is increasing all over the world, with a special reference to the new Asian markets.

4. Quality and postharvest of selected stone fruits

4.1 Peaches (*Prunus persica* L.)

Peaches or *Prunus persica* L. are stone fruits rich in ascorbic acid (vitamin C), carotenoids (provitamin A), minerals, fibre and antioxidant compounds, namely phenolics [33].

Distinct peach varieties were developed and are cultivated in different countries to meet different demands, namely higher yield, disease resistance and higher postharvest quality, among others [34]. This is possible, because of the high metabolic diversity of peaches, which is responsible for the variability in fruit size, texture, flavour, sweetness/acidity ratios and/or skin and flesh colour [33].

Begheldo and co-workers [35] have classified peach cultivars into three different categories according to their fruit texture and firmness, namely: (1) melting flesh, showing soft and juicy fruit when fully ripe; (2) non-melting flesh; and (3) stony hard fruits, which remain firm after harvest.

Peach is a climacteric fruit, meaning that it can ripen fully upon harvest, its ripening being regulated by the production of ethylene. This influences the postharvest strategy of peaches, which must combine maturity with storage conditions [36].

The following ripening parameters are usually considered in peaches: firmness (measured in Newton, as the maximum force needed to penetrate the fruit), skin colour, soluble solids content (SSC), (measured in °Brix), titratable acidity (TA) (measured in %) and the ratio SSC/TA, which is sometimes used as a maturity index [36, 37]. The criteria used for quality evaluation of peaches take into account the satisfaction of consumers. Therefore, the postharvest quality criteria include appearance, firmness and flavour, as well as safety and nutritional value.

Additionally, phenolic compounds can be used to evaluate the postharvest quality of peaches, due to different reasons [26]: (a) they are the main source of antioxidants in peaches [38]; (b) they contribute to the browning of peaches [39]; and (3) they contribute to taste and may be responsible for astringency in peaches [40].

Peaches, as other stone fruits, have a short shelf-life, and thus, both consumers and producers have long been interested in identifying effective measures that can extend peach storage life [41].

Ripening at harvest, storage temperature and postharvest treatment are factors that influence postharvest quality of peaches and their attractiveness to consumers [36].

Cold storage is the most common method employed to delay ripening and increase the postharvest life of peaches [42]. Peaches are ideally kept at around 0°C, but a range of −1 to 1°C is acceptable for up to 2 to 3 weeks [37, 43]. However, prolonged storage at low temperatures may change the ripening processes and result in lack of juice and woolly texture [42].

Chilling injury is a physiological disorder caused by prolonged exposure to the low storage temperatures, triggered at room temperature [43]. It may reduce the ethylene release, and lead to abnormal ripening, causing changes in aroma volatiles (flavour loss), colour changes (flesh browning and flesh bleeding), and changes in texture, namely flesh mealiness, and leatheriness [44].

A recent study evaluated the transcriptomic profile of peaches upon postharvest cold storage [37]. Expression levels of ethylene related genes are correlated with genes involved in cell wall modification, membrane composition, pathogen and stress response, which are all involved in the development of chilling injury [37]. Moreover, early transcriptomic responses to chilling are detectable well before the onset of chilling injury symptoms. Genes that are activated early upon cold storage may provide markers for detecting chilling injury before it can be perceived visually [37].

Several innovative techniques are used to improve the quality of peaches for a long period of time, namely controlled atmosphere storage, modified atmosphere packaging, heat treatment, and use of nanocomposite packaging material and edible coatings [45–49].

It should also be noted that peaches are very susceptible to phytopathogenic fungi, such as *Monilinia laxa*, *Monilinia fructicola* and *Rhizopus stolonifer* [50]. These moulds are the main cause of high postharvest losses in fully mature and ripe peaches [50].

4.2 Plums (*Prunus domestica* L. and *P. salicina* Lindl.)

There are thousands of plums cultivars, and the most commercially important species of plums are generally classified into two groups: European (*Prunus domestica* L.) and Japanese (*Prunus salicina* Lindl.) plums [1, 45]. Plums can be cultivated in a wide range of climatic conditions if cold-hour requirements are suppressed. The Japanese plum has cold-hour requirements in the range of 550 to 800 h, while the requirements in the European plum are higher than 800 h [45].

Plums present an huge diversity of flavour, aroma, texture, colour, form and size [46]. Producers should be encouraged to harvest the fruit at the partial to full ripe stage because the consumers appreciate attributes such as colour, flavour and aroma [47].

The plums maturation leads to an increase in weight, soluble solids content (SSC), sugars and anthocyanins, as well as a decrease in firmness and a darkening of colour. Establishing the ideal ripeness stage at harvest is of extreme importance as it ensures good fruit quality, for both purposes of consumption, fresh or processed. Considering the high variability among plum cultivars, it is necessary to define maturity parameters for each cultivar [47].

The assessment of the correct stage and time-to-harvest plums is based on physical and chemical, methods, physiological evaluation or combinations of them, which allows monitoring the maturation advance by producers [47]. The most common maturation parameters used in plums are fruit exterior colour, pulp firmness, SSC, titratable acidity (TA) and the ratio between the last two [1, 47, 48].

Skin colour is one of the criteria commonly used to determine the ripeness of the plum; however, it should not be used exclusively, as many cultivars develop a misleading pigmentation, while the fruit is still developing [47]. Regarding firmness, an important quality parameter closely related to fruit ripeness is frequently a good indicator of shelf-life potential [47].

Fruit softening is a natural occurrence during storage and radically compromises the market potential, with large volumes of these fruits often being rejected from the market because the firmness values are below acceptable retail standards [47, 49].

In the last years, several works have been developed in plums, employing NIR technology for the rapid evaluation of quality parameters, among them the detection of *Monilia* contamination and prediction of quality parameters (SSC, TA, ratio SSC/TA and firmness) [50, 51].

Plums' quality is rapidly lost after harvest. The high respiratory rates that occur during the transport and marketing process are the main reason why plums do not reach the consumer with the required characteristics [52].

The ideal storage conditions for plums are 0 to 5°C of temperature and 90–95% of relative humidity [53, 54]. However, plums are very sensitive to low temperatures, which cause severe chilling injury, internal browning, translucence of the flesh, loss of flavour and bleeding after being subjected to room temperature [49, 54]. Several research studies have been conducted to improve the postharvest shelf-life. Technologies such as modified atmosphere packaging (MAP), fumigation with ethylene antagonists such as 1-MCP and salicylic acid treatment have been tested in plums [49]. Applications related to prediction and monitoring of temperature induced fruit quality changes. According to Martínez-Romero and colleagues [55], forced air could be a good solution to prevent mechanical damage in plums, and extend shelf-life, specifically if pre-cooling plums immediately after harvest and before moving on to usual handling processes. More recently, new methodologies have emerged. The application of edible coatings of natural origin, such as proteins and polysaccharides, or the use of biodegradable materials, such as starch, has proven to be effective in improving the plum postharvest characteristics [49, 52, 56, 57]. These coating materials are safe and without implications for human health or negative influence on the environment, allowing the preservation of fruit with good quality characteristics for a longer storage period, reduce food waste and offer consumers high-quality plums with a longer shelf-life.

Japanese plums cultivars are used for fresh consumption because they have lower SSC values than European plums [45]. In contrast, most cultivars of European plums are typically used for drying due to high soluble solid content and are usually classified in four groups: Prunes, Reine Claude, Yellow Egg and Lombard with the most commercially important group being Prunes. According to Bahrin et al. [31], 'All prunes are plums, but not all plums are prunes'. The term 'prunes' is commonly used when the plum is dried without removing the stone, while if the stone is removed before product drying is called dried plum [31]. The term 'plums' include every variety and can be considered for fresh consumption, but also for canning, freezing and making jams and jellies [1].

Japanese plums are available on the market in the summer and autumn months, while the European plums are usually processed into dried fruit and can be consumed throughout the year [58].

Plums have a high nutritive value, being rich in vitamins A, B1, riboflavin and are a good source in diet of sugars, proteins, carbohydrates and minerals, such as calcium, phosphorus and iron. It has multiple benefits for human health, presenting anti-inflammatory, antioxidant properties, and may be related to the control of jaundice and can reduce the risk of cardiovascular diseases or cancer [31, 58, 59].

4.3 Sweet cherries (*Prunus avium* L.)

At first glance, consumers look for cherries with a good calibre, attached green stems, shiny, with unblemished skin, and a typical colour [60]. The cherry colour depends on the chosen cultivar: from dark red to mahogany for dark-sweet cherries and in opposition some cherry cultivar should have a yellow background with a slight to full red blush [60]. The firmness is another important quality attribute for cherries. It is desirable that fruits are firm to the touch. At last, when eating cherries consumers valorise flavour, sugar acidity rate and firm flesh with smooth skin. Zoffoli [61] resumed the quality of sweet cherry by referring large fruit size, bright colour, fruit skin without signs of pitting and cracking, firm texture, and high total soluble solids.

The market requires fruits resistant to cracking and diseases, and with an extended ripening season that allows cherries to be available to the consumer for a long period of time. As a particular case of cherry commercialised without stalk, there are 'Picotas', which present lower labour costs in harvesting, less loss of appearance, namely due to dehydration and blackening of the stalk, and less aggression on adjacent fruits, avoiding perforation damage. These varieties of cherry are native to the Valley of Jerte, in northern Extremadura, in Spain where they are highly praised by consumers since the XVII century.

Cherry is a non-climacteric fruit, so it is necessary to harvest the fruit as nearly ripe as possible, because they will not ripe after harvest. Harvest operation generally is spread over 2 weeks and is manually made. After harvest, cherries present a moderate-to-high respiratory activity, and some physical changes occur due to transpiration [62]. Empirically cherries were always harvested according to their external colour. That's why there are so many quality standards for cherry colour, which can vary from light red to black depending on the ripeness stage and variety. But the texture is also considered to define fruit quality and determine storability, taking into account that mechanical resistance of the skin decreases with ripening. A high soluble solid content with minimum softening are the capital conditions used to define perfect time for harvest, considering fruit storage life, fruit quality and acceptability when reaching consumers final fruit value [61].

The respiratory rate is determinant to the longevity of cherry and depends mainly on genetic factors, such as cultivar. Alique et al. [63] evaluated the respiratory rate of 'Ambrunes' and 'Burlat' and found values of 20 to 25 mg CO_2 kg^{-1} h^{-1} and 45 to 50 mg CO_2 kg^{-1} h^{-1}, respectively.

Toivonen and Hampson [64] found that at 0°C there were no differences in the postharvest respiration rate among cultivars, but at higher temperature the cultivars presented different behaviour.

While respiration and transpiration cause loss of quality, the enzymatic changes cause the firmness reducing and so increase the possibility of the invasion of fungus and bacterial diseases. But unfortunately, it is known that more mature fruit is more susceptible to these fungal and bacterial rots.

Generally, sweet cherry presents a short shelf-life, due to severe weight loss, browning, and drying of stems, changes in external colour, pitting, and also softening, with dramatic consequences for stakeholders and consumers. To extend the storage period, the main factor is low temperature, which in cherry should be around 0°C, reducing the physiological activity of the fruit tissues and the growth of microorganisms [65]. The high relative humidity avoids weight loss due to reduced transpiration and maintain turgidity [66]. A period of 15 days is generally considered as the best time to store cherries [67], which limits the commercial aspects and encourages the search for different storage methodologies that can increase this period of time. However, recent technologies used during postharvest period may optimise fruit quality for 45 days at 0°C.

Zoffoli [61] recommend the use of Near Freezing Temperatures (NFT) to store sweet cherries with the goal of maximise storage time. This researcher states that comparing with storage at 0°C, senescence was slowed by reducing the rate of respiration, softening and malondialdehyde accumulation, and so the rate of decay was decreased. Considering that there are a relationship between soluble solids and the freezing temperature, for example cherries with SSC around 22.5% present the freezing temperature at −2.5°C, and those with values of SSC near 16.7% have the freezing point at −1.8°C [68, 69]. For harvesting and handling, the recommended temperature is between 10 and 20°C. A rapid decrease of temperature during the operations is important to slow down fruit respiration rate and stem dehydration and maintain fruit firmness that minimises impact damages [70].

González-Gómez and co-workers [71] tested different postharvest storage conditions in order to find the most appropriate to preserve the overall quality of 'Sweetheart' cherries, harvested in the south of Portugal and Spain. Modified atmosphere (MAP) at 1°C, 95% RH using micro-perforated bags of polypropylene (PPlus® Sidlaw Packaging, Bristol, UK) bags, allows the most appropriate conditions to maintain and the amount of the predominant bioactive compounds such as phenolic compounds doubled their concentration comparing with the amount after harvest.

Nowadays, the use of plastic obtained for petroleum is recognised as an enormous and urgent environmental problem. Therefore, the search for friendly materials for packaging fruits is an update issue. Koutsimanis et al. [72] tested a biomaterial, polylactic acid (PLA) and clamshell containers with a microperforated film, with cherries at 1°C. They found that cherries store in those biodegradable packages, when compared with control with perforated bag, increase their storability up to 27 days, and exhibit less weight loss and higher acceptability.

The use of edible coating is another possible approach to increase shelf-life of fruits mainly cherries. Several studies using Aloe Vera gel to enhance the quality and the shelf-life of cherries demonstrated promising results in reducing respiration rate and decreasing weight loss of coated fruits [73–75].

To take profit of the microbial effect of some natural composts is another line of study to enhance better shelf-life. Afonso et al. [76] found that the use of extracts of *Satureja montana* L. and *Thymus vulgaris* L. could preserve sweet cherry physical and chemical characteristics during 14 days of postharvest storage at 2 ± 1°C and 95% relative humidity in the darkness. Before this, Maghenzani et al. [77] demonstrated that the vapour phase of those essential oils have the ability of controlling some postharvest pathogens.

Tragacanth gum and Eremurus extract were used to made edible coating treatments and being used in cherry 'Takdaneh Mashhad'. The trial was conducted until the 45th days of storage and it was found that the quality was improved allowing larger marketability. One of the treatments (12.5 g $L^{-1}$of tragacanth and eremurus) shows the ability of maintaining fruit firmness and less colour change, and SSC increase and maintain acidity, and simultaneously decrease weight loss [78].

The advances observed in last times in the technologies to avoid decay an enlarge shelf-life are important do reduce waste, and to benefit all, from producers to consumers, and improve a better environmental situation.

4.4 Almonds (*Prunus dulcis* (mill.) D. A. Webb), *Prunus amygdalus* batch, or *Amygdalus communis* L.

Almond trees require a warm climate, such as the one in the Mediterranean region, but also large quantities of water. Water scarcity, a major environmental problem nowadays, could be a limitation for the new intensive almond orchards. Another big environmental problem is the amount of residues produced during the process of peeling almonds [79].

Increasingly health-conscious consumers are driving the rise in demand for almonds [80]. The long shelf-life of the almond itself, with or without shell, along with the various processed products, such as smooth drinks, snacks and toasted almonds, makes its trade reach very important values. Almonds are in high demand around the world and their long shelf-life makes them easy to store and transport.

The almond (*Amygdalus communis* L.) is a drupe, composed by the exocarp, commonly called skin, an outer layer covering its thick, leathery and grey-green coloured mesocarp, called the hull [80]. In the interior of the hull is the endocarp, hoody and reticulated, whose hard shell is called pyrena. Finally, the shell contains the edible part, and the seed, usually called nut. The ripening process can be identified by the natural separation of the hull from the shell when fruits can fall from the tree due to the formation of an abscission layer between the stem and the fruit.

The specific quality parameters for dry fruits, almonds included, are size, colour, texture, flavour (be attentive to the development of stiffness and rancidity), moisture content, and incidence of damaging fungi and insects.

To reach good levels of quality, some storage factors should be considered: controlled moisture content, relative humidity and temperature, oxygen concentration and insect control [81].

The control of water content is essential for the proper storage of dry fruits. For the storage of almonds, the moisture content at harvest based on fresh weight is usually between 5 and 15%. The drying methods used to decrease the value of water activity are sun drying, ambient air drying, two stage drying (first heated-air drying to about 12% moisture and second ambient-air drying to 5–6% moisture), and heated-air drying.

Storage occurs usually at temperatures between 0 and 10°C and low RH, depending on the humidity of the almond, which has normally a low water content. Various potential damages can occur during storage. Loss of quality can be due to darkening of nuts and absorption of off flavours. Wu et al. [82] reported moisture changes during storage, and Kazantzis et al. [83] described that the moisture loss on almonds during 6 months of storage at 20 and 5°C was responsible for their weight loss. The same researchers also reported that storage temperature and relative humidity are determinant for moisture changes during storage [83]. Weight loss was higher at 20°C and 60% RH than at 5 and 80%. Wu et al. [82] affirm that shelled almonds can be stored with good quality at 0°C for 1 year, or at higher temperatures (17.8°C) with a relative humidity between 60 and 75%.

Pleasance et al. [84] found that almonds stored in polypropylene bags had and extended shelf-life, and García-Pascual et al. [81] also showed that the use of plastic bags during storage limits moisture changes. Wu et al. [82] tested roasted and raw almonds for 2 years and observed stabilisation of the measured values (moisture content, firmness and sensory characteristics) with the use of PE film packaging and RH above 50% and temperature above 25°C.

One major problem is rancidification. The controlled atmosphere with O_2 concentrations below 1% delays rancidification and other deterioration symptoms. Hard-shelled varieties are less susceptible to rancidity than soft-shelled varieties [81].

Raise et al. [85] studied the effect of modified atmosphere packaging under vacuum and CO_2 and found that those conditions can allow a shelf-life of at least 10 months for the tested almond kernels, regardless of storage temperature and physical shape of the almonds. To study the oxidation progress, peroxide value (PV) and conjugated trienes (K268) were measured, and sensory evaluation was also performed considering odour and flavour.

Empirical observation has shown that nuts in shell are easier to storage than shelled nuts because shell acts as a protective layer and shelled almonds can absorb odours during storage period. Moreover, according to some producers, whole kernels and halves are easier to storage than broken pieces (F. Galvão, personal communication).

Despite their long shelf-life as a dry fruit, it is important to systematise the available information on the storage of almonds, since there is only limited accessible literature on the topic.

5. Mycotoxins in Prunus fruits: a serious health risk

Fungi are the microorganisms that cause the greatest postharvest losses in *Prunus* spp. *Penicillium*, *Botrytis*, *Rhizopus*, *Mucor*, *Alternaria*, *Cladosporium* and *Monilia* are among the most hazardous genera. Additionally, some undesirable bacteria, such as *Erwinia* and *Pseudomonas*, are also responsible for postharvest losses in stone fruits.

Another noteworthy problem is the presence of mycotoxins. Mycotoxins are secondary metabolites produced by filamentous fungi, mainly *Aspergillus*, *Fusarium*, *Penicillium* and *Alternaria* gender [86–88], and may be present in a large number of substrates, such as nuts, cereals, vegetables, fruits, milk and oilseeds [89, 90].

According to Liu et al. [91], approximately 500 mycotoxins have been identified, and among these, aflatoxins, ochratoxins, fumonisins, patulin, trichothecenes and zearalenone are the more frequent in agricultural goods and those with more adverse health effects on humans [92–94]. Excluding trichothecenes, all abovementioned mycotoxins are carcinogenic [89]. Other adverse health effects of mycotoxin include their teratogenicity, liver and kidney toxicity, neural-tube defects, genotoxicity, immunotoxicity, immunosuppression and cytotoxicity [95–101]. These effects depend on the extent of exposure, amount ingested and sensitivity of the consumer [96]. Nazhand et al. [102] reported that around the world, 4.8 billion people are exposed to mycotoxins at levels that significantly increase morbidity and mortality. Ingestion, inhalation and cutaneous contact can all lead to mycotoxin exposure [89].

Some technics, most of them using thermal mechanisms, have been developed to reduce mycotoxins from food [87]; however, in general, mycotoxins are stable including at temperatures of thermal operations (80–121°C). More recently, non-thermal techniques (cold plasma, pulsed light, pulsed electric fields, high pressure processing, electron beam irradiation) have being studied [87, 93, 94, 103–108].

Among stone fruits, almond has been mostly associated with the presence of mycotoxins [109, 110]. Rodrigues et al. [111] informed about the presence of aflatoxins in almonds, Hidalgo-Ruiz et al. [112] found aflatoxins and zearalenone in almonds, and Ünüsan [113] reported the presence of aflatoxin B1 in almonds. Sadok et al. [110] made reference to the presence of patulin in stone fruits. Suman [87] reported the presence of putalin on olives, and Azaiez et al. [114] the presence of diacetoxyscirpenol in prune samples, while Iqbal et al. [115] showed the presence of ochratoxin A in apricots and plum, and Ünüsan [113] mentioned the presence of patulin in peaches, and the presence of ochratoxins in apricots. Additionally, Vidal et al. [86] mentioned the presence of patulin in apricots, peaches and peach juice, and Sadok et al. [110] reported the presence of patulin in cherries, peaches and plums.

Considering the long-term presence of mycotoxins in foods, also after the elimination of the fungi, an effective control of production, transport and storage conditions is required, as well as the monitorization of the presence of those substances in foods, mainly in fruits and vegetables, which should be both legally enforced and controlled.

6. Conclusions

Prunus is a huge genus of trees and shrubs that comprises the production of different types of stone fleshy fruits, drupes, such as peaches, plums, cherries, among others, and dry fruits as almonds. Some of these drupes are climacteric, while others are non-climacteric fruits. This knowledge is of vital importance for defining the ideal stage of ripeness for harvesting and to take the best decisions about postharvest storage.

Prunus fruits have great relevance in world trade markets. An overall analysis of world production and trade highlights the general increase in *prunus* production, with emphasis on the increase in areas and the upgrading of production techniques for fleshy fruits observed in China and Turkey, while the USA remains a key player in almond production, despite some decline.

The consumers are fond of good quality, which they evaluate looking at fruit size, form, skin and flesh colour, texture, taste, aroma, sweetness, acidity and sometimes easy peeling.

A huge number of new varieties that appear on the market are often very well received by consumers, which justifies the dynamism of this market.

In general, these fleshy fruits present a short shelf-life, and some of them suffer from chilling injury, or other physiological disorders. The emphasis is therefore put on developing new storage technologies that can extend the selling period and increase the distribution distance of the products. The morphological and physiological diversity of fleshy fruits of the *Prunus* genus and the distinct trade requirements have led to the development and implementation of various methods for improving the shelf-life of different species, which cannot be generalised. Ensuring effective storage methods to extend fruit shelf-life, without compromising consumers health and the environment, remains a vital area for future research.

Acknowledgements

This work was funded by National Funds through FCT – Fundação para a Ciência e a Tecnologia – under project number UIDB/05183/2020. Sara Ricardo-Rodrigues acknowledges a PhD fellowship from FCT (Ref. 2021.07663.BD).

Conflict of interest

The authors declare no conflict of interest.

Author details

Sara Ricardo-Rodrigues[1], Marta Laranjo[1], Miguel Elias[1,2] and Ana Cristina Agulheiro-Santos[1,2]*

1 MED-Mediterranean Institute for Agriculture, Environment and Development and CHANGE-Global Change and Sustainability Institute, IIFA-Institute for Advanced Studies and Research, University of Évora, Évora, Portugal

2 Plant Science Department, School of Sciences and Technology, University of Évora, Évora, Portugal

*Address all correspondence to: acsantos@uevora.pt

References

[1] Siddiq M, Sultan MT. In: Sinha NK, Sidhu JS, Barta J, JSB W, Cano MP, editors. Handbook of Fruits and Fruit Processing. Second ed. California, USA: John Wiley & Sons, Ltd.; 2012. pp. 3-679. DOI: 10.1002/9781118352533

[2] Biale JB. Growth, maturation, and senescence in fruits. Science. 1964;**146**(3646):880-888. DOI: 10.1126/science.146.3646.880

[3] Wills RBH, McGlasson WB, Graham D, Joyce DC. Physiology and biochemistry. In: Wills RBH, McGlasson WB, Graham D, Joyce DC, editors. Postharvest: An Introduction to Physiology and Handling of Fruits, Vegetables and Ornamentals. 5th ed. Oxford, UK: University Press; 2007. p. 28-51

[4] Paul V, Pandey R, Srivastava GC. The fading distinctions between classical patterns of ripening in climacteric and non-climacteric fruit and the ubiquity of ethylene-an overview. Journal of Food Science and Technology. 2012;**49**(1):1-21. DOI: 10.1007/s13197-011-0293-4

[5] Fan D, Wang W, Hao Q, Jia W. Do non-climacteric fruits share a common ripening mechanism of hormonal regulation? Frontiers in Plant Science. 2022;**13**:923484. DOI: 10.3389/fpls.2022.923484

[6] El-Sharkawy I, Mila I, Bouzayen M, Jayasankar S. Regulation of two germin-like protein genes during plum fruit development. Journal of Experimental Botany. 2010;**61**(6):1761-1770. DOI: 10.1093/jxb/erq043

[7] Kidd F. Respiration of fruits. Nature. 1934;**766**:66

[8] Dictionary Cambridge. Ripeness. 2023. Available from: https://dictionary.cambridge.org/dictionary/english/ripeness [Accessed: July 10, 2023]

[9] Brizzolara S, Manganaris GA, Fotopoulos V, Watkins CB, Tonutti P. Primary metabolism in fresh fruits during storage. Frontiers in Plant Science. 2020;**11**(February):1-16. DOI: 10.3389/fpls.2020.00080

[10] Fang Y, Wakisaka M. A review on the modified atmosphere preservation of fruits and vegetables with cutting-edge technologies. Agriculture. 2021;**11**(992):1-16. DOI: 10.3390/agriculture11100992

[11] Abdi N, McGlasson WB, Holford P, Williams M, Mizrahi Y. Responses of climacteric and suppressed-climacteric plums to treatment with propylene and 1-methylcyclopropene. Postharvest Biology and Technology. 1998;**14**(1):29-39. DOI: 10.1016/S0925-5214(98)00031-3

[12] Abdi N, Holford P, McGlasson WB, Mizrahi Y. Ripening behaviour and responses to propylene in four cultivars of Japanese type plums. Postharvest Biology and Technology. 1997;**12**(1):21-34. DOI: 10.1016/S0925-5214(97)00041-0

[13] El-Sharkawy I, Kim WS, Jayasankar S, Svircev AM, Brown DCW. Differential regulation of four members of the ACC synthase gene family in plum. Journal of Experimental Botany. 2008;**59**(8):2009-2027. DOI: 10.1093/jxb/ern056

[14] El-Sharkawy I, Sherif S, Mila I, Bouzayen M, Jayasankar S. Molecular characterization of seven genes encoding ethylene-responsive transcriptional factors during plum fruit development and ripening. Journal of Experimental

Botany. 2009;**60**(3):907-922. DOI: 10.1093/jxb/ern354

[15] El-Sharkawy I, Kim WS, El-Kereamy A, Jayasankar S, Svircev AM, Brown DCW. Isolation and characterization of four ethylene signal transduction elements in plums (Prunus salicina L.). Journal of Experimental Botany. 2007;**58**(13):3631-3643. DOI: 10.1093/jxb/erm213

[16] Miller AN, Walsh CS, Cohen JD. Measurement of indole-3-acetic acid in peach fruits (Prunus persica L. Batsch cv Redhaven) during development. Plant Physiology. 1987;**84**(2):491-494. DOI: 10.1104/pp.84.2.491

[17] Abel S, Theologis A. Early genes and auxin action. Plant Physiology. 1996;**111**(1):9-17. DOI: 10.1104/pp.111.1.9

[18] Trainotti L, Tadiello A, Casadoro G. The involvement of auxin in the ripening of climacteric fruits comes of age: The hormone plays a role of its own and has an intense interplay with ethylene in ripening peaches. Journal of Experimental Botany. 2007;**58**(12):3299-3308. DOI: 10.1093/jxb/erm178

[19] Chalmers D, Ende B. A reappraisal of the growth and development of peach fruit. Functional Plant Biology. 1975;**2**(4):623. DOI: 10.1071/pp9750623

[20] García-Gómez BE, Salazar JA, Nicolás-Almansa M, Razi M, Rubio M, Ruiz D, et al. Molecular bases of fruit quality in prunus species: An integrated genomic, transcriptomic, and metabolic review with a breeding perspective. International Journal of Molecular Sciences. 2021;**22**(1):1-38. DOI: 10.3390/ijms22010333

[21] Kapoor L, Simkin AJ, George Priya Doss C, Siva R. Fruit ripening: Dynamics and integrated analysis of carotenoids and anthocyanins. BMC Plant Biology. 2022;**22**:1-22. DOI: 10.1186/s12870-021-03411-w

[22] Seymour GB, Colquhoun IJ, Dupont MS, Parsley KR, R. Selvendran R. Composition and structural features of cell wall polysaccharides from tomato fruits. Phytochemistry. 1990;**29**(3):725-731. DOI: 10.1016/0031-9422(90)80008-5

[23] Brummell DA, Harpster MH. Cell wall metabolism in fruit softening and quality and its manipulation in transgenic plants. Plant Molecular Biology. 2001;**47**(1-2):311-339. DOI: 10.1023/A:1010656104304

[24] Ruiz-May E, Rose JKC. Cell Wall architecture and metabolism in ripening fruit and the complex relationship with softening. In: Seymour GB, Poole M, Giovannoni JJ, Tucker GA, editors. The Molecular Biology and Biochemistry of Fruit Ripening. New Jersey, USA: Blackwell Publishing Ltd.; 2013. p. 163-187. DOI: 10.1002/9781118593714.ch7

[25] Mihaylova D, Popova A, Vrancheva R, Dincheva I. HS-SPME-GC–MS volatile profile characterization of peach (Prunus persica L. Batsch) varieties grown in the eastern Balkan Peninsula. Plants. 2022;**11**(166):1-16. DOI: 10.3390/plants11020166

[26] Lara MV, Bonghi C, Famiani F, Vizzotto G, Walker RP, Drincovich MF. Stone fruit as biofactories of phytochemicals with potential roles in human nutrition and health. Frontiers in Plant Science. 2020;**11**:1-21. DOI: 10.3389/fpls.2020.562252

[27] Silva RFM, Pogačnik L. Polyphenols from food and natural products: Neuroprotection and safety. Antioxidants. 2020;**9**(1):1-13. DOI: 10.3390/antiox9010061

[28] Weng CJ, Yen GC. Chemopreventive effects of dietary phytochemicals against cancer invasion and metastasis: Phenolic acids, monophenol, polyphenol, and their derivatives. Cancer Treatment Reviews. 2012;**38**(1):76-87. DOI: 10.1016/j.ctrv.2011.03.001

[29] USDA. Fresh peaches and cherries: World markets and trade. 2020. Available from: https://downloads.usda.library.cornell.edu/usda-esmis/files/0g354f20t/z029pt97x/2j62st82g/StoneFruit.pdf

[30] FAOSTAT. Statistical data - Crops and livestock products. FAO - Food And Agriculture Organization of the United Nations. 2023. Available from: https://www.fao.org/faostat/en/#data/QCL [Accessed: Mai 10, 2023]

[31] Bahrin AA, Moshawih S, Dhaliwal JS, Kanakal MM, Khan A, Lee KS, et al. Cancer protective effects of plums: A systematic review. Biomedicine and Pharmacotherapy. 2022;**146**:112568. DOI: 10.1016/j.biopha.2021.112568

[32] USDA. Statistical data - Crops. United States Department of Agriculture - National Agricultural Statistics Service. 2023. Available from: https://www.nass.usda.gov/Statistics_by_Subject/index.php?sector=CROPS [Accessed: Mai 10, 2023]

[33] Drincovich MF. Identifying sources of metabolomic diversity and reconfiguration in peach fruit: Taking notes for quality fruit improvement. FEBS Open Bio. 2021;**11**:3211-3217. DOI: 10.1002/2211-5463.13233

[34] Sansavini S, Gamberini A, Bassi D. Peach breeding, genetics and new cultivar trends. Acta Horticulturae. 2006;**713**:23-48. DOI: 10.17660/ActaHortic.2006.713.1

[35] Begheldo M, Manganaris GA, Bonghi C, Tonutti P. Different postharvest conditions modulate ripening and ethylene biosynthetic and signal transduction pathways in stony hard peaches. Postharvest Biology and Technology. 2008;**48**(1):84-91. DOI: 10.1016/j.postharvbio.2007.09.023

[36] Wang Q, Wei Y, Chen X, Xu W, Wang N, Xu F, et al. Postharvest strategy combining maturity and storage temperature for 1-MCP-treated peach fruit. Journal of Food Processing and Preservation. 2020;**44**:1-12. DOI: 10.1111/jfpp.14388

[37] Muto A, Bruno L, Madeo ML, Ludlow R, Ferrari M, Stimpson L, et al. Comparative transcriptomic profiling of peach and nectarine cultivars reveals cultivar-specific responses to chilled postharvest storage. Frontiers in Plant Science. 2022;**13**:1-22. DOI: 10.3389/fpls.2022.1062194

[38] Gil MI, Tomás-Barberán FA, Hess-Pierce B, Kader AA. Antioxidant capacities, phenolic compounds, carotenoids, and vitamin C contents of nectarine, peach, and plum cultivars from California. Journal of Agricultural and Food Chemistry. 2002;**50**:4976-4982. DOI: 10.1021/jf020136b

[39] Lee CY, Kagan V, Jaworski AW, Brown SK. Enzymatic Browning in relation to phenolic compounds and Polyphenoloxidase activity among various peach cultivars. Journal of Agricultural and Food Chemistry. 1990;**38**(1):99-101. DOI: 10.1021/jf00091a019

[40] Tomás-Barberán FA, Cremin P, Waterhouse AL, Hess-Pierce B, Kader AA, Gil MI. HPLC-DAD-ESIMS analysis of phenolic compounds in nectarines, peaches and plums. Journal of Agricultural and Food Chemistry. 2001;**49**(10):4748-4760

[41] Velardo-Micharet B, Pintado CM, Dupille E, Ayuso-Yuste MC, Lozano M, Bernalte-García MJ. Effect of ripening stage, 1-MCP treatment and different temperature regimes on long term storage of 'Songold' Japanese plum. Scientia Horticulturae. 2017;**214**:233-241. DOI: 10.1016/j.scienta.2016.11.043

[42] Pavez L, Hödar C, Olivares F, González M, Cambiazo V. Effects of postharvest treatments on gene expression in Prunus persica fruit: Normal and altered ripening. Postharvest Biology and Technology. 2013;**75**:125-134. DOI: 10.1016/j.postharvbio.2012.08.002

[43] Farcuh M, Hopfer H. Aroma volatiles as predictors of chilling injury development during peach (Prunus persica (L) Batsch) cold storage and subsequent shelf-life. Postharvest Biology and Technology. 2023;**195**(112137):1-12. DOI: 10.1016/j.postharvbio.2022.112137

[44] Lurie S, Crisosto CH. Chilling injury in peach and nectarine. Postharvest Biology and Technology. 2005;**37**(3):195-208. DOI: 10.1016/j.postharvbio.2005.04.012

[45] Roussos PA, Efstathios N, Intidhar B, Denaxa NK, Tsafouros A. Plum (Prunus domestica L. and P. salicina Lindl.). In: Simmonds MSJ, Preedy VR, editors. Nutritional Composition of Fruit Cultivars. Amsterdam, Netherlands: Academic Press, Elsevier Inc.; 2016. p. 639-666

[46] Chavez DJ, Chaparrro JX. The north American plums (Prunus Spp.): A review of the taxonomic and phylogenetic relationships. In: Kuden A, editor. Prunus. London, UK: Intechopen; 2020. DOI: 10.5772/intechopen.91638

[47] Usenik V, Stampar F, Kastelec D. Indicators of plum maturity: When do plums become tasty? Scientia Horticulturae. 2014;**167**:127-134. DOI: 10.1016/j.scienta.2014.01.002

[48] Usenik V, Stampar F, Kastelec D. Phytochemicals in fruits of two Prunus domestica L. plum cultivars during ripening. Journal of the Science of Food and Agriculture. 2013;**93**(3):681-692. DOI: 10.1002/jsfa.5783

[49] Thakur R, Pristijono P, Golding JB, Stathopoulos CE, Scarlett CJ, Bowyer M, et al. Development and application of rice starch based edible coating to improve the postharvest storage potential and quality of plum fruit (Prunus salicina). Scientia Horticulturae. 2018;**237**:59-66. DOI: 10.1016/j.scienta.2018.04.005

[50] Louw ED, Theron KI. Robust prediction models for quality parameters in Japanese plums (Prunus salicina L.) using NIR spectroscopy. Postharvest Biology and Technology. 2010;**58**(3):176-184. DOI: 10.1016/j.postharvbio.2010.07.001

[51] Vitalis F, Nugraha DT, Aouadi B, Bósquez JPA, Bodor Z, Zaukuu JLZ, et al. Detection of *monilia* contamination in plum and plum juice with nir spectroscopy and electronic tongue. Chemosensors. 2021;**9**(12):355-371. DOI: 10.3390/chemosensors9120355

[52] Panahirad S, Naghshiband-Hassani R, Bergin S, Katam R, Mahna N. Improvement of postharvest quality of plum (Prunus domestica L.) using polysaccharide-based edible coatings. Plants. 2020;**9**(1148):1-16. DOI: 10.3390/plants9091148

[53] Crisosto C, Kader A. Plum and Fresh Prune: Postharvest Quality Maintenance Guidelines. Davis, USA: Kearney Agricultural Center & Extension Center. Extension Bulletin; 2000

[54] Pan H, Wang L, Wang R, Xie F, Cao J. Modifications of cell wall pectin in chilling-injured 'friar' plum fruit subjected to intermediate storage temperatures. Food Chemistry. 2018;**242**:538-547. DOI: 10.1016/j.foodchem.2017.09.090

[55] Martínez-Romero D, Castillo S, Valero D. Forced-air cooling applied before fruit handling to prevent mechanical damage of plums (Prunus salicina Lindl.). Postharvest Biology and Technology. 2003;**28**(1):135-142. DOI: 10.1016/S0925-5214(02)00142-4

[56] Basiak E, Linke M, Debeaufort F, Lenart A, Geyer M. Impact of biodegradable materials on the quality of plums. Coatings. 2022;**12**(2):1-15. DOI: 10.3390/coatings12020226

[57] Panahirad S, Naghshiband-Hassani R, Ghanbarzadeh B, Zaare-Nahandi F, Mahna N. Shelf life quality of plum fruits (Prunus domestica L.) improves with carboxymethylcellulose-based edible coating. HortScience. 2019;**54**(3):505-510. DOI: 10.21273/HORTSCI13751-18

[58] Drogoudi P, Pantelidis G. Phenotypic variation and Peel contribution to fruit antioxidant contents in European and Japanese plums. Plants. 2022;**11**(10):1338-1357. DOI: 10.3390/plants11101338

[59] Shamrao BS. Production Technology of Peach, plum and apricot in India. In: Kuden A, editor. Prunus. London, UK: Intechopen; 2020. DOI: 10.5772/intechopen.87486

[60] Ricardo-Rodrigues S, Laranjo M, Agulheiro-Santos AC. Methods for quality evaluation of sweet cherry. Journal of the Science of Food and Agriculture. 2023;**103**(2):463-478. DOI: 10.1002/jsfa.12144

[61] Zoffoli JP. New post harvest technologies for high quality sweet cherries. Italus Hortus. 2022;**29**(1):68-81. DOI: 10.26353/j.itahort/2022.1.C5

[62] Wang Y, Long LE. Respiration and quality responses of sweet cherry to different atmospheres during cold storage and shipping. Postharvest Biology and Technology. 2014;**92**:62-69. DOI: 10.1016/j.postharvbio.2014.01.003

[63] Alique R, Martínez MA, Alonso J. Influence of the modified atmosphere packaging on shelf life and quality of Navalinda sweet cherry. European Food Research and Technology. 2003;**217**(5):416-420. DOI: 10.1007/s00217-003-0789-x

[64] Toivonen PMA, Hampson CR. Respiration rates of sweet cherry cultivars at optimal and abusive temperatures over three growing seasons. Acta Horticulturae. 2017;**1161**:575-579. DOI: 10.17660/ActaHortic.2017.1161.91

[65] Crisosto CH, Crisosto GM, Metheney P. Consumer acceptance of brooks and Bing cherries is mainly dependent on fruit SSC and visual skin color. Postharvest Biology and Technology. 2003;**28**(1):159-167. DOI: 10.1016/S0925-5214(02)00173-4

[66] Nunes C. New development in alternative methods to control postharvest fruit decay. In: Proceedings of the International Conference Environmentally Friendly and Safe Technologies for Quality of Fruit and Vegetables. Faro, Portugal: Universidade do Algarve. 2009. pp. 131-143

[67] Serrano M, Ďiaz-Mula HM, Zapata PJ, Castillo S, Guillén F, Martínez-Romero D, et al. Maturity stage at harvest determines the fruit quality and antioxidant potential after storage of sweet cherry cultivars. Journal of Agricultural and Food Chemistry.

2009;**57**(8):3240-3246. DOI: 10.1021/jf803949k

[68] Whiteman TM. Freezing Points of Fruits, Vegetables and Florist Stocks. Washington, D.C, USA: U.S. Dept. of Agriculture (USDA); 1977. DOI: 10.5962/bhl.title.62920

[69] Jie W, Lite L, Yang D. The correlation between freezing point and soluble solids of fruits. Journal of Food Engineering. 2003;**60**(4):481-484. DOI: 10.1016/S0260-8774(03)00081-5

[70] Crisosto CH, Garner D, Doyle J, Day KR. Relationship between fruit respiration, bruising susceptibility, and temperature in sweet cherries. HortScience. Fevereiro de. 1993;**28**(2):132-135. DOI: 10.21273/HORTSCI.28.2.132

[71] González-Gómez D, Bernalte MJ, Ayuso MC, Fernández-Leon MF, Lozano M, Rato AE, et al. Evaluation of different postharvest conditions to preserve the amount of bioactive compounds, physicochemical quality parameters and sensory attributes of «sweetheart» cherries. Acta Horticulturae. 2017;**1161**:581-586. DOI: 10.17660/ActaHortic.2017.1161.92

[72] Koutsimanis G, Harte J, Almenar E. Freshness maintenance of cherries ready for consumption using convenient, microperforated, bio-based packaging. Journal of the Science of Food and Agriculture. 2015;**95**(5):972-982. DOI: 10.1002/jsfa.6771

[73] Öztürk B. Effects of modified atmosphere packaging and aloe vera treatments on quality traits of cherry Laurel fruit (Prunus laurocerasus L.) during shelf life. Uluslararası Tarım ve Yaban Hayatı Bilimleri Dergisi. 2020;**6**(3):399-406. DOI: 10.24180/ijaws.781564

[74] Öztürk B, Aglar E. Effects of modified atmosphere packaging (MAP) and Aloe vera treatments on quality characteristics of cornelian cherry fruits during cold storage. Akademik Ziraat Dergisi. 2019;**8**(1):1-8. DOI: 10.29278/azd.592897

[75] Öztürk B, Karakaya O, Yıldız K, Saracoglu O. Effects of Aloe vera gel and MAP on bioactive compounds and quality attributes of cherry laurel fruit during cold storage. Scientia Horticulturae. 2019;**249**:31-37. DOI: 10.1016/j.scienta.2019.01.030

[76] Afonso S, Oliveira I, Ribeiro C, Vilela A, Meyer AS, Gonçalves B. Innovative edible coatings for postharvest storage of sweet cherries. Scientia Horticulturae. 2022;**2023**(310):111738. DOI: 10.1016/j.scienta.2022.111738

[77] Maghenzani M, Chiabrando V, Santoro K, Spadaro D, Giacalone G. Effects of treatment by vapour of essential oil from Thymus vulgaris and Satureja montana on postharvest quality of sweet cherry (cv. Ferrovia). Journal of Food and Nutrition Research. 2018;**57**(2):161-169

[78] Esmaeili A, Jafari A, Ghasemi A, Gholamnejad J. Improving postharvest quality of sweet cherry fruit by using Tragacanth and Eremurus. International Journal of Fruit Science. 2022;**22**(1):370-382. DOI: 10.1080/15538362.2022.2043221

[79] Garcia-Perez P, Xiao J, Munekata PES, Lorenzo JM, Barba FJ, Rajoka MSR, et al. Revalorization of almond by-products for the design of novel functional foods: An updated review. Food. 1823;**2021**(10):1-20. DOI: 10.3390/foods10081823

[80] Massantini R, Cristofori V, Frangipane MT. Storage of the early ripe

almonds under modified atmosphere to preserve kernel qualitative and sensory traits. Agriculture (Switzerland). 2022;**12**(7):974. DOI: 10.3390/agriculture12070974

[81] García-Pascual P, Mateos M, Carbonell V, Salazar DM. Influence of storage conditions on the quality of shelled and roasted almonds. Biosystems Engineering. 2003;**84**(2):201-209. DOI: 10.1016/S1537-5110(02)00262-3

[82] Wu J, Lin X, Lin S, Chen P, Huang G, Peng P, et al. California almond shelf life: Changes in moisture content and textural quality during storage. Transactions of the ASABE. 2019;**62**(3):661-671. DOI: 10.13031/trans.12709

[83] Kazantzis I, Nanos GD, Stavroulakis GG. Effect of harvest time and storage conditions on almond kernel oil and sugar composition. Journal of the Science of Food and Agriculture. 2003;**83**(4):354-359. DOI: 10.1002/jsfa.1312

[84] Pleasance EA, Kerr WL, Pegg RB, Swanson RB, Cheely AN, Huang G, et al. Effects of storage conditions on consumer and chemical assessments of raw 'nonpareil' almonds over a two-year period. Journal of Food Science. 2018;**83**(3):822-830. DOI: 10.1111/1750-3841.14055

[85] Raisi M, Ghorbani M, Sadeghi Mahoonak A, Kashaninejad M, Hosseini H. Effect of storage atmosphere and temperature on the oxidative stability of almond kernels during long term storage. Journal of Stored Products Research. 2015;**62**:16-21. DOI: 10.1016/j.jspr.2015.03.004

[86] Vidal A, Ouhibi S, Ghali R, Hedhili A, De Saeger S, De Boevre M. The mycotoxin patulin: An updated short review on occurrence, toxicity and analytical challenges. Food and Chemical Toxicology. 2019;**129**:249-256. DOI: 10.1016/j.fct.2019.04.048

[87] Suman M. Last decade studies on mycotoxins' fate during food processing: An overview. Current Opinion in Food Science. 2021;**41**:70-80. DOI: 10.1016/j.cofs.2021.02.015

[88] Neuenfeldt NH, Silva LP, Pessoa RS, de Oliveira RL. Cold plasma technology for controlling toxigenic fungi and mycotoxins in food. Current Opinion in Food Science. 2023;**52**:101045. DOI: 10.1016/j.cofs.2023.101045

[89] Urugo MM, Teka TA, Berihune RA, Teferi SL, Garbaba CA, Adebo JA, et al. Novel non-thermal food processing techniques and their mechanism of action in mycotoxins decontamination of foods. Innovative Food Science & Emerging Technologies. 2023;**85**(103312):1-17. DOI: 10.1016/j.ifset.2023.103312

[90] Marshall H, Meneely JP, Quinn B, Zhao Y, Bourke P, Gilmore BF, et al. Novel decontamination approaches and their potential application for post-harvest aflatoxin control. Trends in Food Science and Technology. 2020;**106**(November):489-496. DOI: 10.1016/j.tifs.2020.11.001

[91] Liu L, Xie M, Wei D. Biological detoxification of mycotoxins: Current status and future advances. International Journal of Molecular Sciences. 2022;**23**(3):1064. DOI: 10.3390/ijms23031064

[92] Pallarés N, Berrada H, Tolosa J, Ferrer E. Effect of high hydrostatic pressure (HPP) and pulsed electric field (PEF) technologies on reduction of aflatoxins in fruit juices. LWT - Food Science and Technology.

2021;**142**:1-7. DOI: 10.1016/j.lwt.2021.111000

[93] Pallarés N, Sebastià A, Martínez-Lucas V, Queirós R, Barba FJ, Berrada H, et al. High pressure processing impact on emerging mycotoxins (ENNA, ENNA1, ENNB, ENNB1) mitigation in different juice and juice-Milk matrices. Food. 2022;**11**(2):1-12. DOI: 10.3390/foods11020190

[94] Woldemariam HW, Harmeling H, Emire SA, Teshome PG, Toepfl S, Aganovic K. Pulsed light treatment reduces microorganisms and mycotoxins naturally present in red pepper (Capsicum annuum L.) powder. Journal of Food Process Engineering. 2021;**45**(2):1-11. DOI: 10.1111/jfpe.13948

[95] Kamle M, Mahato DK, Devi S, Lee KE, Kang SG, Kumar P. Fumonisins: Impact on agriculture, food, and human health and their management strategies. Toxins. 2019;**11**(6):328. DOI: 10.3390/toxins11060328

[96] Gavahian M, Cullen PJ. Cold plasma as an emerging technique for mycotoxin-free food: Efficacy, mechanisms, and trends. Food Reviews International. 2019;**36**(2):193-214. DOI: 10.1080/87559129.2019.1630638

[97] Wei C, Yu L, Qiao N, Zhao J, Zhang H, Zhai Q, et al. Progress in the distribution, toxicity, control, and detoxification of patulin: A review. Toxicon. 2020;**184**:83-93. DOI: 10.1016/j.toxicon.2020.05.006

[98] Altafini A, Roncada P, Guerrini A, Sonfack GM, Fedrizzi G, Caprai E. Occurrence of ochratoxin a in different types of cheese offered for sale in Italy. Toxins. 2021;**13**(540):1-15. DOI: 10.3390/toxins13080540

[99] Lemos AC, Borba VS de, Badiale-Furlong E. The impact of wheat-based food processing on the level of trichothecenes and their modified forms. Trends in Food Science and Technology. 2021;111:89-99. doi: 10.1016/j.tifs.2021.02.056

[100] de Grave XB, Saltzmann J, Laurain J, Rodriguez MA, Molist F, Dänicke S, et al. Transmission of Zearalenone, Deoxynivalenol, and their derivatives from sows to piglets during lactation. Toxins. 2021;**13**(37):1-13. DOI: 10.3390/toxins13010037

[101] Chen J, Wei Z, Wang Y, Long M, Wu W, Kuca K. Fumonisin B1: Mechanisms of toxicity and biological detoxification progress in animals. Food and Chemical Toxicology. 2021;**149**:1-7. DOI: 10.1016/j.fct.2021.111977

[102] Nazhand A, Durazzo A, Lucarini M, Souto EB, Santini A. Characteristics, occurrence, detection and detoxification of aflatoxins in foods and feeds. Food. 2020;**9**(644):1-26. DOI: 10.3390/foods9050644

[103] Siciliano I, Spadaro D, Prelle A, Vallauri D, Cavallero MC, Garibaldi A, et al. Use of cold atmospheric plasma to detoxify hazelnuts from aflatoxins. Toxins. 2016;**8**(125):1-10. DOI: 10.3390/toxins8050125

[104] Durek J, Schlüter O, Roscher A, Durek P, Fröhling A. Inhibition or stimulation of ochratoxin a synthesis on inoculated barley triggered by diffuse coplanar surface barrier discharge plasma. Frontiers in Microbiology. 2018;**9**(2782):1-9. DOI: 10.3389/fmicb.2018.02782

[105] Wielogorska E, Ahmed Y, Meneely J, Graham WG, Elliott CT, Gilmore BF. A holistic study to understand the detoxification of mycotoxins in maize

and impact on its molecular integrity using cold atmospheric plasma treatment. Food Chemistry. 2019;**301**:1-8. DOI: 10.1016/j.foodchem.2019.125281

[106] Yang K, Li K, Pan L, Luo X, Xing J, Wang J, et al. Effect of ozone and electron beam irradiation on degradation of zearalenone and ochratoxin a. Toxins. 2020;**12**(138):1-10. DOI: 10.3390/toxins12020138

[107] Feizollahi E, Roopesh MS. Mechanisms of deoxynivalenol (DON) degradation during different treatments: A review. Critical Reviews in Food Science and Nutrition. 2021;**62**:1-22. DOI: 10.1080/10408398.2021.1895056

[108] Xue M, Wang T, Sun Q, Qu G, Jia H, Zhu L. Insights into the highly efficient detoxification of the biotoxin patulin in water by discharge plasma oxidation. Chemical Engineering Journal. 2021;**411**:1-9. DOI: 10.1016/j.cej.2021.128432

[109] Owolabi IO, Karoonuthaisiri N, Elliott CT, Petchkongkaew A. A 10-year analysis of RASFF notifications for mycotoxins in nuts. Trend in key mycotoxins and impacted countries. Food Research International. 2023;**172**:112915. DOI: 10.1016/j.foodres.2023.112915

[110] Sadok I, Szmagara A, Krzyszczak A. Validated QuEChERS-based UHPLC-ESI-MS/MS method for the postharvest control of patulin (mycotoxin) contamination in red-pigmented fruits. Food Chemistry. 2023;**400**(134066):1-10. DOI: 10.1016/j.foodchem.2022.134066

[111] Rodrigues P, Venâncio A, Lima N. Mycobiota and mycotoxins of almonds and chestnuts with special reference to aflatoxins. Food Research International. 2012;**48**(1):76-90. DOI: 10.1016/j.foodres.2012.02.007

[112] Hidalgo-Ruiz JL, Romero-González R, Martínez Vidal JL, Garrido FA. Determination of mycotoxins in nuts by ultra high-performance liquid chromatography-tandem mass spectrometry: Looking for a representative matrix. Journal of Food Composition and Analysis. 2019;**82**(103228):1-7. DOI: 10.1016/j.jfca.2019.05.011

[113] Ünüsan N. Systematic review of mycotoxins in food and feeds in Turkey. Food Control. 2019;**97**:1-14. DOI: 10.1016/j.foodcont.2018.10.015

[114] Azaiez I, Font G, Mañes J, Fernández-Franzón M. Survey of mycotoxins in dates and dried fruits from Tunisian and Spanish markets. Food Control. 2015;**51**:340-346. DOI: 10.1016/j.foodcont.2014.11.033

[115] Iqbal SZ, Mehmood Z, Asi MR, Shahid M, Sehar M, Malik N. Co-occurrence of aflatoxins and ochratoxin A in nuts, dry fruits, and nuty products. Journal of Food Safety. 2018;**38**(4):1-7. DOI: 10.1111/jfs.12462

Chapter 14

The Role of Some Pre and Postharvest Applications on Storage Behavior and Protein Pattern of Date Palm Fruits *Phoenix dactylifera* L. cvs. Berhi and Breim

Dhia Ahmed Taain, Hamzah Abbas Hamzah
and Abbas Mahdi Jasim

Abstract

Storage experiment was conducted to study the effect of some pre and postharvest natural control treatments which include ultraviolet light (UV) at the concentration of 1KGY for periods of (0, 5, 10) minutes, 1-MCP at the concentration of (0,0.5,1) ppm for 24 hour at 0°c, ozone (O3) at the concentration of 5 ppm for the periods of (0, 0.5, 1) hour, soaking in chitosan at the concentratons of (0,1,2) % and control treatment in addition to field-treated chitosan at the concentrations of 0%, 1%, and 2% by the aim of improving storage behavior of date palm fruits cvs. Berhi and Breim and determination the protein pattern of fruits after six months of storage at -10 ± 2°C. Results of the protein pattern showed that protein bundles on polyacrylamide gel differed by their molecular weights, the number of protein bundles, and Breim cultivar fruits treated with chitosan recorded the highest number of bundles of seven bundles and the highest molecular weight of (173.857) kDa for the first bundle.

Keywords: date palm fruits, Berhi and Breim, chitosan, ultraviolet light (UV), ozone (O3)

1. Introduction

The date palm, *Phoenix dactylifera* L., belongs to the family Arecaceae and is one of the subtropical fruit trees cultivated and spread in Iraq and some regions of the Middle East [1]. It is considered one of the most important fruit trees in Iraq because of its great nutritional and economic value. It is a sacred tree mentioned in all monotheistic religions. The evidence available at present indicates that the Sumerians were the first to be interested in the cultivation of the date palm, and used its fruits as basic food in the Tigris and Euphrates valleys for more than four thousand years BC [2].

The number of date palms in Iraq has decreased significantly in recent years due to the wars, the high salinity rate in soil and irrigation water, and the problem of the housing crisis that led to the bulldozing of many palm orchards, which led to a severe shortage of dates production. Therefore, it is important to take the necessary measures to develop the production of dates by increasing the number of planted palm trees, as well as increasing the yield, improving the qualities of fruits, and reducing the percentage of spoilage, especially the desired cultivars, such as Berhi and Breim, through conducting some of the pre- and post-harvest treatments and avoiding treatment of fruits with chemicals that have negative impact on consumer health, as consumer demand has increased recently for fruits whose production has bio-safety factors [3, 4].

Chitosan is a vital polymer, the second largest biomaterial after cellulose, which is found in the outer structure of crustaceans, insects, and fungal cell walls. It is also characterized by no toxicity and biological decay and has no local effects on living tissue. It is a compound with vital functions [5], which has attracted the interest of researchers in the last few years for its commercial uses. Chitosan is composed of glucosamine units, which are associated with each other with a type of beta-type (1–4) cyclic bonds. It possesses many free hydroxyl and amino acids that enable it to form ionic, hydrogen, and hydrofluidic bonds with other molecules such as fats and proteins [6, 7].

Cold storage of date palm fruits is one of the important means at present, which is used to try to keep those fruits in the rutab phase as long as possible, thus prolonging the display period of those fruits in the local markets in the rutab phase, as cold storage reduces pathogens and the vital activities of the fruits, especially the process of respiration and the production of ethylene [8]. The activity of ethylene can also be inhibited through the use of the compound 1-methylcyclopropene, which symbolizes (1-MCP) commercially called smart fresh, which is in the form of a white powder that can be dissolved with water and releases the active substance 1-methylcyclopropene in the form of gas, which prevents contact ethylene with its receptors in the cells, which leads to inhibiting the formation of ethylene, and this, in turn, delays the natural ripening processes in the fruit, which keeps it fresh and of good quality for a longer time [9]. Studies have shown that treatment with the compound (1-MCP) limits the rate of ethylene production in fruits, reduces their respiration rate, and delays their entry into the ripening phase compared to untreated fruits [10–13].

Ultraviolet treatment is one of the alternative methods that has spread to be effective in inactivating bacteria, protozoa, algae, and viruses. Ultraviolet rays have the ability to destroy microorganisms, as is the case with heat treatment, but it has better advantages than heat treatment, as it does not affect the sensory properties of fruits, and it has a lower cost than heat treatment from an economic point of view. The radiation treatment process makes foods free of dangerous substances that are commonly used to kill insects by fumigation, such as ethyl dibromide, methyl bromide, and phosphine. Ultraviolet radiation in the range of 250–260nm is lethal to most microorganisms and acts as a strong bactericide [14], in addition that the treatment is certified and approved by various international health organizations.

Ozone (O3) is one of the powerful disinfectants against a wide range of microorganisms [15]. Ozone has number of features that make it suitable as an ideal post-harvest treatment, quickly decomposes into oxygen without leaving any residue and is applied either as a gas or it is soluble in water, so it can effectively reduce post-harvest losses during storage for several crops [16–18].

The current study aims to improve the storage behavior of date palm fruits cvs. Berhi and Breim and determination the protein pattern of fruits after six months of storage at $-10 \pm 2°C$. Increasing the display period of the fruits of the two cultivars in the rutab phase for the longest possible period and improving their storage qualities and marketable of fruits through the use of some natural control treatments before and after harvest, especially chitosan, which is used according to the available references for the first time in the field of improving the yield, qualitative characteristics, and storage ability of the fruits of the two date palm cultivars. In addition to treat the fruits with some post-harvest treatments, which include the use of 1-MCP, ozone, and ultraviolet rays, and studying the physical and chemical changes of fruits during storage under the influence of these treatments.

2. Materials and methods

The storage experiment which was conducted, where the fruits of Berhi and Breim cultivars that treated with chitosan at the concentration of (0, 1, 2%) could be summarized as follows:

2.1 Chitosan extraction process

The shrimps were obtained from the local fish market in Basrah and the crusts were washed with water and dried by leaving them exposed to the sun. The method mentioned in [19], was followed for the extraction of chitin from the shrimp. The shrimp crust was crushed into small pieces using an electric mill then, the process of removal of the proteins (deprotienization) has been done by treating the crusts with sodium hydroxide solution at a concentration of 3.5% for two hours at a temperature of 65°c. by 1:10 (weight/volume). The mineral elements are removed in the process called demineralization by using a solution of hydrochloric acid at the concentration of 1 N for 1/2hour at room temperature by 1:15 (weight/volume). Crusts were washed well with water several times, then the pigment was removed by acetone and then by sodium hypochlorite solution at 0.315% for 5 minutes at room temperature by 1:10 (weight/volume). Finally, the white product was washed with distilled water and dried in an oven at 60° C for 24 hours to obtain the chitin.

Chitosan was prepared according to the method mentioned by [20] by removing the acetyl groups (Deacetylation) by treating with 50% sodium hydroxide at 1:10 (weight/volume) at 100°C for 20 hours to obtain chitosan with low molecular weight, and then dried at 110° C for 6 hours. The resulting chitosan is a white powder.

The viscosity was determined by the use of the Ostwald viscometer. After the preparation of the solution, the amount of time required to flow it at a certain distance at 25°. Molecular weight was determined depending on the viscosity of the solution according to [21]. The degree of removal of acetyl groups was determined by mixing 40 mg chitosan with 120 mg potassium bromide and then pressing and dried, then determined by using the Fourier Transform Infrared Spectroscopy (FTIR) Instrument.

The parameters of the product were measured as follows: Viscosity = 64.16 Centi Boyz, molecular weight 720 K, dalton, and the degree of removal of acetyl groups 87.6%.

Fruits have been brought after 18 weeks of pollination to the cold store in the early morning immediately after picking, then cleaned and each part of the three parts was divided into four parts for each cv. The first part was treated with ultraviolet rays at an amount of 1 kgY at intervals (0, 5, 10) minutes, and the second part was treated with ozone at 5 ppm at intervals of (0, 0.5, 1) an hour, and the third part was treated with the compound (1-MCP) at a concentration of (0, 0.5, 1) ppm for 24 hours under 0°C, and the fourth part was immersed in chitosan at the three concentrations (0, 1,

2%). Fruits were packed in the transparent plastic container automatically with six replicates for each concentration, then three replicates were stored at a temperature of -10 ± 2°C for six months and after that, the following characteristics were determined:

2.2 The protein pattern was determined according to the following method

Electrophoresis for proteins

1. Lyophilization of samples

   The samples were lyophilized by the freeze-dryer lyophilization technique, where the samples to be lyophilized were placed in plastic containers and then placed in a lyophilization device at a temperature of −26°C until almost most of the water was removed, after which powdered was used in protein electrophoresis on polyacrylamide gel in the presence of sodium dodecyle sulfate (SDS) by SDS-PAGE method according to [22].

2. Identification and quantification of proteins:

   A. Sample preparation:

   2g of lyophilized samples were crushed with 14 ml of cooled acetone three times, then the powder was thoroughly mixed with the extraction solution consisting of 0.2M sodium phosphate, 5% SDS, and 4 molar urea pH 7.0. The extraction solution was prepared by dissolving 3.12g sodium phosphate dissolving NaH_2PO_4, 5 g SDS, and 24.024 g urea in a volume of distilled water, the pH was adjusted to 7.0 and the volume was filled to 100 ml with distilled water, and then centrifuged at 4000 cycles min-^{1} for 15 min. The protein was precipitated using acetone in a ratio of 1:4 (volume:volume) and centrifuged at a speed of 10,000 cycles min-^{1}, the filter was neglected, and the precipitate was taken and dissolved in the buffer solution of the sample.

   B. Electrophoresis: Protein electrophoresis was carried out on a Polyacrylamide gel using the Slab–electrophoresis method in the presence of SDS according to the method of [23] and described by [24].

3. The solutions used:

   A. Resolving gel buffer prepared at a concentration of 1.5 M by dissolving 18.2 g of Tris (hydroxymethyl) methylamine in 80 ml of distilled water, the pH adjusted to 8.8 using 1 M of hydrochloric acid and the volume completed to 100 ml with distilled water.

   B. Stacking gel buffer (pH = 6.8) was prepared at 0.5 M concentration by dissolving 6 g of Tris (hydroxymethyl) methylamine in 40 ml of distilled water, the pH adjusted to 6.8 using 1 M HCl, and the volume completed to 100 ml with distilled water.

   C. 10% SDS solution was prepared by dissolving 10 g of sodium dodecyl sulfate in a volume of distilled water and then, the volume completed to 100 ml with distilled water.

D. Electrode buffer was Prepared by dissolving 1.5 g of Tris (hydroxymethyl) methylamine and 2.7 g of glycine in an amount of distilled water and the volume completed to 500 ml with distilled water with the addition of 5 ml of 10% SDS solution.

E. Acryl amide stock solution was prepared by adding 29.2 g of acrylamide with 0.8 g of Bis-acryl amide in 60 ml of distilled water and the volume completed to 100 ml with distilled water. The solution is filtered through filter paper no. 1 and 4 ml of a 10% SDS solution is added to it.

F. Ammonium persulfate (Aps) solution was prepared immediately at a concentration of 1.5% by dissolving 0.15 g of Ammonium Persulfate (Aps) in 10 ml of distilled water.

G. TEMED (N,N,N,N-tetra methyl ethylene diamine).

H. Staining Solution (0.1) was prepared by dissolving 0.25 g of Coomassie brilliant blue R-250 in 250 ml of a mixture consisting of acetic acid: methyl alcohol: distilled water in a 1:4:5 ratio, respectively.

I. Detaining solution consisted of a mixture of acetic acid: methyl alcohol: and distilled water in a ratio of 1:4:5, respectively.

J. A solution of bromophenol blue (0.25%) was prepared by dissolving 0.25 g of bromophenol blue dye in a 50% solution of glycerol.

K. sample buffer consisted of SDS at a concentration of 10%, bromophenol blue at a concentration of 0.5%, bromoethanol at a concentration of 0.5%, and sucrose at 20%.

4. Method

A. Sample preparation: It was prepared by dissolving the precipitated protein after the precipitation treatment in the buffer solution of the sample, and then it was placed in a water bath for 5 minutes at the boiling point and left to cool to the laboratory temperature to transfer the sample later.

B. Preparation of the gel:

1. Preparation of the separation gel: Separation gel 7.5% acrylamide was prepared by mixing 14.55 ml of distilled water, 7.5 ml of acrylamide solution, 7.5 ml of buffer solution for separation gel, 0.3 ml of SDS solution, 150 μl of ammonium persulfate solution, and 15 μl of TEMED, leave to harden for an hour and a half.

Finally, removal of the gel: Carefully remove the gel from the two glass plates by adding a little water with a syringe to avoid tearing the gel. Then, the dyeing solution was added and left for a whole day. After that, the gel was removed from the dyeing basin and the dye removal solution was added to it, and the process of washing the gel continued until the bands appeared. It was photographed with an English-origin Gel Documentation Device.

2. Total soluble solids (T.S.S.) were measured by hand refractometer and the results were corrected to 20°C according to [25].

3. Total and reducing sugars (%) of fruits were determined according to Lane and Eynon method outlined in [26].

4. Total titratable acidity%): Total titratable acidity was determined according to the method outlined in [26].

2.3 Statistical analysis

A completely randomized design (CRD) was used for a factorial storage experiment with three factors: the first factor is field-treated with chitosan, the second factor is storage treatments with three concentrations for each treatment, and the third factor is different storage periods that include six months at −10°C. The analysis was done using the statistical program (SPSS), and the mean values were compared using the least significant difference test (R.L.S.D) at the level of significance (5%) [27].

3. Results and discussion

3.1 The protein pattern

As shown in **Figure 1**, the electrophoresis in acrylamide gel of fruits proteins of Berhi and Breim cultivars field-treated with 2% chitosan and postharvest treated with UV rays for 10 minutes, 1-MCP at a concentration of 1 ppm, ozone for 1 hour, and chitosan at a concentration of 2% in addition to control treatment, respectively.

Breim cultivar treated with chitosan recorded the highest height of bundle (180) for the second bundle, while Berhi cultivar treated with the compound (1-MCP) recorded the lowest bundle height of (104) for the first bundle. Berhi cultivar treated with the compound (1-MCP) recorded the largest bundle area of (14,112) for the fourth bundle, while Breim cultivar treated with ultraviolet rays recorded the smallest bundle area of (48) for the first bundle (**Figure 2 a–d**).

As shown in **Figure 3**, the number and sites of protein bundles and **Table 2** showed the changes in the number of protein bundles and their molecular weights (kilodalton). It is clear that Breim cultivar fruits treated with chitosan recorded the highest number of bundles of seven bundles, while the control treatments of the two cultivars as well as Breim cultivar fruits treated with ultraviolet rays recorded five bundles each, while all treatments of Berhi cultivar except the control treatment recorded four bundles, as well as the ozone-treated Breim cv. fruits, while the Breim cv. fruits treated with the compound (1-MCP) recorded the lowest number of bundles, which amounted to only three bundles.

The Breim cultivar treated with chitosan recorded the highest molecular weight of (173.857) kDa for the first bundle, while the lowest molecular weight was (32.00) kDa for the Breim dipped in chitosan for the seventh bundle, see **Table 1**.

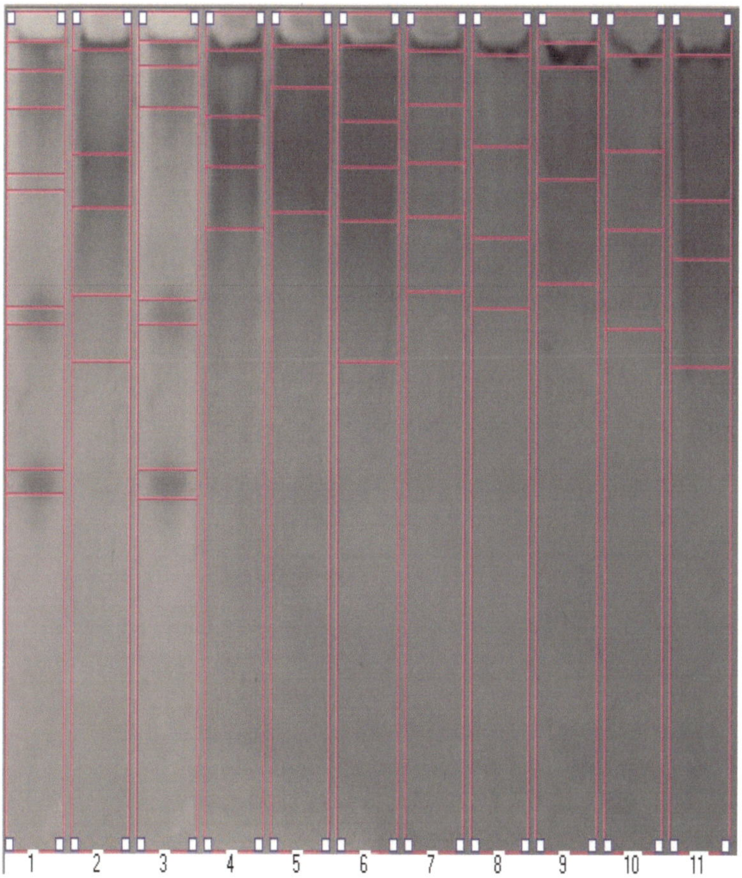

Figure 1.
The electrophoresis.

Through the results obtained from the gel-electrophoresis of the proteins of the date palm fruits of the Berhi and Breim cultivars, it is noted that there are significant differences in the number of protein bundles as well as the sites of their appearance between the control treatment and other treatments. There is no doubt that the dependence on the physical and chemical characteristics of the fruits is no longer sufficient to identify and distinguish among date cultivars and to detect commercial fraud for dates, especially after the processes of pressing them. Therefore, the recent trend is to use techniques such as electrophoresis to identify the protein patterns of dates and determine their behavior during storage. These differences in the protein pattern of the fruits mean that the fruits have differed in the process of gene expression.

It is well known in recent years that changes in the process of gene expression played an important role in regulating the process of fruit growth and ripening, and scientific development in the field of molecular biology has led to a significant increase in our knowledge of the mechanisms in which the genes responsible for the ripening of fruits are regulated, and thus, there may be gene expression of heat shock proteins made fruits to bear low temperatures when freezing [28].

3.2 Total soluble solids

The results of **Tables 2** and **3** showed the effect of spraying chitosan, storage treatments, and storage periods, and the interaction among them on the percentage of total soluble solids in the fruits of the Berhi and Breim cultivars stored at a temperature of (−10 ± 2) °C for the two seasons 2014 and 2015. It is noted that spraying

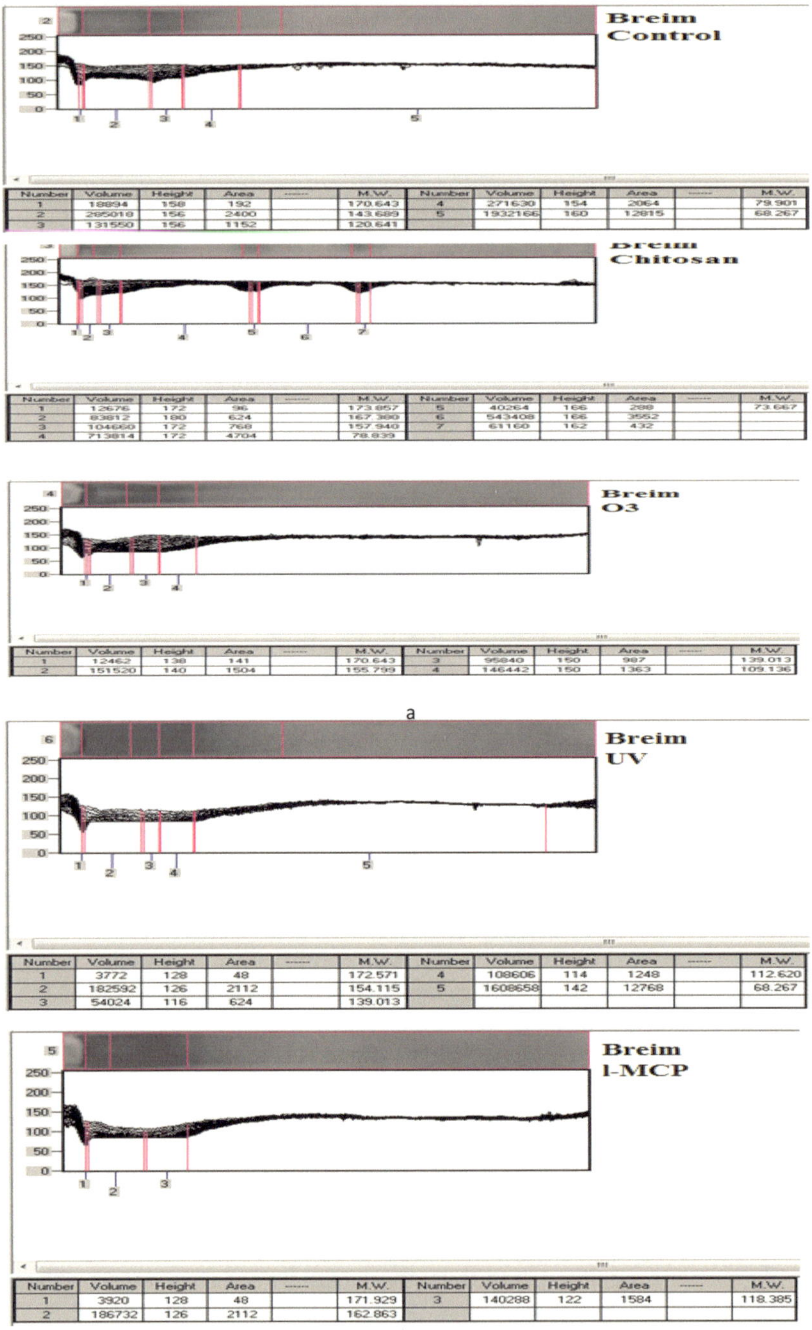

b

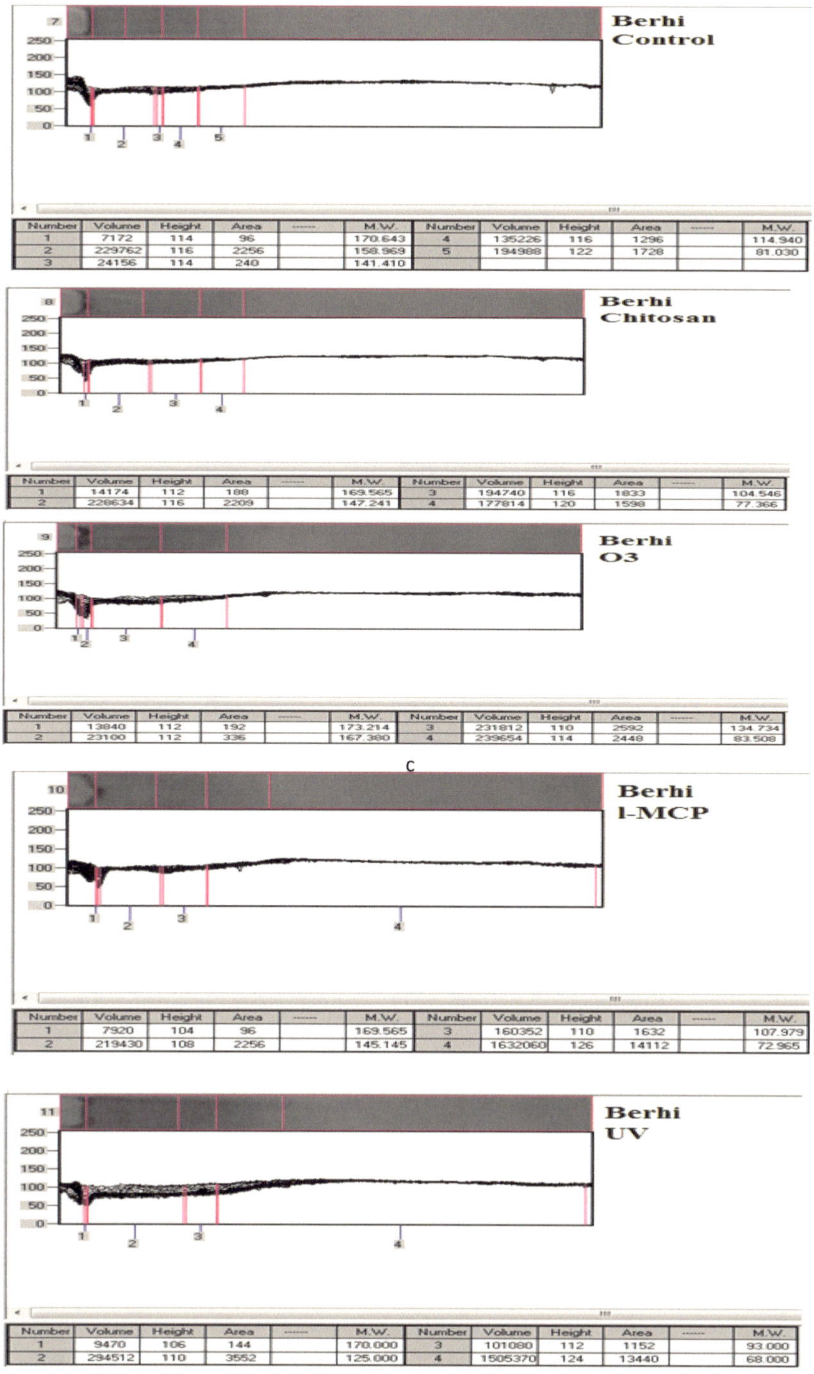

Figure 2.
a-d. some specifications of protein bundles for study parameters.

chitosan had a significant effect in decreasing the percentage of total soluble solids, the lowest percentage of total soluble solids was (36.12 and 35.72), (33.61 and 33.48)% for the fruits of the Berhi and Breim cultivars treated in the field with 2%chitosan for

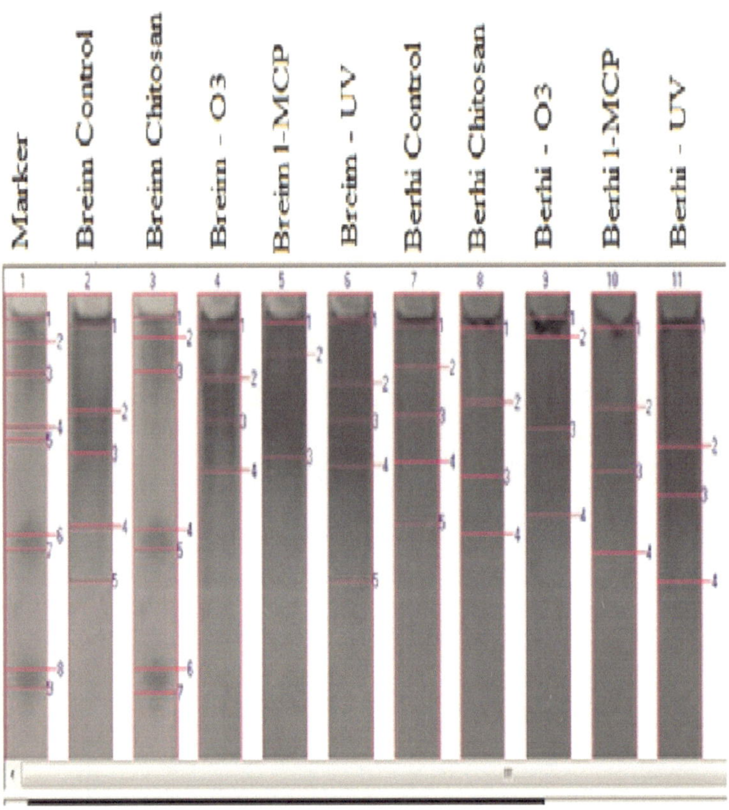

Figure 3.
The number and sites of protein bundles for the study parameters.

the two seasons, respectively, with a significant difference from the rest of the treatments, while it reached the highest percentage of total soluble solids (42.06 and 41.93)% and (37.38 and 44.98)% for the control fruits of Berhi and Breim cultivars for the two seasons, respectively. The results are consistent with [29], which referred to the effect of pre-harvest chitosan treatment in decreasing the percentage of soluble solids, and the untreated fruits had a higher percentage of soluble solids.

As for the effect of 1-MCP at a concentration of 1 ppm, it caused a reduction in the percentage of total soluble solids, which amounted to (37.64 and 37.38), (34.67 and 34.36)% for the fruits of Berhi and Breim cultivars for the two seasons, respectively, with a significant difference from the untreated fruits of the two cvs. For the same seasons, which amounted to (39.94 and 39.85), (36.24 and 35.87)%.

It was noted from the same table that the percentage of total soluble solids mild increased with the increment of the storage periods reached (42.03 and 41.89), (37.19 and 37.00)% for the fruits of the two cvs. For the two seasons, respectively after six months of storage, while the lowest percentage of total soluble solids was (35.55 and 34.88), (34.15 and 33.94%) for the fruits of the two cvs. For the two seasons, respectively, after one week of storage. This is may be due to that, the percentage of total soluble solids increasing by decreasing the percentage of the water content of the fruits.

As for the interaction effect between spraying chitosan and storage treatments, the results indicated that the fruits treated with 2% chitosan and (1-MCP) at a

Berhi cultivar					No.bandles	Breim cultivar					No.bandles
UV	1-mcp	O3	chitosan	control		UV	1-mcp	O3	chitosan	control	
170.00	169.565	173.214	169.565	170.643	1	172.929	171.643	170.643	173.857	170.643	1
125.00	145.145	167.380	147.241	158.969	2	154.115	162.863	155.799	167.380	143.689	2
93.00	107.979	134.734	104.546	141.410	3	139.013	118.385	139.013	157.940	120.641	3
68.00	72.965	83.508	77.366	114.940	4	112.620		109.136	78.839	79.901	4
				81.030	5	68.267			73.667	68.267	5
					6				41.00		6
					7				32.00		7

Table 1.
Changes in the number of protein bundles and their molecular weights (kilodaltons) for the study parameters.

Field chitosan treatment	Storage treatments	2014						Field chitosan x Storage treatments	2015						Field chitosan x Storage treatments
		Storage period (month)							Storage period (month)						
		1	2	3	4	5	6		1	2	3	4	5	6	
0%	Control	38.13	40.23	42.40	44.4	46.66	49.00	43.47	40.03	40.23	42.86	44.60	46.33	48.26	43.72
	UV 5 min.	37.13	39.70	41.30	43.43	45.50	46.86	42.32	37.46	38.70	41.63	43.10	44.83	46.86	42.10
	UV 10 min.	37.03	39.63	41.13	43.30	45.53	46.73	42.22	38.36	39.70	41.70	43.30	45.20	46.13	42.40
	(1-MCP) 0.5 ppm	35.93	36.56	40.03	42.20	44.50	46.20	40.90	35.70	36.26	39.63	41.86	44.16	45.86	40.58
	(1-MCP) 1 ppm	35.70	37.63	39.90	42.06	44.46	45.73	40.91	35.60	37.63	39.23	41.80	44.23	45.46	40.66
	Ozone half an hour	37.40	39.86	41.46	43.56	45.80	47.00	42.51	37.63	39.96	41.56	43.66	45.86	47.26	42.66
	Ozone one hour	37.80	40.06	41.83	43.80	46.10	47.40	42.83	37.90	40.06	41.86	43.80	46.10	47.53	42.87
	(Chitosan) 1%	36.60	38.70	41.23	43.03	44.83	46.73	41.85	35.93	38.36	40.90	42.63	44.36	46.23	41.40
	(Chitosan) 2%	35.70	37.63	39.90	42.06	43.33	46.30	41.35	35.63	38.30	40.86	42.10	43.16	46.00	41.01
1%	Control	35.96	37.16	38.33	40.13	41.13	42.13	39.14	35.63	36.50	37.86	39.56	41.46	41.80	38.80
	UV 5 min.	35.30	36.00	37.50	39.30	40.23	41.13	38.24	35.16	35.26	37.16	38.96	40.06	40.80	37.90
	UV 10 min.	34.96	35.96	37.40	39.33	40.23	40.23	38.02	34.36	35.76	37.06	38.66	40.23	40.43	37.75
	(1-MCP) 0.5 ppm	33.86	35.00	36.40	38.20	39.60	40.13	37.20	33.66	34.66	36.23	37.90	39.36	39.80	36.93
	(1-MCP) 1 ppm	33.86	34.93	36.36	38.30	39.53	39.60	37.10	33.53	34.80	35.96	38.20	39.20	39.13	36.80
	Ozone half an hour	35.50	36.06	37.73	39.43	40.53	41.43	38.45	35.76	36.23	37.80	39.60	40.86	41.43	38.61
	Ozone one hour	35.73	36.20	38.00	39.70	40.73	41.70	38.67	36.06	36.30	38.33	39.76	40.73	41.70	38.81
	(Chitosan) 1%	34.80	35.33	36.86	38.50	39.80	40.73	37.67	34.33	35.03	36.46	38.16	39.33	40.53	37.31
	(Chitosan) 2%	34.23	35.10	37.30	38.36	39.50	40.40	37.48	33.96	35.10	37.06	38.06	39.16	40.03	37.23
2%	Control	34.06	35.20	36.60	38.06	39.06	40.26	37.21	34.73	34.43	36.26	38.06	38.06	39.93	37.02
	UV 5 min	32.93	33.93	35.60	36.36	38.70	39.06	36.10	32.60	33.60	35.26	36.36	38.70	38.40	35.82
	UV 10 min	32.83	33.76	35.46	36.43	38.60	38.70	35.96	32.50	33.43	35.46	36.43	38.26	38.70	35.80

DOI: http://dx.doi.org/10.5772/intechopen.109899

Field chitosan treatment	Storage treatments	2014						Field chitosan x Storage treatments	2015						Field chitosan x Storage treatments
		Storage period (month)							Storage period (month)						
		1	2	3	4	5	6		1	2	3	4	5	6	
	(1-MCP) 0.5 ppm	31.93	32.93	34.60	35.36	37.33	38.06	35.03	31.60	32.60	34.26	35.20	37.00	37.93	34.76
	(1-MCP) 1 ppm	31.83	32.76	34.46	35.43	37.23	37.70	34.90	31.63	32.43	34.33	35.23	36.90	37.50	34.67
	Ozone half an hour	33.20	34.26	35.76	36.53	38.83	39.30	36.31	33.43	34.60	36.10	36.53	38.86	39.46	36.50
	Ozone one hour	33.36	34.66	36.06	36.80	39.03	39.60	36.58	33.70	34.66	36.33	36.80	39.30	39.66	36.74
	(Chitosan) 1%	32.93	33.86	35.00	35.80	38.16	37.76	35.58	32.60	33.46	34.83	35.46	37.83	37.36	35.26
	(Chitosan) 2%	32.66	33.50	35.30	35.43	37.86	37.66	35.40	32.46	33.26	34.96	35.10	37.53	36.96	35.05
								Mean of field chitosan treatment							Mean of field chitosan treatment
Field chitosan x Storage period	0%	37.28	38.90	41.30	43.12	45.04	46.73	42.06	37.13	38.8	41.13	42.98	44.91	46.62	41.93
	1%	34.79	35.57	37.19	38.85	40.16	40.78	37.89	34.71	35.51	37.10	38.7	40.04	40.62	37.794
	2%	34.58	33.71	35.35	36.24	38.24	38.57	36.12	32.80	33.60	35.31	36.13	38.04	38.43	35.72
								Mean of storage treatments							Mean of storage treatments
Storage treatments x Storage period	control	36.05	37.53	39.11	40.86	42.28	43.80	39.94	36.80	37.05	39.00	40.74	42.17	43.33	39.85
	UV 5 min.	35.12	36.54	38.13	39.70	41.47	42.35	38.88	35.07	35.85	38.02	39.47	41.20	42.02	38.60
	UV 10 min.	34.94	36.45	38.00	39.68	41.45	41.88	38.73	35.07	36.30	38.07	39.46	41.23	41.75	38.65
	(1-MCP) 0.5 ppm	33.91	34.83	37.01	38.58	40.47	41.46	37.71	33.65	34.51	36.71	38.32	40.17	41.20	37.43
	(1-MCP) 1 ppm	33.80	35.11	36.91	38.60	40.41	41.01	37.64	33.58	34.95	36.51	38.41	40.11	40.70	37.38
	Ozone half an hour	35.36	36.73	38.32	39.84	41.72	42.57	39.09	35.61	36.93	38.48	39.93	41.86	42.72	39.25
	Ozone one hour	35.63	36.97	38.63	40.10	41.95	41.90	39.36	35.88	37.01	38.84	40.12	42.04	42.96	39.48
	(Chitosan) 1%	34.77	35.96	37.70	39.11	40.93	41.74	38.37	34.28	35.62	37.40	38.75	40.51	41.37	37.99
	(Chitosan) 2%	34.40	35.74	37.90	38.74	40.23	41.45	38.08	34.02	35.55	37.63	38.42	39.95	41.00	37.76

Field chitosan treatment	Storage treatments	2014						Field chitosan x Storage treatments	2015						Field chitosan x Storage treatments
		Storage period (month)							Storage period (month)						
		1	2	3	4	5	6		1	2	3	4	5	6	
Mean of Storage period		35.55	36.06	37.95	39.40	41.14	42.03		34.88	35.97	37.85	39.29	41.02	41.89	

R.L.S.D. 5%	Field chitosan treatment	Storage treatments	Storage period	Field chitosan x Storage treatments	Storage treatments x Storage period	Storage treatments x Storage period	Triple interaction
2014	0.09	0.17	0.14	0.14	0.24	0.42	0.72
2015	0.05	0.10	0.08	0.17	0.14	0.25	0.43

Table 2.
Effect of field chitosan spraying, storage treatments, storage period, and interaction between them in the percentage of total soluble solids in the fruits of the Berhi cultivar stored at a temperature of (−10 ± 2) °C for the 2014 and 2015 seasons.

DOI: http://dx.doi.org/10.5772/intechopen.109899

Field chitosan treatment	Storage treatments	2014						Field chitosan x Storage treatments	2015						Field chitosan x Storage treatments
		Storage period (month)							Storage period (month)						
		1	2	3	4	5	6		1	2	3	4	5	6	
0%	control	36.20	36.53	36.73	37.80	40.83	42.76	38.47	35.73	36.20	36.46	37.46	40.50	42.43	38.13
	UV 5 min.	36.03	36.33	36.73	37.40	39.10	41.23	37.80	35.70	36.00	36.03	36.93	38.73	40.90	37.38
	UV 10 min.	36.00	36.30	36.53	37.40	39.43	41.10	37.79	35.73	36.10	36.03	37.03	39.06	40.50	37.41
	(1-MCP) 0.5 ppm	35.66	36.06	36.20	36.40	37.23	38.56	36.68	35.20	35.73	35.86	36.06	36.90	38.30	36.34
	(1-MCP) 1 ppm	35.63	36.10	36.20	36.40	37.23	38.63	36.70	35.46	35.76	36.16	36.06	36.90	38.30	36.44
	Ozone half an hour	36.06	36.26	36.53	36.80	39.13	40.53	37.55	36.26	36.43	36.53	36.96	39.13	40.73	37.67
	Ozone one hour	36.03	36.06	36.46	36.86	38.86	40.23	37.42	36.23	36.26	36.46	36.86	38.86	40.56	37.54
	(Chitosan) 1%	35.76	35.80	36.46	36.56	38.40	39.20	37.03	35.46	35.03	35.96	36.16	37.93	38.93	36.58
	(Chitosan) 2%	35.60	35.66	36.40	36.60	38.20	39.23	36.95	34.90	35.10	36.16	36.26	38.13	38.90	36.57
1%	Control	34.03	34.33	34.66	36.00	37.06	38.06	35.69	33.70	34.00	34.33	34.86	36.40	37.73	35.17
	UV 5 min.	33.96	34.16	34.50	35.30	36.70	35.86	35.08	33.70	34.13	33.86	35.20	35.70	35.53	34.68
	UV 10 min.	34.03	34.33	34.53	35.20	36.60	36.60	35.21	33.70	33.33	34.20	34.83	35.93	36.26	34.71
	(1-MCP) 0.5 ppm	33.60	33.63	34.06	34.40	35.36	35.93	34.50	33.26	33.30	33.73	34.06	35.03	35.60	34.16
	(1-MCP) 1 ppm	33.50	33.60	33.96	33.93	35.30	35.76	34.34	33.16	33.20	33.63	33.60	35.10	35.43	34.02
	Ozone half an hour	34.03	34.36	34.43	34.70	35.06	36.36	34.82	34.06	34.50	34.60	34.83	35.26	36.53	34.96
	Ozone one hour	34.00	34.20	34.36	34.83	35.10	36.23	34.78	34.06	34.30	34.56	34.90	35.23	36.43	34.91
	(Chitosan) 1%	33.60	33.86	33.80	34.36	35.30	36.66	34.60	33.23	33.53	33.40	34.10	34.96	36.20	34.23
	(Chitosan) 2%	35.30	33.96	33.63	34.23	35.13	36.50	34.51	33.26	33.56	33.30	33.96	34.96	36.26	34.22
2%	Control	33.00	33.23	33.70	34.66	35.40	37.33	34.55	32.80	32.90	33.36	34.53	35.30	37.00	34.31
	UV 5 min.	33.06	33.30	33.46	33.73	34.63	35.56	33.96	32.73	33.30	33.13	33.46	34.30	35.23	33.69
	UV 10 min.	33.10	33.13	33.36	33.70	34.63	35.26	33.86	32.76	32.80	33.36	33.36	34.63	34.93	33.68

Field chitosan treatment	Storage treatments	2014						Field chitosan x Storage treatments	2015						Field chitosan x Storage treatments
		Storage period (month)							Storage period (month)						
		1	2	3	4	5	6		1	2	3	4	5	6	
	(1-MCP) 0.5 ppm	32.50	33.06	32.73	32.80	34.10	34.00	33.20	32.16	32.73	32.40	32.46	33.76	33.50	32.83
	(1-MCP) 1 ppm	32.50	32.63	32.63	32.76	33.43	33.93	32.98	32.16	32.30	32.30	32.36	33.10	33.53	32.62
	Ozone half an hour	33.00	33.26	33.56	34.06	34.53	35.06	33.91	33.16	33.30	33.63	34.20	34.50	35.27	34.01
	Ozone one hour	32.96	33.23	33.46	33.86	34.73	34.80	33.84	33.06	33.40	33.60	33.90	34.80	34.93	33.95
	(Chitosan) 1%	32.63	32.70	32.83	33.16	34.60	35.10	33.50	32.36	32.36	32.50	33.00	34.26	34.40	33.15
	(Chitosan) 2%	32.66	32.66	32.73	33.00	34.43	35.10	33.43	32.43	32.36	32.43	32.83	33.90	34.90	33.14
								Mean of field chitosan treatment							Mean of field chitosan treatment
Field chitosan x Storage period	0%	35.88	36.12	36.474	36.915	38.715	40.167	37.38	35.63	35.84	36.18	36.64	38.46	39.95	37.12
	1%	33.82	34.05	34.219	34.774	35.700	36.370	34.82	33.57	33.76	33.95	34.48	35.39	36.219	34.56
	2%	32.75	32.97	33.063	33.415	34.437	35.044	33.61	32.62	32.82	32.96	33.3	34.28	34.85	33.48
								Mean of storage treatments							Mean of storage treatments
Storage treatments x Storage period	Control	34.41	34.70	35.03	36.15	37.76	39.38	36.24	34.07	34.36	34.72	35.62	37.40	39.05	35.87
	UV 5 min.	34.35	34.60	34.90	35.47	36.81	37.55	35.61	34.04	34.47	34.34	35.20	36.24	37.22	35.25
	UV 10 min.	34.37	34.58	34.81	35.43	36.88	37.65	35.62	34.06	34.07	34.53	35.16	36.54	37.23	35.27
	(1-MCP) 0.5 ppm	33.92	34.25	34.33	34.53	35.56	36.16	34.79	33.54	33.92	34.00	34.20	35.23	35.80	34.45
	(1-MCP) 1 ppm	33.87	34.11	34.26	34.36	35.32	36.11	34.67	33.60	33.75	34.03	34.01	35.03	35.75	34.36
	Ozone half an hour	34.36	34.63	34.84	35.18	36.24	37.32	35.43	34.50	34.74	34.92	35.33	36.30	37.51	35.55
	Ozone one hour	34.33	34.50	34.76	34.86	35.83	37.07	35.35	34.45	34.65	34.87	35.22	36.30	37.31	35.47
	(Chitosan) 1%	34.00	34.12	34.36	34.70	36.10	36.98	35.04	33.68	33.64	33.95	34.42	35.72	36.51	34.65
	(Chitosan) 2%	33.96	34.10	34.25	34.61	35.92	36.94	34.96	33.53	33.67	33.96	34.35	35.66	36.68	34.6

Field chitosan treatment	Storage treatments	2014						Field chitosan x Storage treatments	2015						Field chitosan x Storage treatments
		Storage period (month)							Storage period (month)						
		1	2	3	4	5	6		1	2	3	4	5	6	
Mean of Storage period		34.15	34.38	34.58	35.03	36.28	37.19		33.94	34.14	34.36	34.83	36.04	37.00	

R.L.S.D. 5%	Field chitosan treatment	Storage treatments	Storage period	Field chitosan x Storage treatments	Storage treatments x Storage period	Storage treatments x Storage period	Triple interaction
2014	0.05	0.10	0.08	0.17	0.14	0.25	0.434
2015	0.05	0.10	0.08	0.17	0.14	0.25	0.434

Table 3.
Effect of field chitosan spraying, storage treatments, storage period and interaction between them in the percentage of total soluble solids in the fruits of the Breim cultivar stored at a temperature of (−10 ± 2) °C for the 2014 and 2015 seasons.

concentration of 1 ppm significantly decreased the percentage of total soluble solids, as it was the lowest percentage of total soluble solids (34.90 and 34. 67), (32.98 and 32.83%) was in the fruits of the Berhi and Breim cultivars for the two seasons, respectively, while the highest percentage of total soluble solids was (43.47 and 43.72%) and (38.47 and 38.13%) for the untreated fruits of the two seasons, respectively and this is in the same line with [30]. The results also showed that the effect of the interaction between spraying with chitosan and the storage periods had a significant effect, as the lowest percentage of total soluble solids was (38.57 and 38.43), (35.04 and (34.85%) for the fruits of the two cultivars treated with chitosan at a concentration of 2% at the end of the storage period for the two seasons, respectively. The highest percentage of total soluble solids was (46.73 and 46.62), (40.16 and 39.95%) for the untreated fruits of the two cultivars for the two seasons, respectively, at the end of the storage periods. Findings are in agreement with the results obtained by [8].

The interaction between storage treatments and storage periods had a significant effect, as the lowest percentage of total soluble solids was (41.01 and 40.70), (36.11 and 35.75%) for the fruits of the Berhi and Breim treated with (1-MCP) at a concentration of 1 ppm at the end of the storage periods for the two seasons, respectively. The highest percentage of total soluble solids was (43.80 and 43.33%) and (39.38 and 39.05%) for the control fruits of the Berhi and Breim cultivars at the end of the storage periods for the two seasons, respectively. Regarding the effect of the interaction among the spraying chitosan, postharvest treatments, and storage periods, the lowest percentage of total soluble solids was (37.66 and 36.96), (33.93 and 33.50%) for the Berhi fruits sprayed with 2% chitosan and dipped in 2% chitosan and for the Breim fruits treated with (1-MCP) at a concentration of 1 ppm for the two seasons at the end of the storage period of (six months), respectively, while, the highest percentage of total soluble solids was (49.00 and 48.26%) and (42.76 and 42.43)% in the fruits of the two cultivars sprayed with 0% chitosan and control treatment at the end of the storage period, respectively.

3.3 Total sugars

The results of **Tables 4** and **5** showed the effect of spraying chitosan in the field, storage treatments and storage period, and the interaction between them on the percentage of total sugars in the fruits of the Berhi and Breim cultivars stored at a temperature of (-10 ± 2) °C for the two seasons 2014 and 2015. It is noted that spraying with field chitosan had a significant effect in reducing the percentage of total sugars where the lowest percentage of total sugars was (49.92 and 49.49), (47.21 and 47.07%) for the fruits of the mentioned cultivars that were field-treated with 2% chitosan for the two seasons, respectively, with a significant difference from the rest of the treatments, while the highest percentages of total sugars were (55.86 and 55.73), (50.98 and 50.70%) in the control fruits of mentioned cultivars for the two seasons, respectively. As for the effect of storage treatments, it was noted that it worked to reducetotal sugars percentage, which reached (51.44 and 51.18), (48.27 and 47.96%) for the fruits of the Berhi cultivar for the first season and Breim cultivar for the two seasons treated with the compound 1-MCP at a concentration of 1 ppm and at a concentration of 0.5 ppm for the Berhi cv. in the second season, respectively, with a significant difference from the control treatment, which amounted to (53.65 and 53.56%), (49.73 and 52.65%)for Berhi and Breim cultivars for the two seasons, respectively.

Field chitosan treatment	Storage treatments	2014						Field chitosan x Storage treatments	2015						Field chitosan x Storage treatments
		Storage period (month)							Storage period (month)						
		1	2	3	4	5	6		1	2	3	4	5	6	
0%	Control	53.83	54.03	56.66	58.40	60.13	62.06	57.52	53.83	54.03	56.66	58.40	60.13	62.06	57.52
	UV 5 min.	51.26	52.50	55.43	56.90	58.63	60.66	55.90	51.26	52.50	55.43	56.90	58.63	60.66	55.90
	UV 10 min	52.16	53.50	55.50	57.10	59.00	59.93	56.20	52.16	53.50	55.50	57.10	59.00	59.93	56.20
	(1-MCP) 0.5 ppm	49.73	50.36	53.83	56.00	58.30	60.00	54.70	49.50	50.06	53.43	55.66	57.96	59.66	54.38
	(1-MCP) 1 ppm	49.50	51.43	53.70	55.86	58.26	59.53	54.71	49.40	51.43	53.03	55.60	58.03	59.26	54.46
	Ozone half an hour	51.20	53.66	55.26	57.36	59.60	60.80	56.31	51.43	53.76	55.36	57.46	59.66	61.06	56.46
	Ozone one hour	51.60	53.86	55.63	57.60	59.90	61.20	56.63	51.70	53.86	55.66	57.60	59.90	61.33	56.67
	(Chitosan) 1%	50.40	52.50	55.03	56.83	58.63	60.53	55.65	49.73	52.16	54.70	56.43	58.16	60.03	55.20
	(Chitosan) 2%	50.10	52.43	54.90	56.23	57.13	60.10	55.15	49.43	52.10	54.66	55.90	56.96	59.80	54.81
1%	Control	49.76	50.96	52.13	53.93	54.93	55.93	52.94	49.43	50.30	51.66	53.36	55.26	55.60	52.60
	UV 5 min.	49.10	49.80	51.30	53.10	54.03	54.93	52.04	48.96	49.06	50.96	52.76	53.86	54.60	51.70
	UV 10 min	48.76	49.76	51.20	53.13	54.03	54.03	51.82	48.16	49.56	50.86	52.46	54.03	54.23	51.5
	(1-MCP) 0.5 ppm	47.66	48.80	50.20	52.00	53.40	53.93	51.00	47.46	48.46	50.03	51.70	53.16	53.60	50.73
	(1-MCP) 1 ppm	47.66	48.73	50.16	52.10	53.33	53.40	50.90	47.33	48.60	49.76	52.00	53.00	52.93	50.60
	Ozone half an hour	49.30	49.86	51.53	53.23	54.33	55.23	52.25	49.56	50.03	51.60	53.40	54.66	55.23	52.41
	Ozone one hour	49.53	50.00	51.80	53.50	54.53	55.50	52.47	49.86	50.10	52.13	53.56	54.53	55.50	52.61
	(Chitosan) 1%	48.60	49.13	50.66	52.30	53.60	54.53	51.47	48.13	48.83	50.26	51.96	53.13	54.33	51.11
	(Chitosan) 2%	48.03	48.90	51.10	52.16	53.30	54.20	51.28	47.76	48.90	50.86	51.86	52.96	53.83	51.03
2%	Control	47.86	49.00	50.40	51.86	52.86	54.06	51.01	48.53	48.23	50.06	51.86	52.53	53.73	50.82
	UV 5 min.	46.73	47.73	49.40	50.16	52.50	52.86	49.90	46.40	47.40	49.06	50.16	52.50	52.20	49.62
	UV 10 min	46.63	47.56	49.26	50.23	52.40	52.50	49.76	46.30	47.23	49.26	50.23	52.06	52.50	49.62

Field chitosan treatment	Storage treatments	2014						Field chitosan x Storage treatments	2015						Field chitosan x Storage treatments
		Storage period (month)							Storage period (month)						
		1	2	3	4	5	6		1	2	3	4	5	6	
	(1-MCP) 0.5 ppm	45.73	46.73	48.40	49.16	51.13	51.86	48.83	45.40	46.40	48.06	49.00	50.80	51.73	48.56
	(1-MCP) 1 ppm	45.63	46.56	48.26	49.23	51.03	51.50	48.70	45.43	46.23	46.23	49.03	50.70	51.30	48.47
	Ozone half an hour	47.00	48.06	49.56	50.33	52.63	53.10	50.11	47.23	48.40	49.90	50.33	52.66	53.26	50.30
	Ozone one hour	46.63	48.46	49.86	50.60	52.83	53.40	50.38	47.50	48.46	50.13	50.60	53.10	53.46	50.54
	(Chitosan) 1%	46.73	47.66	48.80	49.60	51.96	51.56	49.38	46.40	47.26	48.63	49.26	51.63	51.16	49.06
	(Chitosan) 2%	46.46	47.30	49.10	49.23	51.66	51.46	49.20	46.26	47.06	48.76	48.90	51.33	50.76	48.85
								Mean of field chitosan treatment							Mean of field chitosan treatment
Field chitosan x Storage period	0%	51.08	52.700	55.107	56.922	58.844	60.537	55.86	50.93	52.6	54.93	56.78	58.71	60.42	55.73
	1%	48.59	49.374	50.996	52.656	53.963	54.581	51.69	48.51	49.31	50.90	52.56	53.8	54.42	51.59
	2%	48.38	47.519	49.156	50.048	52.041	52.370	49.92	46.60	47.40	48.89	49.93	51.92	52.23	49.49
								Mean of storage treatments							MEAN of field chitosan treatment
Storage treatments x Storage period	control	50.60	50.856	52.800	54.544	55.978	57.60	53.65	50.60	50.80	52.80	54.54	55.97	57.13	53.65
	UV 5 min	48.87	49.656	51.822	53.278	55.000	56.15	52.40	48.87	49.65	51.82	53.27	55.00	55.82	52.40
	UV 10 min	48.87	50.100	51.878	53.267	55.033	55.68	52.45	48.87	50.10	51.87	53.267	55.03	55.55	52.45
	(1-MCP) 0.5 ppm	47.71	48.633	50.811	52.389	54.278	55.26	51.51	47.45	48.31	50.51	52.12	53.97	55.00	51.23
	(1-MCP) 1 ppm	47.60	48.911	50.711	52.400	54.211	54.81	51.44	47.38	48.75	50.31	52.21	53.91	54.50	51.18
	Ozone half an hour	49.16	50.533	52.122	53.644	55.522	56.37	52.89	49.41	50.73	52.28	53.73	55.66	56.52	53.05
	Ozone one hour	49.43	50.778	52.433	53.900	55.756	56.70	53.16	49.68	50.81	52.64	53.92	55.84	56.76	53.28

Field chitosan treatment	Storage treatments	2014						Field chitosan x Storage treatments	2015						Field chitosan x Storage treatments
		Storage period (month)							Storage period (month)						
		1	2	3	4	5	6		1	2	3	4	5	6	
	(Chitosan) 1%	51.17	49.767	51.500	52.911	54.733	55.54	52.60	48.08	49.42	51.20	52.55	54.31	55.17	51.79
	(Chitosan) 2%	50.77	49.544	51.700	52.544	54.033	55.25	52.30	47.82	49.35	51.43	52.22	53.75	54.80	51.56
mean of Storage period		49.35	49.86	51.75	53.20	54.94	55.83		48.68	49.76	51.65	53.09	54.82	55.69	
R.L.S.D. 5%	Field chitosan treatment	Storage treatments			Storage period			Field chitosan x Storage treatments	Storage treatments x Storage period			Storage treatments x Storage period			Triple interaction
2014	0.09	0.17			0.14			0.29	0.24			0.42			0.72
2015	0.05	0.10			0.08			0.17	0.14			0.25			0.434

Table 4.
Effect of field chitosan spraying, storage treatments, storage period and interaction between them in the percentage of total sugars in the fruits of the Berhi cultivar stored at a temperature of (−10 ± 2) °C for the 2014 and 2015 seasons.

Field chitosan treatment	Storage treatments	2014						Field chitosan x Storage treatments	2015						Field chitosan x Storage treatments
		Storage period (month)							Storage period (month)						
		1	2	3	4	5	6		1	2	3	4	5	6	
0%	Control	49.80	50.13	50.33	51.40	54.43	56.36	52.07	49.33	49.80	50.06	50.06	54.10	56.03	51.73
	UV 5 min	49.63	49.93	50.33	51.00	52.70	54.83	51.40	49.30	49.60	49.63	50.53	52.33	54.50	50.98
	UV 10 min	49.60	49.90	50.13	51.00	53.03	54.70	51.39	49.33	49.70	49.63	50.63	52.66	54.10	51.01
	(1-MCP) 0.5 ppm	49.80	49.26	49.66	49.80	50.00	52.16	50.28	48.80	49.33	49.46	49.66	50.50	51.90	49.94
	(1-MCP) 1 ppm	49.23	49.70	49.80	50.00	50.83	52.23	50.30	49.06	49.36	49.76	49.66	50.50	51.90	50.04
	Ozone half an hour	49.66	49.86	50.13	50.40	52.73	54.13	51.15	49.87	50.03	50.13	50.56	52.73	54.33	51.27
	Ozone one hour	49.63	49.66	50.06	50.46	52.46	53.83	51.02	49.83	49.86	50.06	50.46	52.46	54.16	51.14
	(Chitosan) 1%	49.36	49.40	50.06	50.16	52.00	52.80	50.63	49.06	48.63	49.56	49.76	51.53	52.53	50.18
	(Chitosan) 2%	49.20	49.26	50.00	50.20	51.80	52.83	50.55	48.50	48.70	49.76	49.86	51.73	52.50	50.17
1%	Control	47.63	47.93	48.26	49.60	50.66	51.66	49.29	47.30	47.60	47.93	48.46	50.00	51.33	48.77
	UV 5 min	47.56	47.76	48.10	48.90	50.30	49.46	48.68	47.30	47.73	47.46	47.80	49.30	49.13	48.28
	UV 10 min	47.63	47.93	48.13	48.80	50.20	50.20	48.81	47.30	46.93	47.80	48.43	49.53	49.86	48.31
	(1-MCP) 0.5 ppm	47.20	47.23	47.66	48.00	48.96	49.53	48.10	46.86	46.90	47.33	47.66	48.63	49.20	47.76
	(1-MCP) 1 ppm	47.10	47.20	47.56	47.53	48.90	49.36	47.94	46.76	46.80	47.23	47.20	48.70	49.03	47.62
	Ozone half an hour	47.63	47.96	48.03	48.30	48.66	49.96	48.42	47.66	48.10	48.20	48.43	48.87	50.13	48.56
	Ozone one hour	47.60	47.80	47.96	48.43	48.70	49.83	48.38	47.67	47.90	48.16	48.50	48.83	50.03	48.51
	(Chitosan) 1%	47.20	47.46	47.40	47.96	48.90	50.26	48.20	46.83	47.13	47.00	47.70	48.56	49.80	47.83
	(Chitosan) 2%	47.23	47.56	47.23	47.83	48.73	50.10	48.11	46.86	47.16	46.90	47.56	48.56	49.86	47.82
2%	Control	46.60	46.83	47.30	48.26	49.00	50.93	48.15	46.40	46.50	46.96	48.13	48.90	50.60	47.91
	UV 5 min	46.66	46.90	47.06	47.33	48.23	49.16	47.56	46.33	46.90	46.73	47.06	47.90	48.83	47.29
	UV 10 min	46.70	46.73	46.96	47.30	48.23	48.86	47.46	46.36	46.40	46.96	47.23	48.23	48.53	47.28

Field chitosan treatment	Storage treatments	2014						Field chitosan x Storage treatments	2015						Field chitosan x Storage treatments
		Storage period (month)							Storage period (month)						
		1	2	3	4	5	6		1	2	3	4	5	6	
	(1-MCP) 0.5 ppm	46.10	46.66	46.33	46.40	47.70	47.60	46.80	45.76	46.33	46.00	46.06	47.36	47.10	46.43
	(1-MCP) 1 ppm	46.10	46.23	46.23	46.36	47.03	47.53	46.58	45.76	45.90	45.90	45.96	46.70	47.13	46.22
	Ozone half an hour	46.60	46.86	47.16	47.66	48.13	48.66	47.51	46.76	46.90	47.23	47.00	48.10	48.86	47.61
	Ozone one hour	46.56	46.83	47.06	47.46	48.33	48.66	47.44	46.66	47.00	47.20	47.50	48.40	48.53	47.55
	(Chitosan) 1%	46.23	46.30	46.43	46.76	48.20	48.70	47.10	45.96	45.96	46.10	46.60	47.86	48.00	46.75
	(Chitosan) 2%	46.26	46.26	46.33	46.60	48.03	48.70	47.03	46.03	45.96	46.03	46.43	47.50	48.50	46.74
								Mean of field chitosan treatment							Mean of field chitosan treatment
Field chitosan x Storage period	0%	49.48	49.72	50.07	50.51	52.31	53.76	50.98	49.23	49.44	49.78	50.13	52.06	53.55	50.70
	1%	47.42	47.65	47.81	48.37	49.30	49.97	48.42	47.17	47.36	47.55	47.971	48.99	49.81	48.14
	2%	46.35	46.57	46.66	47.01	48.03	48.64	47.21	46.22	46.42	46.56	46.88	47.88	48.45	47.07
								Mean of storage treatments							Mean of storage treatments
Storage treatments x Storage period	control	47.90	48.19	48.52	49.64	51.33	52.81	49.73	47.67	47.96	48.32	49.22	51.00	52.65	49.47
	UV 5 min	47.93	48.16	48.41	48.98	50.26	51.07	49.13	47.64	48.07	47.94	48.80	49.84	50.82	48.85
	UV 10 min	47.91	48.18	48.30	48.90	50.38	51.03	49.11	47.67	47.67	48.13	48.76	50.14	50.83	48.87
	(1-MCP) 0.5 ppm	47.52	47.85	47.93	48.13	49.16	49.76	48.39	47.14	47.52	47.60	47.80	48.83	49.40	48.05
	(1-MCP) 1 ppm	47.47	47.71	47.86	47.96	48.92	49.71	48.27	47.20	47.35	47.63	47.61	48.63	49.35	47.96
	Ozone half an hour	47.96	48.23	48.44	48.78	49.84	50.92	49.03	48.10	48.34	48.52	48.93	49.90	51.11	49.15
	Ozone one hour	47.93	48.10	48.36	48.78	49.83	50.68	48.95	48.05	48.25	48.47	48.82	49.90	50.91	49.07
	(Chitosan) 1%	47.60	47.72	47.96	48.30	49.70	50.58	48.64	47.28	47.24	47.55	48.02	49.32	50.11	48.25
	(Chitosan) 2%	47.56	47.70	47.85	48.21	49.52	50.54	48.56	47.13	47.27	47.56	47.95	49.26	50.28	48.24

Field chitosan treatment	Storage treatments	2014						Field chitosan x Storage treatments	2015						Field chitosan x Storage treatments
		Storage period (month)							Storage period (month)						
		1	2	3	4	5	6		1	2	3	4	5	6	
Mean of Storage period		47.75	47.98	48.18	48.63	49.88	50.79		47.54	47.74	47.96	48.43	49.64	50.60	

R.L.S.D. 5%	Field chitosan treatment	Storage treatments	Storage period	Field chitosan x Storage treatments	Storage treatments x Storage period	Storage treatments x Storage period	Triple interaction
2014	0.05	0.10	0.08	0.17	0.14	0.25	0.43
2015	0.05	0.10	0.08	0.17	0.14	0.25	0.43
2015	0.03	0.06	0.05	0.11	0.09	0.15	0.27

Table 5.
Effect of field chitosan spraying, storage treatments, storage period and interaction between them in the percentage of total sugars in the fruits of the Breim cultivar stored at a temperature of (-10 ± 2) *°C for the 2014 and 2015 seasons.*

The storage period had a clear effect, as it was noted from the two mentioned tables that the percentage of total sugars increases with the increment in the storage period, where the highest percentages of total sugars reached (55.83 and 55.69), (50.79 and 50.60%) for the fruits of the two seasons after six months of storage, while the lowest percentages of total sugars were (49.35 and 48.68), (47.75 and 47.54%) for the fruits of the mentioned cultivars for the two seasons, respectively, after one week of storage. The reason may be due to that the percentage of total sugars increases by decreasing the percentage of water content in the fruits [31]. As for the effect of the interaction between spraying with chitosan in the field and storage treatments, the results indicated that the fruits treated with 2% chitosan and stored with the compound (1-MCP) at a concentration of 1 ppm worked significantly in reducing the percentage of total sugars, where it was the lowest percentage of total sugars (48.70 and 48.47), (46.58 and 46.22%) for the fruits of the studied cultivars for the two seasons respectively, while the highest percentages of total sugars were (57.52 and 57.52), (52.07 and 51.73%) for the control fruits of the studied cultivars.

The results also showed that the interaction between spraying chitosan in the field and the storage period had a significant effect, as the lowest percentage of total sugars was (52.37 and 52.23), (48.64 and 48.45%) for the fruits of the Berhi and Breim cultivars treated in the field with 2% chitosan at the end of the storage period for the two seasons respectively. The highest percentage of total sugars was (60.53 and 60.42) (53.76 and 53.55%) for the control fruits of the two cultivars for the two seasons, respectively after six months of storage. The results showed that the interaction between storage treatments and storage period had a significant effect, as the lowest percentage of total sugars reached (54.81 and 54.50), (49.71 and 49.35%) for the fruits of the studied cultivars treated with the compound (1-MCP) at a concentration of 1 ppm at the end of the storage period for the two seasons respectively. The highest percentage of total sugars was (57.60 and 57.13%) and (52.81 and 52.65%) for the control fruits of the studied cultivars, for the two seasons, respectively after six months of the storage.

The interaction among the three factors (spraying chitosan in the field, storage treatments, and storage period) had a significant effect. It was noted that the highest percentage of total sugars were (62.06 and 62.06%), (56.36 and 56.03%) in the pre and post-untreated fruits of the Berhi and Breim cultivars after six months of storage for the two seasons respectively.

3.4 Total titratable acidity

The results of **Tables 6** and 7 showed the effect of spraying chitosan in the field, storage treatments and storage period, and the interaction among them on the percentage of total titratable acidity in the fruits of the Berhi and Breim cultivars stored at a temperature of $-10 \pm 2°C$ for the two seasons 2014 and 2015. It is noted that field chitosan spraying had a significant effect on the preservation of the total titratable acidity percentage, where the highest percentage of total titratable acidity was (0.293 and 0.275), (0.287 and 0.313)% for the fruits of the two cultivars Berhi and Breim, field-treated with 2% chitosan for the two seasons, respectively, with a significant difference from the rest of the treatments, while the lowest percentage reached to (0.244 and 0.230), (0.246 and 0.246%) in the control fruits of Berhi and Breim cultivars for the two seasons, respectively. The results are consistent with [29], which indicated that the effect of pre-harvest chitosan spraying in all treatments led to an increase in acidity compared to the control treatment, except for the concentration of

Field chitosan treatment	Storage treatments	2014						Field chitosan x Storage treatments	2015						Field chitosan x Storage treatments
		Storage period (month)							Storage period (month)						
		1	2	3	4	5	6		1	2	3	4	5	6	
0%	Control	0.203	0.206	0.196	0.193	0.190	0.186	0.196	0.170	0.166	0.156	0.153	0.150	0.146	0.157
	UV 5 min	0.216	0.210	0.213	0.206	0.206	0.196	0.208	0.176	0.170	0.173	0.166	0.166	0.156	0.168
	UV 10 min	0.230	0.220	0.226	0.220	0.213	0.210	0.220	0.190	0.180	0.186	0.180	0.173	0.170	0.180
	(1-MCP) 0.5 ppm	0.267	0.253	0.240	0.235	0.230	0.223	0.241	0.290	0.285	0.275	0.270	0.250	0.243	0.269
	(1-MCP) 1 ppm	0.250	0.248	0.241	0.233	0.236	0.220	0.238	0.287	0.270	0.266	0.253	0.256	0.240	0.262
	Ozone half an hour	0.240	0.330	0.323	0.250	0.230	0.223	0.266	0.220	0.310	0.303	0.230	0.210	0.203	0.246
	Ozone one hour	0.170	0.340	0.320	0.240	0.226	0.216	0.252	0.150	0.320	0.300	0.220	0.206	0.196	0.232
	(Chitosan) 1%	0.206	0.196	0.290	0.240	0.230	0.216	0.230	0.186	0.176	0.270	0.220	0.210	0.196	0.210
	(Chitosan) 2%	0.203	0.206	0.253	0.206	0.226	0.216	0.218	0.183	0.186	0.233	0.186	0.206	0.196	0.198
1%	Control	0.223	0.220	0.216	0.213	0.210	0.203	0.214	0.183	0.180	0.176	0.173	0.170	0.163	0.174
	UV 5 min	0.250	0.240	0.236	0.236	0.230	0.220	0.235	0.210	0.200	0.196	0.196	0.190	0.180	0.195
	UV 10 min	0.253	0.243	0.240	0.240	0.240	0.230	0.241	0.213	0.203	0.200	0.200	0.200	0.190	0.201
	(1-MCP) 0.5 ppm	0.320	0.300	0.236	0.270	0.236	0.236	0.266	0.340	0.320	0.256	0.290	0.256	0.256	0.286
	(1-MCP) 1 ppm	0.280	0.286	0.300	0.283	0.243	0.213	0.273	0.300	0.306	0.320	0.303	0.263	0.270	0.293
	Ozone half an hour	0.306	0.300	0.236	0.276	0.233	0.236	0.265	0.286	0.280	0.216	0.256	0.213	0.216	0.245
	Ozone one hour	0.280	0.286	0.290	0.280	0.250	0.243	0.271	0.260	0.266	0.270	0.260	0.230	0.223	0.251
	(Chitosan) 1%	0.306	0.300	0.236	0.276	0.233	0.236	0.265	0.286	0.280	0.216	0.256	0.213	0.216	0.245
	(Chitosan) 2%	0.280	0.286	0.300	0.280	0.250	0.243	0.273	0.260	0.266	0.280	0.260	0.230	0.223	0.253
2%	Control	0.246	0.243	0.236	0.223	0.216	0.213	0.230	0.206	0.203	0.196	0.183	0.176	0.173	0.190
	UV 5 min	0.276	0.270	0.266	0.263	0.250	0.250	0.262	0.236	0.230	0.226	0.223	0.210	0.210	0.222
	UV 10 min	0.276	0.273	0.266	0.266	0.270	0.260	0.268	0.236	0.233	0.226	0.226	0.230	0.220	0.228

DOI: http://dx.doi.org/10.5772/intechopen.109899

Field chitosan treatment	Storage treatments	2014						Field chitosan x Storage treatments	2015						Field chitosan x Storage treatments
		Storage period (month)							Storage period (month)						
		1	2	3	4	5	6		1	2	3	4	5	6	
	(1-MCP) 0.5 ppm	0.340	0.360	0.350	0.283	0.250	0.280	0.310	0.360	0.380	0.370	0.303	0.270	0.300	0.330
	(1-MCP) 1 ppm	0.340	0.350	0.356	0.293	0.273	0.290	0.317	0.360	0.370	0.376	0.313	0.293	0.310	0.337
	Ozone half an hour	0.333	0.360	0.350	0.283	0.250	0.280	0.309	0.313	0.340	0.330	0.263	0.230	0.260	0.289
	Ozone one hour	0.340	0.350	0.350	0.290	0.273	0.273	0.312	0.320	0.330	0.330	0.270	0.253	0.253	0.292
	(Chitosan) 1%	0.346	0.360	0.350	0.283	0.250	0.280	0.317	0.326	0.340	0.330	0.263	0.230	0.260	0.291
	(Chitosan) 2%	0.340	0.366	0.350	0.286	0.280	0.283	0.317	0.320	0.346	0.330	0.266	0.260	0.263	0.297
								Mean of field chitosan treatment							Mean of field chitosan treatment
Field chitosan x Storage period	0%	0.221	0.245	0.256	0.225	0.221	0.212	0.230	0.205	0.229	0.240	0.208	0.203	0.194	0.213
	1%	0.277	0.273	0.254	0.261	0.2361	0.228	0.255	0.259	0.255	0.236	0.243	0.218	0.215	0.238
	2%	0.315	0.325	0.319	0.274	0.256	0.267	0.293	0.297	0.308	0.301	0.256	0.239	0.249	0.275
								Mean of storage treatments							Mean of storage treatments
Storage treatments x Storage period	Control	0.224	0.223	0.216	0.210	0.205	0.201	0.213	0.186	0.183	0.176	0.170	0.165	0.161	0.173
	UV 5 min	0.247	0.240	0.238	0.235	0.228	0.222	0.235	0.207	0.200	0.198	0.195	0.188	0.182	0.195
	UV 10 min	0.253	0.245	0.244	0.242	0.241	0.233	0.243	0.213	0.205	0.204	0.202	0.201	0.193	0.203
	(1-MCP) 0.5 ppm	0.360	0.331	0.305	0.267	0.238	0.246	0.291	0.380	0.351	0.325	0.287	0.258	0.266	0.311
	(1-MCP) 1 ppm	0.442	0.327	0.324	0.270	0.251	0.253	0.311	0.462	0.347	0.344	0.344	0.271	0.273	0.331
	Ozone half an hour	0.293	0.330	0.303	0.270	0.237	0.246	0.280	0.273	0.310	0.283	0.250	0.217	0.226	0.260
	Ozone one hour	0.263	0.325	0.320	0.270	0.250	0.244	0.278	0.243	0.305	0.300	0.250	0.230	0.224	0.258
	(Chitosan) 1%	0.286	0.285	0.292	0.266	0.237	0.244	0.268	0.266	0.265	0.272	0.246	0.217	0.224	0.248
	(Chitosan) 2%	0.274	0.286	0.301	0.257	0.252	0.247	0.270	0.254	0.267	0.281	0.237	0.232	0.227	0.250

Field chitosan treatment	Storage treatments	2014						Field chitosan x Storage treatments	2015						Field chitosan x Storage treatments
		Storage period (month)							Storage period (month)						
		1	2	3	4	5	6		1	2	3	4	5	6	
Mean of Storage period		0.293	0.288	0.282	0.254	0.237	0.237		0.276	0.270	0.264	0.242	0.219	0.219	
R.L.S.D. 5%	Field chitosan treatment		Storage treatments		Storage period			Field chitosan x Storage treatments		Storage treatments x Storage period			Storage treatments x Storage period		Triple interaction
2014	0.005	0.009	0.008	0.017	0.013	0.024	0.041								
2015	0.005	0.009	0.008	0.017	0.013	0.024	0.041								

Table 6.
Effect of field chitosan spraying, storage treatments, storage period and interaction between them in the percentage of total titratable acidity in the fruits of the Berhi cultivar stored at a temperature of (−10 ± 2) °C for the 2014 and 2015 seasons.

Field chitosan treatment	Storage treatments	2014						Field chitosan x Storage treatments	2015						Field chitosan x Storage treatments
		Storage period (month)							Storage period (month)						
		1	2	3	4	5	6		1	2	3	4	5	6	
0%	Control	0.243	0.253	0.253	0.240	0.220	0.213	0.237	0.263	0.273	0.273	0.260	0.240	0.233	0.257
	UV 5 min	0.280	0.260	0.196	0.230	0.196	0.196	0.226	0.340	0.320	0.256	0.290	0.256	0.256	0.286
	UV 10 min	0.240	0.246	0.260	0.243	0.203	0.210	0.233	0.300	0.306	0.320	0.303	0.263	0.270	0.293
	(1-MCP) 0.5 ppm	0.320	0.300	.236	0.270	0.236	0.236	0.266	0.360	0.340	0.276	0.310	0.276	0.276	0.306
	(1-MCP) 1 ppm	0.280	0.286	0.300	0.283	0.243	0.225	0.273	0.320	0.326	0.340	0.323	0.283	0.290	0.313
	Ozone half an hour	0.290	0.270	0.206	0.240	0.206	0.206	0.236	0.260	0.240	0.176	0.210	0.176	0.176	0.206
	Ozone one hour	0.250	0.256	0.270	0.253	0.213	0.220	0.243	0.220	0.226	0.240	0.223	0.183	0.190	0.213
	(Chitosan) 1%	0.300	0.280	0.216	0.250	0.216	0.216	0.246	0.340	0.320	0.256	0.290	0.256	0.256	0.286
	(Chitosan) 2%	0.260	0.266	0.280	0.263	0.223	0.230	0.253	0.300	0.306	0.320	0.303	0.263	0.270	0.293
1%	Control	0.250	0.306	0.273	0.246	0.203	0.213	0.248	0.270	0.326	0.293	0.266	0.223	0.233	0.268
	UV 5 min	0.380	0.293	0.290	0.210	0.190	0.183	0.257	0.440	0.440	0.350	0.270	0.250	0.243	0.317
	UV 10 min	0.366	0.306	0.276	0.193	0.196	0.180	0.303	0.726	0.366	0.336	0.253	0.256	0.240	0.363
	(1-MCP) 0.5 ppm	0.273	0.333	0.330	0.250	0.230	0.223	0.273	0.313	0.373	0.370	0.290	0.270	0.263	0.313
	(1-MCP) 1 ppm	0.270	0.346	0.316	0.233	0.236	0.220	0.270	0.310	0.386	0.356	0.273	0.276	0.260	0.310
	Ozone half an hour	0.223	0.303	0.300	0.220	0.200	0.193	0.240	0.193	0.273	0.270	0.190	0.170	0.163	0.210
	Ozone one hour	0.353	0.316	0.286	0.203	0.206	0.190	0.259	0.323	0.286	0.256	0.173	0.176	0.160	0.229
	(Chitosan) 1%	0.286	0.313	0.310	0.230	0.210	0.203	0.258	0.326	0.353	0.350	0.270	0.250	0.243	0.298
	(Chitosan) 2%	0.350	0.326	0.296	0.213	0.216	0.200	0.267	0.390	0.366	0.336	0.253	0.256	0.240	0.307
2%	Control	0.280	0.276	0.290	0.223	0.230	0.243	0.257	0.300	0.296	0.310	0.243	0.250	0.263	0.277
	UV 5 min	0.300	0.320	0.310	0.243	0.210	0.240	0.270	0.360	0.380	0.370	0.303	0.270	0.300	0.330
	UV 10 min	0.300	0.310	0.316	0.253	0.233	0.250	0.277	0.360	0.360	0.376	0.313	0.293	0.310	0.337

Field chitosan treatment	Storage treatments	2014						Field chitosan x Storage treatments	2015						Field chitosan x Storage treatments
		Storage period (month)							Storage period (month)						
		1	2	3	4	5	6		1	2	3	4	5	6	
	(1-MCP) 0.5 ppm	0.340	0.360	0.350	0.283	0.250	0.280	0.310	0.380	0.400	0.390	0.323	0.290	0.320	0.350
	(1-MCP) 1 ppm	0.340	0.350	0.356	0.293	0.273	0.290	0.317	0.380	0.390	0.396	0.333	0.313	0.330	0.357
	Ozone half an hour	0.310	0.330	0.320	0.253	0.220	0.250	0.280	0.280	0.300	0.290	0.223	0.190	0.220	0.250
	Ozone one hour	0.310	0.320	0.326	0.263	0.243	0.260	0.287	0.280	0.290	0.296	0.233	0.213	0.230	0.257
	(Chitosan) 1%	0.320	0.340	0.330	0.263	0.230	0.260	0.290	0.360	0.380	0.370	0.303	0.270	0.300	0.330
	(Chitosan) 2%	0.320	0.330	0.336	0.273	0.253	0.270	0.297	0.360	0.370	0.376	0.313	0.293	0.310	0.337
								Mean of field chitosan treatment							Mean of field chitosan treatment
Field chitosan x Storage period	0%	0.273	0.268	0.246	0.252	0.217	0.220	0.246	0.300	0.295	0.273	0.279	0.244	0.246	0.273
	1%	0.339	0.316	0.297	0.222	0.210	0.200	0.264	0.365	0.352	0.324	0.248	0.236	0.227	0.292
	2%	0.313	0.326	0.326	0.261	0.2381	0.260	0.287	0.34	0.351	0.352	0.287	0.264	0.287	0.313
								Mean of storage treatments							Mean of storage treatments
Storage treatments x Storage period	Control	0.257	0.278	0.272	0.236	0.217	0.223	0.247	0.277	0.298	0.292	0.256	0.237	0.243	0.267
	UV 5 min	0.320	0.291	0.265	0.227	0.198	0.206	0.251	0.380	0.351	0.325	0.287	0.258	0.266	0.311
	UV 10 min	0.402	0.287	0.284	0.230	0.211	0.213	0.271	0.462	0.347	0.344	0.290	0.271	0.273	0.331
	(1-MCP) 0.5 ppm	0.311	0.331	0.305	0.267	0.238	0.246	0.283	0.351	0.371	0.345	0.307	0.278	0.286	0.323
	(1-MCP) 1 ppm	0.296	0.327	0.324	0.270	0.251	0.253	0.287	0.336	0.367	0.364	0.310	0.291	0.293	0.327
	Ozone half an hour	0.274	0.301	0.275	0.237	0.208	0.216	0.252	0.244	0.271	0.245	0.207	0.178	0.186	0.222
	Ozone one hour	0.304	0.297	0.294	0.240	0.221	0.223	0.263	0.274	0.267	0.264	0.210	0.191	0.193	0.233
	(Chitosan) 1%	0.302	0.311	0.285	0.247	0.218	0.226	0.265	0.342	0.351	0.325	0.287	0.258	0.266	0.305
	(Chitosan) 2%	0.310	0.307	0.304	0.250	0.231	0.223	0.272	0.350	0.347	0.344	0.290	0.271	0.273	0.312

Field chitosan treatment	Storage treatments	2014						Field chitosan x Storage treatments	2015						Field chitosan x Storage treatments
		Storage period (month)							Storage period (month)						
		1	2	3	4	5	6		1	2	3	4	5	6	
Mean of Storage period		0.308	0.303	0.290	0.245	0.222	0.227		0.335	0.330	0.316	0.271	0.248	0.253	

R.L.S.D. 5%	Field chitosan treatment	Storage treatments	Storage period	Field chitosan x Storage treatments	Storage treatments x Storage period	Storage treatments x Storage period	Triple interaction
2014	0.005	0.009	0.008	0.017	0.013	0.024	0.041
2015	0.005	0.009	0.008	0.017	0.013	0.024	0.041

Table 7.
Effect of field chitosan spraying, storage treatments, storage period and interaction between them in the percentage of total titratable acidity in the fruits of the Breim cultivar stored at a temperature of (-10 ± 2) *°C for the 2014 and 2015 seasons.*

1% chitosan, which decreased this treatment with no significant differences in both seasons. Li and Yu (2000) found a decrease in the acidity of peach fruits during the storage period and at the end of the storage period, and an increase in acidity on fruits treated with chitosan, while in other fruits such as mango, the acidity decreased slowly, and linked this decrease with loss of quality [32, 33].

As for the effect of storage treatments, the highest value of acidity percentage was (0.311 and 0.3931), (0.287 and 0.327%) for the fruits of the two cultivars Berhi and Breim treated with the compound (1-MCP) at a concentration of 1 ppm for the two seasons respectively with a significant difference from the control treatment which amounted to (0.213, 0.161%) and (0.247, 0.222%) for Berhi and Breim cultivars for the two seasons respectively, except for Breim cultivar for the second season. The storage period had a clear effect, as it was noted from the mentioned table that the percentage of total titratable acidity decreased, with the increment of the storage period, where the lowest percentage of total titratable acidity reached (0.222, 0.219) and (0.227, 0.248%) for Berhi and Breim fruits after 5 months of storage for the two seasons, respectively. As for the effect of the interaction between spraying chitosan in the field and storage treatments, the results indicated that the fruits treated in the field with 2% chitosan and stored with the compound (1-MCP) at a concentration of 1 ppm have worked to maintain the highest percentage of total titratable acidity (0.317, 0.337) and (0.317, 0.357%), while the lowest percentage of total titratable acidity was (0.196, 0.157) and (0.226, 0.213)% for the fruits of the Berhi cultivar treated with 0% chitosan in the field for the control treatment for the two seasons and for the fruits of the Breim cultivar treated with ultraviolet rays for (5) minutes for the first season and with the compound (1-MCP) at a concentration of 1 ppm for the second season respectively.

The results also showed that the effect of the interaction between spraying chitosan in the field and the storage period had a significant effect, as the highest percentage of total titratable acidity reached (0.260, 0.287) and (0.267, 0.249%) for the fruits of the Berhi and Breim cultivars treated in the field with 2% chitosan at the end of the storage period for the two seasons respectively. As for the lowest percentage of total acidity, it was (0.222, 0.227) and (0.211, 0.194%) for the fruits of the Berhi cultivar field-treated with chitosan at a concentration of 1% and for the fruits of the Breim cultivar for the comparison treatment at the end of the storage period. The results also showed that the effect of the interaction between the storage treatments and the storage period had a significant effect, as the highest percentage of total titratable acidity was (0.253, 0.273) and (0.253, 0.293%) for the fruits of the Berhi cultivar treated with the compound (1-MCP) at a concentration of 1 ppm at the end of the storage period for the two seasons, respectively.

The effect of the interaction between the three factors was spraying chitosan in the field, storage treatments, and storage period. It was noted that the highest percentage of total titratable acidity was (0.290, 0.310) and (0.290, 0.330%) for the fruits of the Berhi and Breim cultivars treated with 2% chitosan and with the compound 1-MCP0 at a concentration of 0.5 ppm at the end of the storage period for the two seasons respectively, while the lowest percentage of total acidity was (0.186, 0.146) for the fruits of the Berhi cultivar treated in the field with chitosan at a concentration of 0% for the comparison treatment at the end of the storage period for the two seasons respectively, and (0.180, 0.176%) for the fruits of Breim cultivar treated in the field with chitosan at a concentration of 1% and UV rays for 10 minutes for the first season and field treated with chitosan at a concentration of 1% and ozone for one hour for the second season at the end of the storage period.

The results of the present study indicate the role of the treatments in improving the qualitative characteristics of date palm fruits of the two cultivars, Berhi and Breim, which were stored by freezing. No doubt that preserving the palm fruits in the rutab stage (fresh stage) after harvesting is one of the priorities of the technology of storing these fruits, especially the soft ones such as Berhi and Breim, which are characterized by an excellent flavor as well as price is higher compared to other cultivars. Refrigerated storage of fruits, in principle, aims to reduce the vital activities that occur [31] in fruits, especially the process of respiration [2]. In addition to limit the growth of microorganisms, especially fungi. The studies showed that the high temperatures after harvesting, and during storage lead to an acceleration of physiological processes, increase affection of pathogens, and the speed of consumption of food stored in the fruis, thus storage ability decreases.

The process of ripening fruits as mentioned by [8] is a series of changes in the color, taste, and composition making fruits in an edible state, as is known, the process

a

b

Figure 4.
(a, b) Berhi fruits after six months of storage at -10 ± 2 *°C.*

of ripening is a complex process in which many factors interaction, making the fruits finally edible. Concerning the date palm fruits, the changes that occur at maturity are identical to those that occur in the climacteric fruits, which is closely related to changes in respiratory rate. Khalal stage has been considered as the maturity stage (completeness of growth, while the rutab stage is the stage of ripening. Undoubtedly, controlling the ripening process requires first lowering the temperature, as low temperatures slow down respiration, ethylene production, and vital activity of fruits, especially the enzymatic activity [2]. Results in the same line with [34] who mentioned that the low temperature (0 °C) led to a decrease in the respiration rate of date palm fruits, cv. Breim, and no climacteric rise was observed in them climatically, while it was observed in the stored fruits at room storage temperature.

c

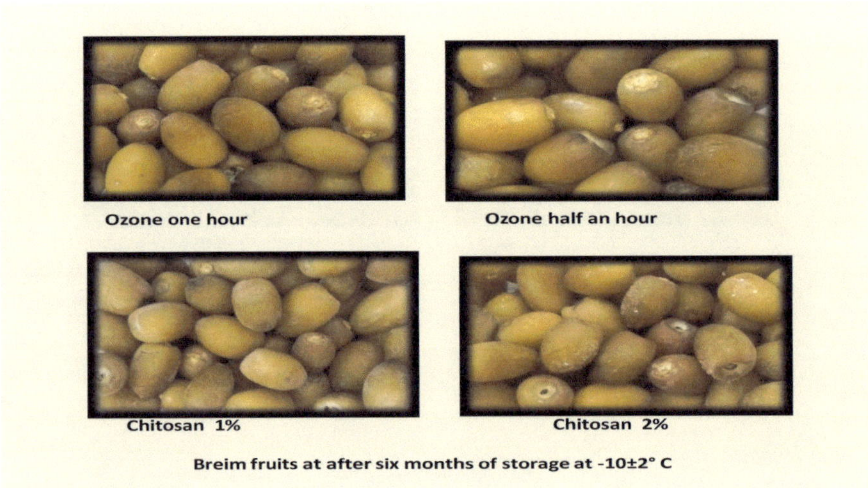

d

Figure 5.
(c, d). Breim fruits after six months of storage at -10 ± 2 *°C.*

4. Conclusions and recommendations

1. The spraying of pre-harvest chitosan improved the storage ability of the Berhi and Breim fruits.

2. The storage treatments (ultraviolet rays, ozone, 1-MCP and immersion in chitosan) improved the storage ability of the fruits.

3. Through the study of the protein pattern, new proteins were identified during storage, as well as a distinction between the two cultivars after pressing and packing the fruits, so we recommend using this technique for the purpose of detecting fraud cases in palm fruits of different cultivars, especially after pressing and packing dates.

4. The treatments that were used in the experiment are considered natural alternatives, so we recommend using them before and after harvest to increase the yield, improve its qualitative characteristics, and improve the storage ability of the fruits, as shown in **Figures 4** and 5.

Author details

Dhia Ahmed Taain*, Hamzah Abbas Hamzah and Abbas Mahdi Jasim
Department of Horticulture and Landscape Design, College of Agriculture, University of Basrah, Iraq

*Address all correspondence to: golden_fruitb@yahoo.com

References

[1] Kahramanoglu I, Usanmaz S. Preharvest and postharvest treatments for increasing the rate of ripening of date palm fruit (Phoenix dactylifera L.) cv. Medjool Progress in Nutrition. 2019; **21**(1):215-224

[2] Taain DA. Study on physico-chemical and physiological characteristics of date palm fruits (Phoeni dactylifera L.) CV. Um-Aldehn. Pakistan Journal of Agricultural Science. 2013;**50**(1):1-5

[3] Dessouki IM, Algizawi AM, Abdel Azim MM, Ahmed S. Technology of Storage and Export of Horticultural Crops. Egypt: College of Agriculture, Ain Shams University; 2001

[4] De Costa DM, Erabadupitiya HRUT. An integrated method to control postharvest diseases of banana using a member of the Burkholderia cepacia complex. Postharvest Biology and Technology. 2005;**36**:31-39

[5] Rudrapatnam NT, Farooqahmed SK. Chitin – The undisputed biomolecules of great potential. Critical Reviews in Food Science and Nutrition. 2003;**43**: 61-87

[6] Shahidi F, Arachchi JKV, Jeon Y. Food application of chitin and chitosans. Trends in Food Science and Technology. 1999;**10**:37-51

[7] Dutta PK, Dutta J, Tripathi VS. Chitin and chitosan: Chemistry, properties and applications. Journal of Science & Industrial Research. 2004;**63**: 20-31

[8] Taain DA. Effect of the package kind and storage temperature on qualitative characteristics and storage behavior of date fruits cv. Barhi. Basrah. Journal of Date Palm Researches. 2005;**4**:54-70

[9] Blankenship SM, Dole JM. 1-Methylcyclopropene: A review. Postharvest Biology and Technology. 2003;**28**:1-25

[10] Jiang Y, Joyce DG. 1-methylcyclopropene treatment effects on intact and fresh apple. The Journal of Horticultural Science and Biotechnology. 2002;**77**:19-21

[11] Moran RE, McManus P. Firmness retention and prevention of core line browning and senescence in, Macoun, apples with 1- methylcyclopropene. Horticultural Science. 2005;**40**:161-163

[12] Pre-Aymard C, Fallik E, Wesler LS. Sensory analysis and instrumental measurements of, Anna, apple treated with 1- methylcyclopropene. Postharvest Biology and Technology. 2005;**36**:135-142

[13] Toivonen PM, Lu CW. Studies on elevated temperature, short- term storage of, sunrise, summer apple using 1-MCP to maintain quality. Horticultural Science and Biotechnology. 2005;**80**: 43-446

[14] Bintsis T, Tzanetaki EL, Robinson RK. Existing and potential applications of ultraviolet light in the food industry – A critical review. Journal of the Science of Food and Agriculture. 2000;**80**:637-645

[15] Khadre MA, Yousef AE. Sporicidal action of ozone and hydrogen peroxide: A comparative study. International Journal of Food Microbiology. 2001;**71**: 131-138

[16] Langlais B, Reckhow DA, Brink DR. Practical applications of ozone: Principle and case study. In: Ozone in Water Treatment. Michigan: Lewis Publishers; 1991

[17] Graham DM, Pariza M, Glaze WH, Newell GW, Erdman JW, Borzelleca JF. Use of ozone for food processing. Food Technology. 1997;**51**:72-75

[18] Xu L. Use of ozone to improve the safety of fresh fruit and vegetable. Food Technology. 1999;**53**:58-62

[19] Salman DD, Abadi EMK. Applications of chitosan in food processing. Iraqi Journal of Agricultural Sciences. 2009;**40**(3): 136-153

[20] Kamil JYVA, Jeon Y, Shahidi F. Antioxidant activity of chitosans of different viscosity in cooked comminuted flesh of herring(Clupea herengus). Food Chemistry. 2002;**79**: 69-77

[21] No HK, Lee S, Meyers SP. Correlation between physicochemical characteristics and binding capacities of chitosan products. Journal of Food Science. 2000;**65**(7):1134-1137

[22] Shaheen MA, El-Meleigi MA. Identification of date palm mail cultivars by the electrophoresis pattern of pollen soluble proteins. JKAU. Meteorology Environment Arid Land Agricultural Science. 1991;**2**:97-103

[23] Laemmli UK. Cleavage of structural proteins during the assembly of the heat of bacteriophage T4. Nature. 1970;**227**: 680-685

[24] Carffin DE. Purification procedures electrophoretic methods. In: Murray ED, Dentscher PJ, editors. Methods in Enzymology. Vol. 182. 1990. pp. 425-441

[25] Shirokov EP. Partical Course in Storage and Processing of Fruit and Vegetable USDA|NSF. Washington. D.C, USA; 1968. p. 161

[26] Helrich K. Official Method of Analysis. Association of Officical Analytical Chemists. Washington, D. C; 1990. p. 910

[27] Al-Rawi KM, Khalf AM. Design and Analysis of Agricultural Experiments. Iraq: Mosul Univ; 1980

[28] Taain DA, Al-Najar MA, El-Qatarani NA. Ivestigation the protein pattern of leaves and roots of Barhi and Khalas date palm (Phoenix dactylifera L.) cultivars propagated by offshoots and tissue culture techiques. Plant Cell Biotechnology and Molecular Biology. 2021;**22**(1&2):9-17

[29] El-Badawy HEM. Effect of chitosan and calcium chloride spraying on fruits quality of Florida prince peach under cold storage. Research Journal of Agricultural & Biology Science. 2012; **8**(2):272-281

[30] Taain DA, Hamza HA, Jaber FN. The effect of cultivar, chitosan and storage period on qualitative charactetristics of date palm fruits (*phoenix dactylifera* L.) and their infection with the saw-toothed grain beetle. IOP Conf. Series: Journal of Physics: Conf. Series. 2019;**1294**(9)

[31] Taain DA. The role of some plant extracts and storage temperature in improving storageability of date palm fruits cv. Dayri (*Phoenix dactylifera* L. cv. Dayri). AAB Bioflux. 2014;**6**(1):26-32

[32] Jiang Y, Li Y. LiY. Effects of chitosan coating on postharvest life and quality of longan fruit. Food Chemistry. 2001;**73**: 139-143

[33] Srinivasa PC, Baskaran R, Armes MN, Harish KV, Tharanathan RN. Storage studies of mango packed using biodegradable chitosan film. European Food Research and Technology. 2002;**215**:504-508

[34] Taain DA. Effect of Heat Treatments and Packing Method on the Storage Ability of Date Palm Fruits (Phoenix Dacdylifera L.) cv. Braim. Master's Thesis. Basra - Iraq: College of agriculture - University of Basra; 1997

Chapter 15

Smallholder Maize Farmers Need Better Storage for Food Security: An Exploratory Study over the Storage Types Used in Uganda

Anthony Tibaingana, Godswill Makombe and Tumo Kele

Abstract

Storage is a crucial link in the food supply chain. It helps to even-out fluctuations in food demand and supply. This ensures food availability during the lean periods. Despite the immense contribution of storage, a knowledge gap exists on the storage types used by smallholder maize farmers, how they are acquired, used, and their cost in Uganda. Storage affects the social and economic well-being of smallholder maize farmers. In this study, smallholder maize farmers in three districts of eastern Uganda (Iganga, Manafwa, and Katakwi) were interviewed during the maize storage season of 2014/2015. The aim was to: describe the different storage types; find out how they were acquired and used; the length of storage and the cost. The findings show that sacks were the most used storage type. Storage types were acquired through purchase; however, some were constructed by the smallholder maize farmers. Affordability and accessibility determined the storage type used. Some storage types were not used across all the districts; for example, the granary was used in two out of the three sampled districts. Thus, the findings show that maize storage is a challenge. We recommend that maize storage facilities should be improved with affordable to the farmers.

Keywords: maize, storage, storage characteristics, smallholder maize farmers

1. Introduction

Maize (*Zea mays L.*) is among the most important cereal crops produced in the world [1–3]. Maize is the most popular temporary crops grown in Uganda besides, beans and cassava, with more than 50% of the agricultural households involved in their cultivation. Small scale farmers who constitute the bulk (80%) of the rural poor also account for the largest share of maize production. It is grown in every part of the country and a direct source of livelihood to over 2 million households. It is the most popular contributor to income and food for smallholder farmers and for institutions like the police, the army, prisons, schools, and hospitals. Thus, maize act as the main source of food for such institutions and contributes over 11% of the caloric intake [4–6].

In sub-Saharan Africa (SSA), maize forms an important part of the staple food supplies [3, 7]. In this regard, maize is a strategic crop [8]. Although important, its production is dominated by smallholder farmers [9]. As a popular food, maize is eaten by humans as well as used to brew local alcohol, used as feeds for animals, and birds, it is prudent that we understand its storage. Storage has been a challenge to smallholder farmers and greatly impacts on their income and food security. Hence, its storage becomes crucial. Therefore, the core of the thesis of this chapter stems from the dearth of descriptions of storage systems used by smallholder maize farmers, how they are acquired and used, and at what cost [10].

Udo and Vorotnikova [11, 12] argue that the challenge of grain storage affects farmers and merchants simultaneously. Consequently, grain storage is among the key constraints to food and income security in SSA [13]. These challenges sometimes compel smallholder maize farmers to harvest late [8]. After harvest, maize is dried and stored at the household level to provide food, cash, and seed for the next planting season [14].

Although maize is a subsistence crop which is produced seasonally, its demand is throughout the year [15]. Moreover, in some seasons, the harvest is poor because production in the tropics depends mainly on climatic conditions [16]. As a result, maize supply fluctuates widely and production hardly meets the high demand [9]. In some parts of the tropics, there is only one production season, which makes storage important for ensuring food supply during the off-season [17]. Maize storage at household level is critical due to its dual purpose of food and income security [18, 19]. Because little is written about the maize storage types in SSA, it became paramount to describe them [20]. This exploration was therefore necessitated by the need to describe the storage types used, how they are acquired, and why they are used in the Ugandan context.

Brennan [21] posits that "since production is not stable for all commodities, consumers demand that the storage function be performed so that the flow of commodities for sale will be made relatively stable". Wright [22] argues that storage is important in balancing consumption, supply and stock. In other words, storage occupies a crucial position in the socio-economic development of smallholder maize farmers. Therefore, maize storage is desired in guaranteeing domestic income and food supply [23].

In Uganda, as in other SSA countries, maize is a staple food [24] whose storage greatly moderates supply fluctuations from one season to another [8]. In addition, it makes a considerable input to diets of many rural and urban populations [25]. Govender et al. [23] posit that maize is a major source of energy and protein in the human diet. Therefore, maize needs to be stored well to preserve it against thieves, pests and diseases. Storage may, therefore, lead to quantity equalisation and market stabilisation [26]. Moreover, good storage of the seeds is critical to smallholder maize farmers because it determines the quality and quantity of maize to be produced in the subsequent seasons.

However, when maize is improperly stored, numerous challenges are encountered; for example, it can be destroyed by pests such as *Sitophiluszeamais*, it can rot due to rat damage, or it can be stolen. Maize storage among smallholder farmers is affected by temperature, moisture, carbon dioxide, oxygen, grain characteristics, micro-organisms, insects, mites, rodents, birds, geographical location, and the storage structure [23]. Consequently, smallholder farmers are often forced to sell off their maize immediately after harvest and later buy it back at higher prices for household consumption because of the lack of adequate and effective storage facilities [27].

All such risks affect the quality and quantity of maize produced. Prevention of pest damage during storage is critical for the availability of maize and its continued supply during the off-season [3]. Therefore, smallholder maize farmers struggle to ensure that maize is kept well for future use [26]. Proctor [15] argues that the only way the costs of storage may be reduced is by having a full understanding of the different storage types.

Although maize storage has evolved over the years and is considered a key component of economic development, storage is still poor and traditional in most developing countries [13]. Consequently, insect pests like *Sitrotogacerealella* (Olivier), *Plodiainterpunctella* (Hübner, 1813), *Sitophiluszeamais* (Motschulsky), *Rhyzoperthadominica* (F.) and larger grain borer, *Prostephanustruncatus* (Horn), cause enormous maize loss during storage. The same consequences are also reported by [8, 25]. The poor storage exacerbates post-harvest maize losses to levels that require attention [28]. The losses lead to a reduction in the quantity and quality of maize available on the market [29]. Maize scarcity is therefore not caused by low production per say, but mainly by wastage during storage.

Although smallholder farmers are the main producers of maize, they still use traditional storage methods. Hence, it is not always economically feasible for them to keep their maize long enough and in sufficient quantities to take full advantage of price changes in the marketplace [28]. Indeed, [25, 30] estimated that post-harvest loss due to pests in three months of storage stands at 20–30% in SSA. While fulfilling food demand is a global challenge, the loss in storage is still very high and has made it harder to meet the demand [7]. In Uganda, 63% of post-harvest maize losses are related to storage [31].

The losses of maize grain have far-reaching effects on the smallholder farmers leading to reduction in maize available for family consumption, deterioration in nutritional maize quality, and disruption of maize supply at household level, thus forcing farmers to buy maize later at higher prices. It also impacts on local maize processing and cross-border trade, leading to losses in food, revenue and profit, especially in rural areas. A study conducted in Uganda by [32] established that 83% of smallholder maize farmers sell off their maize within two months after harvest. This has forced some smallholder farmers in Katakwi district to eat tree leaves due to famine.

Although evidence of maize losses is glaring, little is known of the storage types, acquisition processes, and the reasons for use of particular storage methods among smallholder maize farmers. Indeed, in spite of the unprecedented contribution of storage, the description of storage types used has received little attention. No study so far has described the storage types used by smallholder maize farmers in Uganda. This is the void that this study intends to fill.

2. Types of maize storage

Globally, smallholder maize farmers use different storage methods. Post-harvest losses resulting from poor storage are considered highest in grain [2]. In a bid to address the loss of maize due to poor storage, smallholder maize farmers have adopted different storage methods. Across the African continent, storage methods tend to vary even within the same country [28, 33]. Specifically, the variation in storage types among smallholder maize farmers are due to social and economic conditions, cultural traditions, production scale, and climatic zones [28]. The preference of one storage type over another is based on the perceived relative safety of the storage type. Safety means that the storage system is capable of preserving the quality and quantity of the grain against damage by pests and by poor storage conditions [34].

In this study, some storage types are considered traditional, while others are seen as modern. Traditional storage is the predominant type used by smallholder maize farmers [15, 34, 35]. These traditional methods include the granary, the basket, the crib, the house roof, the house corner, above-the-fire, clay pots, old jerry-cans, tins, and sacks. However, some smallholder maize farmers who are slightly better-off financially have adopted modern storage types such as metal silos, triple bags, modern cribs, and warehouses.

Stored maize is used for various purposes: selling, domestic consumption, planting, and brewing beer which is consumed by the household or sold. Of all the storage types used, sacks were predominantly found to be used for storage of maize for consumption and selling, while above-the-fire storage was mainly used for storage of maize seed for planting (because above-the-fire maize does not get weevilled). However, the effectiveness of storage depends on various ecological conditions of the storage, the nature and structure of the storage facility, how long the maize will stay in storage, the pesticides used (if any), and the characteristics of the maize to be stored [17].

Despite the various efforts directed towards storage, smallholder maize farmers still face significant losses resulting from inappropriate and/or ineffective storage practices [2]. Coupled with the inappropriate and or ineffective storage practices is the high cost of pesticides, which are necessary for the protection of maize from damage. In most developing countries, pesticides, besides being potentially adulterated, are prohibitively expensive for smallholders, most of whom earn less than USD1.00 per day, the internationally accepted minimum to keep one above the poverty line [36]. Unfortunately, solutions to post-harvest losses require modest financial investment, which the majority of smallholder maize farmers cannot afford [7].

There is a dearth of literature describing the storage types used by smallholder maize farmers [20]. However, some studies show that the dominant form of storage used by smallholder maize farmers include: baskets, clay pots, jute bags, polyethylene bags, open and closed cribs, house corners, tins, open-fields, platforms, trees, jerry-cans, granaries, and roofs of houses [2, 13, 25, 26, 32, 33]. The storage types used by most smallholder farmers in SSA are traditional and crude; they cannot enable smallholder maize farmers to store and then sell when prices are most attractive [26].

Maximum losses in maize are known to occur during storage due to inappropriate storage infrastructure [7]. Moreover, good hygiene in storage is known to limit insect infestation [8]. Although in the studies mentioned above some storage types are identified, there was no attempt to describe how they are made and/or acquired. Also, the reasons for the preference of one storage type over another were not reported, and yet it is pertinent.

Indeed, post-harvest maize storage losses resulting from traditional storage types is reported to range between 9% and 40% globally and 20% and 40% in Africa [2]. The storage structures used are susceptible to damage by natural calamities thereby causing considerable loss every season. When maize is lost in storage, other resources such as water, land, nutrients, inputs and labour that were used to produce the maize are lost too [7]. Therefore, [2] argue that effective storage is critical in order to keep adequate quantities of good quality maize grain until the next harvest. Accordingly, it is important for us to understand the characteristics of the storage types if we are to improve on maize storage by smallholder farmers in Uganda. It is envisaged that reduced post-harvest grain losses could increase smallholder food availability, subdue pressure on natural resources, and improve smallholder maize farmers' livelihoods [7].

More important to note is the fact that as food becomes scarce due to global warming, deliberate effort should be directed to storage if we are to reduce excessive losses. For example, the Famine Early Warning System Network (FEWS NET) in its report on Uganda's food security outlook June 2016 to January 2017, indicated that poor households would spend nearly 100% of their income on food, compared to 55% spent in the previous year.

2.1 Justification of the study

Knowledge about the characteristics of storage types is still limited specifically in Uganda and generally in SSA. This study focused on household storage because it is what most smallholder maize farmers in Uganda practice. In addition, there are very few studies on storage and only 5% of research funding in the previous years has been directed to post-harvest storage [7]. Moreover, about USD one trillion is lost in post-harvest food losses every year globally, while about USD four billion is lost in SSA. In east and southern Africa, the postharvest grain loss are valued at US$1.6 billion/year, or about 13.56% of the total value of grain production in the region. This could potentially reach nearly US$4 billion/year in SSA out of an estimated annual value of US$27 billion. Also, most studies on post-harvest storage have focused on losses and the issues and challenges surrounding the processes; only few studies have focused on describing storage characteristics.

Hence, the working hypothesis is that this knowledge is required to develop strategies that will improve maize storage at the household level. Therefore, the aim of this study is to describe the storage types used with the view of creating a clear understanding of the different storage characteristics. The central argument is that despite the importance of storage as a basis of food and income security for smallholder maize farmers, the storage structures used have not received the attention they deserve as shown by the few studies have attempted to describe them [20]. The dearth of information on the description, acquisition and why, how often and for how long the particular storage types are used, merits investigation in the Ugandan context.

3. Material and methods

3.1 Study area

Uganda is divided into four administrative regions, namely: Northern, Eastern, Western and Central [37]. The study was conducted in three districts of Eastern Uganda between January and May 2014/2015. The area covers the major maize producing region of Uganda.

3.2 Sampling

In a cross-sectional survey, interviews were conducted to collect comprehensive data on maize storage among smallholder maize farmers in three districts of Eastern Uganda. The Eastern region was purposively selected because, based on Uganda Census of Agriculture (UCA) 2008/2009 [38], it produces the highest volume of maize in the country. Three high (Iganga), medium (Manafwa) and low (Katakwi) ranking maize producing districts were selected for the study based on national ranking of maize production by volume (see coloured areas in **Figure 1**). While the highest and lowest maize producing districts were easily identified, for the medium producing district we chose the one whose production volume was closest to the average production of all the districts in the region.

Using maize production volume, we selected three sub-counties from each district. Thereafter, we contacted agricultural extension workers from each sub-county to provide registers of all smallholder maize farmers in their respective sub-counties. For the purpose of achieving gender balance in the sample, the lists were rewritten separating male from female smallholder maize farmers in each sub-county. Thereafter, the names were written on small pieces of paper which were then folded and put into two separate containers for each sub-county

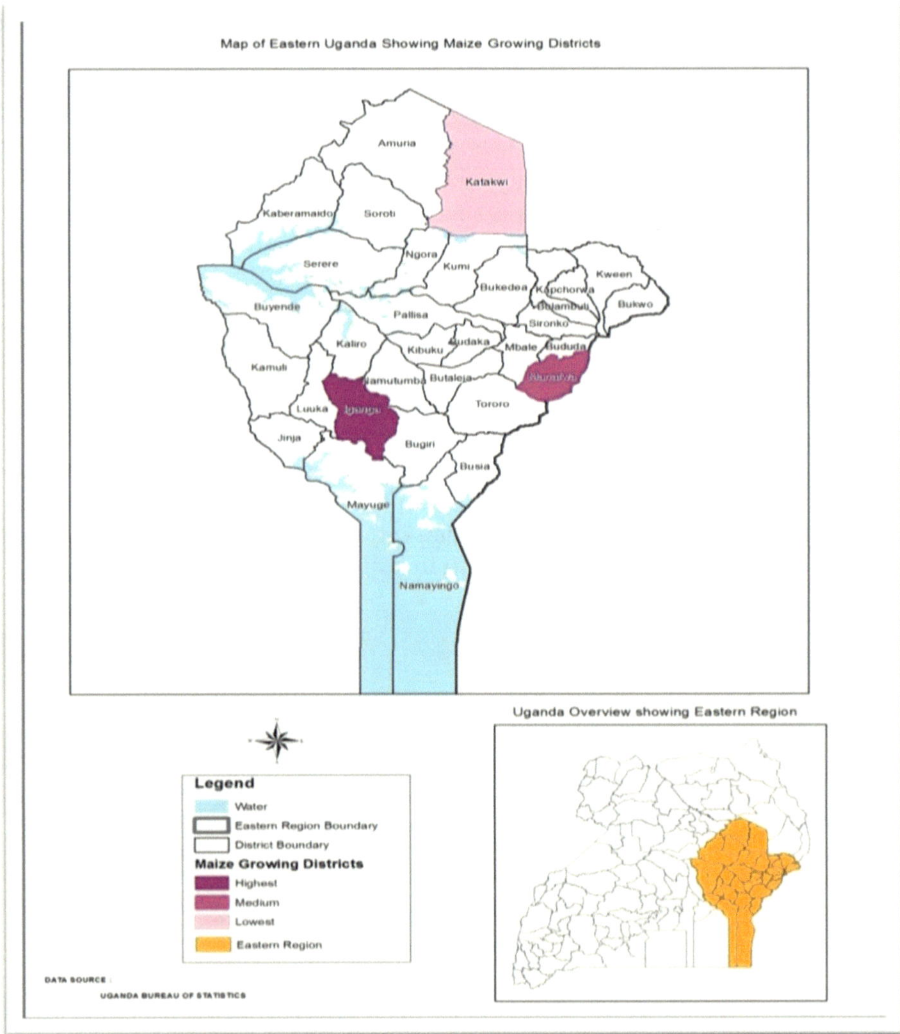

Figure 1.
Map of eastern region showing sampled districts.

containing names of male and female smallholder maize farmers. These were shaken to ensure proper mix. Then the researcher closed his eyes and used his hand to randomly pick a folded paper from each of the containers. This was done several times to align the respondents with the sample. Then the papers picked from the two containers were unfolded and the names were read to identify the respondents. Upon completing the reading of the names, the agriculture extension worker for each sub-county was asked for their contacts. The key respondents were later called, and appointments were fixed for interviews. This was done to avoid selection bias for interviewees.

In total, there were twenty-four interviews: six per district, and two per sub-county. Since there is no agreed sample size for qualitative data collection [39], the study adopted [40] view of 12-and-above participants. These were contacted for appointment and later briefed about the aim of the study and consent sought for their participation. The interviews were conducted at the smallholder maize farmers' homes. Where a farmer scheduled for an interview was absent from his or her home, the interview was rescheduled to another agreed day and time. This was important for courtesy and to increase the response rate.

3.3 Data collection procedure

Prior to the interview, each respondent was given a consent form to complete to confirm their willingness to participate in the interview [41]. Permission to record the conversation was sought and obtained. Respondents were also informed that they could stop the interview if they felt uncomfortable with the questions being asked. Each interview session started with an interviewee confirming that the respondent was responsible for maize storage in the household. An interview guide with open-ended questions was used, in order to avoid limiting the farmer's response and to stimulate the discussion [42]. The interviews were conducted in English and, where it became complicated, respondents were free to talk in their local languages. The researcher or a research assistant would then translate and record the responses in English. This was crucial if farmers were to adequately express their views. All translations were cross-checked with a language expert to ensure accuracy. However, most of the time, the researcher's translations accurately captured the meaning. Responses were voice-recorded and notes taken simultaneously. Information sought included farmers' demographic characteristics, characteristics of storage types used, process of acquisition and durability of the storage type.

3.4 Data analysis

This started in the field where quotes that were clearly stated were recorded verbatim. Themes were identified using Nvivo software. This helped to organise the data. Then, a Microsoft Word Document file was created. Codes were generated by reading each excerpt several times to ensure familiarity with the content [42, 43]. Content analysis was done based on the identified themes [41]. The themes were based on the different storage types because we needed to describe them. Descriptive analysis was done by gleaning from the interviews and recording the quotes. To understand the storage types and their cost, farmers were asked to describe how they were acquired and to give their local names. Pictures of the storage types were also taken.

4. Results

4.1 Social-economic characteristic of smallholder maize farmers

The farmer's ages ranged from 16 to 90 years with an average age of 41 years. Over half of the smallholder maize farmers interviewed had attained formal primary level of education. Many farmers were growing more than one crop and keeping a few domesticated animals. All of them depended on agriculture for their livelihood, and mainly employed family labour. The famers interviewed mainly used rudimentary tools such as hand hoes. Many farmers owned the land, and a few were renting. The majority of the families, which were extended, lived in small mud-and-wattle grass-thatched huts. Some families had six to ten members in the household.

4.2 Perception of the smallholder farmers of storage types

The study captured smallholder maize farmers' general experience of storage. The study was carried out in the second harvest season of 2014/2015. All respondents stated that they experienced storage challenges arising from the storage methods they used. Since the majority of them rely on the maize produced for food and income, storage challenges affected their food and income security. The poor nature of storage negatively affected the quantity and quality of the maize. Sadly, the smallholder maize farmers reported that when they ran out of maize for food, they were forced to buy back at inflated prices the maize that they had sold off cheaply earlier in the year.

Table 1 shows the different storage types used by smallholder maize farmers.

The findings show that ten different storage types were used by smallholder maize farmers in the sampled districts of Eastern Uganda. As can be seen from **Table 1**, storage types have different names in the different dialects. Sometimes, even within the same district they are referred to differently; for example, *edula* and *etuuja* to mean granary in Katakwi. Some storage types like the jerry-can had no local name in Ateso because it was not made locally. Respondents explained that some storage types were not used in their districts for various reasons. For example, while the granary storage type was common in Manafwa and Katakwi districts, it was not used in Iganga because of the local people had limited knowledge of how to construct it; and the materials used for making it were also scarce.

4.3 Description, acquisition and usage of storage types

In SSA, post-harvest losses contribute substantially to the total crop losses incurred by smallholder maize farmers. Most often, maize prices are very low just after the harvest, when there is plenty of maize on sale. Thus, farmers have to store grain for family consumption, for sale, and for seed in the next planting season. These, and many more factors, call for proper storage in order to serve these purposes when there is no production. The subsequent paragraphs present the description of the different storage types, how they are acquired, and why they are used.

4.3.1 Granary storage type

The granary has a conical roof made of poles, reeds, and grass, and the bottom is round and made of mud. It is normally raised above the ground to avoid damage from termites, rats, and ground moisture. Granaries are built for specific crops. For

Storage type used	Local names for the different storage types in the sample districts			Percentage usage of each storage type[1]
	Lusoga in Iganga	Lumasaba in Manafwa	Ateso in Katakwi	
Granary	*Ekyagi*	*Sirara*	*Edula/Etuuja*	9.9
Cribs	*Ekyagi*	*Hayu*	*Etekati*	4.4
Tins	*Endebe*	*Kolokolo*	*Edebe*	0.3
Basket	*Ekibo*	*Sisibbo*	*Edita*	0.4
Above fire	*Ekibanyi*	*Inunga*	*Aitoola/Aruodo*	0.4
House corner	*Munsonda*	*Mukona*	*Esoodaloekai*	8.9
Jerry-can (old used)	*Ekidomola*	*Kupila*	*Jerikan*	0.1
House roof	*Akasolya*	*Sisoli*	*Atuluru/Ebibiru*	0.7
Sacks	*Ekutiya/Kavera*	*Isawu*	*Ipukoi*	74.4
Pot	*Ensuwa*	*Inyungu*	*Amot*	0.5
Total				**100**

[1]*The percentages are calculated to give the extent of use within the sample. However, because of the small sample size, this does not reflect the prevalence of the storage type.*

Table 1.
Name in local dialect and percentage usage by storage type.

example, a maize granary is different from a millet granary. Although at first glance it seems easy to construct, some smallholder maize farmers admitted that they do not use granaries because they are hard to construct and require expertise. The granary has an advantage of being cheap and made from locally available materials and can store relatively large amount of grain. However the disadvantage of such storage is that they are prone to fires and rodent attack which may destroy the stored maize grain (**Figure 2**).

On the construction of the granary, one participant had this to say:

> *Depending on where one gets the materials for granary construction from, one may not need money to construct it because all the materials are locally available free of charge in some villages. However, if a farmer lacks the skills for granary construction, he/she has to hire someone to construct it. Since granaries are constructed outside the house one, must choose a strategic place where water flow is limited to protect it from floods. More so, the location should be safe to prevent easy access by thieves. Thus, granaries are normally constructed close to the main house for security purposes and are mainly used by male farmers (Male respondent, Omodoi sub-county, Katakwi district).*

The cost of a granary varies from farmer to farmer, village to village and district to district. This is because most smallholder maize farmers use locally available materials for construction (where these exist). The cost is difficult to estimate because the owners hardly know how to monetise the materials used. This is because materials are

Figure 2.
A farmer explaining why maize granary is never smeared with mud.

obtained freely from within their surroundings. In addition, the cost also varies with the size, shape and with the season. During dry spells, the grass and water become scarce, which increases the cost. However, by triangulating information from different farmers, the granary was estimated to cost on average about 67 USD.

A granary built for maize is not smeared with mud to allow for continuous flow of air, which is useful in maize drying, while the one for millet must be smeared because the grain is tiny. One participant had this to say:

> *When making a granary, one starts by creating a sign of plus {+} by joining two short sticks of the desired bottom size which are either tied or nailed to keep them stuck together. Here, you start making it with small sticks or reeds. The length is determined by the size, in that if it is very tall and big, the granary can easily collapse. The granary is normally placed on four stones to prevent damage by flooding water. The stones help in raising the granary above the ground. The stones must be at the same level to avoid tilting the granary. The poles are supposed to support the granary and*

> *the roof. Stones are used to prevent storm water from damaging the stored maize (Male farmer, Butiru sub-county, Manafwa district).*

If the granary is well-kept, it can be used for two or more years, but it must be roofed properly and protected from termites. The main reason for using the granary is that it can keep a lot of maize outside the house, which reduces inconvenience to the inhabitants. However, the demerit of using the granary is its susceptibility to pests, floods, rats and thieves.

The use of traditional granaries to store maize is attributed to inadequate resources; even farmers who knew about modern storage types said their price was prohibitive. Besides, the modern storage types are scarce.

4.3.2 Sack storage type

Sacks were the most commonly used storage type (**Figure 3**).

From **Table 1**, 74.4% of the sampled smallholder maize farmers stored their maize in sacks. Even where a farmer used more than one storage type, a sack would be among them. Farmers used sacks because of their affordability, accessibility, and flexibility. Sacks were bought from the shops around the villages. They cost between 0.27USD and 0.40USD. The capacity of each sack is 100 kilogrammes. The advantages of this method is quick access and it could easily be re-used if it is not destroyed by rats. A participant said:

> *Sacks can also keep the maize in one place and enable lifting for sun-drying easily. One would also know how much maize he/she has by just counting the sacks. The drawback of sacks is that it is susceptible to mould, pests, rats and termites (Female farmer, Ibulanku sub-county, Iganga district).*

Although sacks are used by the majority of the farmers, they absorb moisture when put on bare ground and offer little protection against pests. If they are not

Figure 3.
Maize stored in sack.

damaged by rats, sacks can be used for one year. Well-to-do farmers have started adopting modern sacks referred to as hermetic triple bags, which are airtight [8]. However, they are expensive; each costs 1.88 USD to 2.14 USD, hence few smallholder maize farmers can afford them. Therefore, they are very scarce. This is demonstrated by the fact that it took 18 years for hermetic bags that were used in central America in 1990 to reach Africa in 2008 [10].

4.3.3 Pot storage type

Pots are made of clay and are burnt to make them strong and impermeable. Large and small pots with storage capacities ranging from 5 to 20 kilogrammes were used. The maize stored in pots would remain clean and safe, but it requires regular sun-drying to kill off the pests and to avoid mould. One participant noted that:

> *The challenges of using pots are that they are fragile, require continuous sun-drying, have limited capacity, and are very difficult to make. Pots are either purchased or made locally. Smallholder maize farmers with pot-making skills would make them for either home use or sell (Female farmer, Katakwi sub-county, Katakwi district).*

A big pot as shown in **Figure 4** costs 10.72 USD. The maize kept in a pot cannot be attacked by rats and other pests.

The top of the big pot is covered with small pot. Cow-dung mixed with soil or ash is used to seal off the top, thus preventing access by rats and other pests as shown in **Figure 4**. Air permeation is limited, but in case of excessive moisture around the pot, the maize may start to germinate. This is an ancient traditional storage type which is slowly getting extinct. However, smallholder maize farmers who use pots said that they did so for ancestral attachment. Pots are safe and can be stored in the house, are locally available, and are relatively cheap compared to cribs. Due to its fragile nature, the pot can be easily damaged, even by children. Hence, their durability depends on how careful the users are.

4.3.4 Above-the-fire storage type

This storage type was used by few of the smallholder maize farmers. Above-the-fire storage type takes various forms. Some farmers use strings, while others use ropes or wire to tie the maize together. The tied maize cobs are then hung above the cooking fireplace. However, some farmers construct platforms above the cooking fireplace. The platform raised above the cooking fireplace is constructed by use of poles, nails and ropes. After harvest, maize is placed on this platform. When maize is off-season the platform may be used to keep other household items like firewood (**Figure** 5).

When ropes and wires are used, the maize is tied on the rope or wire and is hung above the cooking fireplace. The rope or wire is hung above the fire, which makes it very difficult for rats to attack the maize. The smoke from the cooking fireplace helps to protect the maize from damage by insects and rodents. This storage type is cheap and is easy to use, provided the cooking fireplace is safe. It also enables continuous drying of the maize. The maize kept above the fireplace was reported to germinate easily. Generally, this storage type is used mainly to preserve the seeds. A participant stated that:

Figure 4.
Pot used to keep maize.

> *The disadvantage with this storage type is that it cannot store big amounts of maize. Hence, it stores mainly for planting. The use of smoke makes our maize to turn brown or black and so we cannot eat it because it looks dirty (Male farmer, Omodoi sub-county, Katakwi district).*

The storage type is constructed with poles, strings, and nails, which are locally available. Smallholder maize farmers with cooking fire places do not incur extra cost in acquiring this storage type. It is a storage type that has been used for a long time and those who use it copied it from past generations. A participant noted that:

> *Another important advantage of this storage type is that it can be used for many things in that when maize is not available it can store other household items (Female farmer, Ngariam sub-county, Katakwi district).*

Figure 5.
Above-the-fire storage type.

This storage type was very difficult to monetise because the users normally got the materials free and used their own labour to construct it. Nevertheless, the estimated cost is about 26.81 USD.

4.3.5 Basket storage type

A basket is the other type of storage used by smallholder maize farmers in Uganda (**Figure 6**).

These are mainly used to store maize temporarily during transitory periods. **Table 1** showed that few of smallholder maize farmers used baskets. It is a round-shaped open bowl made locally with banana and papyrus fibre. This storage type has ancestral attachments. Baskets are made locally and are affordable. They vary in size and may keep between five and twenty kilogrammes of maize. A participant noted that:

> *Those who cannot make the baskets buy them from the communities because some experts make baskets for sell. They are less expensive compared to granaries and cribs. The maize is kept clean as it undergoes drying. The basket is durable in that it can last for two or more years if properly kept. The danger with basket storage is that it is highly susceptible to pests, rats, and mould, which cause high losses. When purchased, baskets cost between 2.68 UDS and 5.36 USD depending on the size (Female farmer, Bubutu sub-county, Manafwa district).*

Respondents explained that the baskets are normally used every season but for a short period because, like sacks, they are easy to carry around when sun-drying maize.

4.3.6 The jerry-can storage type

Although jerry-cans are bought to fetch water, they have a dual purpose among smallholder maize farmers (**Figure 7**). When the jerry-cans get old, they are turned

Figure 6.
Basket storage type.

Figure 7.
Jerry-can storage type.

into maize storage containers. Unlike other storage types, this particular one is not locally made. In this storage type, maize is put into the jerry-can and then sealed with a lid. One participant noted that:

> *This storage type is better because the maize is protected from rats, pests and birds. When the jerry-can is put in a dry place, the maize can stay safe for one to two months. The jerry-can should not be placed on the cold wet ground because it will mould. The challenge with this kind of storage is that it cannot store maize for long and stores little in quantities. You also have to keep sun-drying (Female farmer, Makuutu sub-county, Iganga district).*

Although jerry-cans are readily available, they are not cheap. A new one costs about 1.88 USD; hence, farmers have to find damaged ones that can no longer be used to fetch water.

4.3.7 The crib storage type

Cribs are either open or closed. During the interview, smallholder maize farmers explained that cribs are constructed using locally available materials. They are built like a house, roofed with iron sheets, or thatched with grass. **Figure 8** shows a crib made of eucalyptus tree poles, timber, nails and iron sheet without wire mesh.

The well-to-do farmers use iron sheets, timber, nails and wire mesh to construct the crib. However, the poor farmers (unfortunately the majority) use grass, papyrus, tarpaulins, reeds, poles and nails to construct cribs. A participant noted that:

> *To protect maize from pilferage, we add chicken mesh on the crib, which makes it impossible for one to pick the maize cob. Sometimes we do not put a door but just leave a space towards the roof and use a ladder to put and remove the maize. Smallholder maize farmers construct cribs with doors when one has security in the home (Male farmer, Butiru sub-county, Manafwa district).*

This storage type is important because it enables continuous drying of maize and offers a large storage capacity compared to granaries. Farmers with no expertise in crib-making hire experts to undertake the construction. A crib is the most expensive storage type costing approximately 268.06 USD.

The crib may be made with two small doors to aid in adding new stock or taking out the maize. If the first part of the crib up to the first door level is full, the farmer uses the higher door to put in more maize. Similarly, in removing the maize one can use the first-in first-out (FIFO) method.

The use of crib storage type depends on the amount of maize produced. Smallholder maize farmers producing less than one tonne can hardly ever use them. The nature or type of a crib depends on the farmer's financial ability and expertise. Besides being expensive, the cribs are susceptible to attack by rats, pests, birds and thieves.

Figure 8.
Traditional crib.

4.3.8 The plastic bucket and iron container storage type

Another storage type used by smallholder maize farmers is the plastic bucket and iron container. This storage type is an open round- or square-shaped container made of plastic or iron. The iron containers today are quite rare because plastic containers are more available. Plastic buckets are more common today. Buckets are either bought directly from the market for storage or obtained after emptying buckets containing products like washing detergent packaged in these containers. Buckets are hermetic (airtight) and limit many pests. However, they are susceptible to fire, besides being more expensive than sacks (**Figure 9**).

However, to prevent moulding, the container must be kept in a dry place and the maize must be clean and dry. The tins are of varying capacities ranging from two kilogrammes to 20 kilogrammes of grain. Tins are not locally made and hence have to be purchased at 0.54 USD for five kilogrammes to 8.04 USD for twenty kilogrammes. Thus, the majority of smallholder maize farmers find them expensive compared to sacks.

4.3.9 The house-corner storage type

The house-corner is another storage type used by smallholder maize farmers. Quite often, the farmers spread their harvested maize in the house-corners. Depending on whether the house is cemented or not, they turn one corner of the house into a storage place. This space is normally in their sitting room because it enables them to easily take the maize out for sun-drying.

However, the drawback with this storage type is that maize is susceptible to destruction by rats, birds and pests, which exacerbates grain losses. A participant noted that:

> *This storage type is acquired through house construction. Once one has built a house then the maize seeds are just spread in one of the house corners. (Female farmer, Bulamagi sub-county, Iganga district).*

This storage was very difficult to monetise. Nevertheless, in this study, it is treated as a zero cost because the house is built for accommodation and not for storage.

Figure 9.
Sample of plastic bucket.

4.3.10 The house-roof storage type

This storage type is transitional in that it does not keep maize for long. When maize is harvested, it can be stored on the rooftop of the house to help with continued drying. Maize cobs with or without husks are deposited on rooftops. It prevents domestic birds from destroying the maize because they cannot easily climb to the roof of the house. This storage type also limits pilferage and rat damage but is susceptible to mould and pests. One participant said:

> *When the roof is thatched with grass, then the maize remains on the cob and with husks; but for an iron-roofed house, the maize may be removed from the cob (Female farmer, Ibulanku sub-county, Iganga district).*

House-rooftops store maize for a short period. They are also affected by weather changes, particularly during the rainy season. The storage type is difficult to monetise; first, because some rooftops are made of grass and others of iron sheets. Secondly, because the rooftops are part-and-parcel of their accommodation, farmers did not see it as an extra cost to acquire the storage type and hence it was difficult for them to monetize it.

5. Discussion

Household storage presents complex challenges for smallholder maize famers in Uganda. It leads to loss of grain in storage [11]. The description of the various storage types used reveals that there is an urgent need to tackle storage inadequacies among smallholder maize farmers [12]. This study concurs with [33] findings in his study of grain storage in Africa. He revealed that storage methods differ even within the same country. This study shows that, to a large extent, all the storage types described present a degree of risk. And yet maize has the potential of expanding the income and food security of the smallholder maize farmers in Uganda [19, 23, 24]. The storage challenge has contributed to escalating food and income insecurity, thereby condemning many smallholder maize farmers to perpetual poverty and famine [31, 33, 34].

Some storage types, such as sacks, are susceptible to pests and yet cribs that are less susceptible are quite expensive to construct. The findings show that smallholder maize farmers are still trapped in the traditional storage types that alarmingly exacerbate their income and food insecurity (**Table 1**). In developing countries, losses in traditional storage facilities account for a significant proportion of all post-harvest losses in cereals [7, 30]. In Uganda, storage losses of maize after harvest are about 6% of the quantity harvested, on average [31]. The findings lend some credence on how storage types are acquired and used and the reason for their use. This is pertinent to researchers and policy-makers because it adds anecdotal evidence on the description of storage types. This is crucial for policy formulation in the fight against food and income insecurity.

Purchasing was the most common method of acquiring the storage facilities used. However, while some farmers would construct their storage structures, others hired experts to construct the storage facilities for them. Given the limited financial ability of most smallholder maize farmers, it is hard for them to acquire better storage types.

This, therefore, constrains their effort to increase food supply since much is lost during storage. Farmers purchase the storage types that are available and affordable, but also remain committed to traditional attachments.

Taking the examples of the storage types used by smallholder maize farmers, the findings demonstrate that storage requires urgent attention because it impacts negatively on food and income security of the farmers. It is safe to say that deliberate efforts need to be directed towards improving maize storage at household level. Attention should be directed towards improving the existing storage technologies at household level, improving the storage efficiency, facilitating more storage capacity and perhaps extending the storage length, which is critical for income and food security. The emphasis on household storage is premised on the dire need for safe storage and the fact that much of the grain produced is for human consumption. Given the level at which the storage types currently operate, it seems the gains from such an approach would be high.

It is arguable that the current state of storage at household level cannot eradicate food and income insecurity. The storage types elucidated above clearly demonstrate the urgent need to improve on the structures if safety of the grain is to be achieved. When safety is threatened, smallholder maize farmers' participation in economic activity through the sale of maize is compromised. Unfortunately, the farmers' ability to improve the storage on their own seems limited given their continued use of traditional storage methods, despite the losses experienced. This has left them in a predicament because the rate of damage to the grain in storage by rats and other pests is alarming.

Consequently, this limits economic development because the majority of the farmers derive their livelihood from agriculture. And yet protection of the grains in the current storage types is almost impossible [25]. Besides, smallholder maize farmers predominantly use traditional protection methods like *neem* tree leaves, sun-drying, and smoking, which are not very effective in protecting the grains.

In conclusion, despite the predominant use of traditional storage types among smallholder maize farmers in Uganda, its safety is still very poor. It also causes high losses to the farmers. Clearly, these storage types need urgent improvement to enable the famers to become food and incomes secure [28]. Overall, the present storage types described suggest a need for change, which must be integrated in the way maize is stored at smallholder level. Apart from being traditional, the storage types have no ability to protect the maize from waste and, therefore, the farmers will certainly continue to suffer if no change is instituted. As a key component of family security, every effort should be directed at improving storage for better income and food security.

It can also be seen that farmers use different types of storage. Two aspects that need further research are:

1. Farmers were not asked their preferred storage type (irrespective of cost or local material availability, but just based on their perceptions of storage efficiency). Such knowledge would assist in directing the efforts to improve local storage performance.

2. The variation and prevalence of storage types in Uganda and indeed in SSA needs to be established. The effort to improve storage efficiency can be directed based on farmer efficiency perceptions and prevalence.

6. Conclusions

This book chapter is novel in that it explicates an important aspect of maize storage and acquisition among the vulnerable smallholder farmers. It is demonstrated that they use rudimentary that cannot protect their maize grain from damage by rodents, insects, rats etc. that greatly affects their ability to be income and food secure moreover maize is a duo purpose crop for these smallholder.

Acknowledgements

The authors are grateful to IDRC, Gordon Institute of Business Science, University of Pretoria and the College of Business and Management Sciences (CoBAMS) Makerere University for the support.

Conflict of interest

There is no conflict of interest to declare.

Notes/thanks/other declarations

I take this opportunity to thank three institution Makerere University, University of Pretoria- Gordon institute of Business Sciences and IDRC. In addition, I thank my co-authors and mentors as well as the respondents.

Author details

Anthony Tibaingana[1,2]*, Godswill Makombe[2] and Tumo Kele[3]

1 Department of Marketing and Management, School of Business, Makerere University College of Business and Management Sciences, Kampala, Uganda

2 The University of Pretoria, Gordon Institute of Business Science, Johannesburg, South Africa

3 Tshwane School for Business and Society, Tshwane University of Technology, Pretoria, South Africa

*Address all correspondence to: atibaingana@gmail.com

References

[1] di Domenico AS, Christ D, Hashimoto EH, Busso C, Coelho SRM. Evaluation of quality attributes and the incidence of Fusarium sp. and Aspergillus sp. in different types of maize storage. Journal of Stored Products Research. 2015;**61**:59-64

[2] Paudyal S, Opit GP, Osekre EA, Arthur FH, Bingham GV, Payton ME, et al. Field evaluation of the long-lasting treated storage bag, deltamethrin incorporated, (ZeroFly® Storage Bag) as a barrier to insect pest infestation. Journal of Stored Products Research. 2017;**100**(70):44-52

[3] Williams SB, Murdock LL, Baributsa D. Storage of maize in Purdue improved crop storage (PICS) bags. PLoS One. 2017;**12**(1):e0168624

[4] Agona A, Nabawanuka J, Muyinza H. An overview of maize in Uganda. 2001. Retrieved from Kampala

[5] Bwambale J, Mourad KA. Modelling the impact of climate change on maize yield in Victoria Nile Sub-basin, Uganda. Arabian Journal of Geosciences. 2022;**15**(1):1-19

[6] Okoboi G, Muwanga J, Mwebaze T. Use of improved inputs and its effect on maize yield and profit in Uganda. African Journal of Food, Agriculture, Nutrition and Development. 2012;**12**(7):6931-6944

[7] Kumar D, Kalita P. Reducing postharvest losses during storage of grain crops to strengthen food security in developing countries. Food. 2017;**6**(1):8

[8] Anankware P, Fatunbi A, Afreh-Nuamah K, Obeng-Ofori D, Ansah A. Efficacy of the multiple-layer hermetic storage bag for biorational management of primary beetle pests of stored maize. Academic Journal of Entomology. 2012;**5**(1):47-53

[9] Chigoverah AA, Mvumi BM. Efficacy of metal silos and hermetic bags against stored-maize insect pests under simulated smallholder farmer conditions. Journal of Stored Products Research. 2016;**100**(69):179-189

[10] Ndegwa MK, De Groote H, Gitonga ZM, Bruce AY. Effectiveness and economics of hermetic bags for maize storage: results of a randomized controlled trial in Kenya. Crop Protection. 2016;**90**:17-26

[11] Udo IO. Potentials of Zanthoxylum xanthoxyloides (LAM.) for the control of stored product insect pests. Journal of Stored Products and Postharvest Research. 2011;**2**(3):40-44

[12] Vorotnikova E. Optimal storage capacity allocation in grain merchandizing. Paper presented at the 2016. In: Annual Meeting of Southern Agricultural Economics Association. Texas: San Antonio; 2016

[13] Mdangi M, Mulungu L, Massawe A, Eiseb S, Tutjavi V, Kirsten F, et al. Assessment of rodent damage to stored maize (Zea mays L.) on smallholder farms in Tanzania. International Journal of Pest Management. 2013;**59**(1):55-62

[14] Ouma DO. Intra-regional agricultural exports in the East African Community. African Journal of Economic Review. 2017;**5**(1):14-31

[15] Proctor D. Grain Storage Techniques: Evolution and Trends in Developing Countries. Rome: Food and Agriculture Organisation; 1994

[16] Stathers T, Lamboll R, Mvumi BM. Postharvest agriculture in changing climates: Its importance to African smallholder farmers. Food Security. 2013;**5**(3):361-392

[17] Garbaba CA, Denboba LG, Ocho FL, Hensel O. Nutritional deterioration of stored Zea mays L. along supply chain in southwestern Ethiopia: Implication for unseen dietary hunger. Journal of Stored Products Research. 2017;**100**(70):7-17

[18] Kagoda F, Gidoi R, Isabirye BE. Status of maize lethal necrosis in eastern Uganda. African Journal of Agricultural Research. 2016;**11**(8):652-660

[19] Owach C, Bahiigwa G, Elepu G. Factors influencing the use of food storage structures by agrarian communities in Northern Uganda. Journal of Agriculture, Food Systems, and Community Development. 2017;**7**(2):127-144

[20] Udoh JM, Cardwell KF, Ikotun T. Storage structures and aflatoxin content of maize in five agroecological zones of Nigeria. Journal of Stored Products Research. 2000;**36**(2):187-201

[21] Brennan MJ. The supply of storage. The American Economic Review. 1958;**48**(1):50-72

[22] Wright BD. The economics of grain price volatility. Applied Economic Perspectives and Policy. 2011;**33**(1):32-58

[23] Govender V, Aveling T, Kritzinger Q. The effect of traditional storage methods on germination and vigour of maize (Zea mays L.) from northern KwaZulu-Natal and southern Mozambique. South African Journal of Botany. 2008;**74**(2):190-196

[24] Midega CA, Murage AW, Pittchar JO, Khan ZR. Managing storage pests of maize: Farmers' knowledge, perceptions and practices in western Kenya. Crop Protection. 2016;**90**:142e149

[25] Baoua I, Amadou L, Ousmane B, Baributsa D, Murdock L. PICS bags for post-harvest storage of maize grain in West Africa. Journal of Stored Products Research. 2014;**58**:20-28

[26] Sekumade A, Akinleye S. Comparative analysis of maize storage technologies in North Central Nigeria. Research Journal of Social Sciences. 2009;**4**(1):7-14

[27] Stephens EC, Barrett CB. Incomplete credit markets and commodity marketing behaviour. Journal of Agricultural Economics. 2011;**62**(1):1-24

[28] Jones M, Alexander C, Lowenberg-DeBoer J. An initial investigation of the potential for hermetic Purdue Improved Crop Storage (PICS) bags to improve incomes for maize producers in sub-Saharan Africa. In: Department of Agricultural Economics. West Lafayette: Purdue University; 2011. pp. 11-03

[29] Affognon H, Mutungi C, Sanginga P, Borgemeister C. Unpacking postharvest losses in sub-Saharan Africa: a meta-analysis. World Development. 2015;**66**(C):49-68

[30] Tapondjou L, Adler C, Bouda H, Fontem D. Efficacy of powder and essential oil from Chenopodium ambrosioides leaves as post-harvest grain protectants against six-stored product beetles. Journal of Stored Products Research. 2002;**38**(4):395-402

[31] Omotilewa OJ, Ricker-Gilbert J, Ainembabazi H, Shively G. Impacts of improved storage technology among smallholder farm households in Uganda. Paper presented at the 2016. In: Fifth

International Conference. Ethiopia: Addis Ababa; 2016

[32] Jones M, Alexander C, Widmar NO, Ricker-Gilbert J, Lowenberg-DeBoer JM. Do insect and mold damage affect maize prices in Africa? Evidence from Malawi. Modern Economy. 2016;7(11):1168-1185

[33] Shepherd A. Grain storage in Africa: Learning from past experiences. Food Chain. 2012;**2**(2):149-163

[34] Thamaga-Chitja JM, Hendriks SL, Ortmann GF, Green M. Impact of maize storage on rural household food security in Northern Kwazulu-Natal. Journal of Family Ecology and Consumer Sciences. 2004;**32**(1):8-15

[35] Jenkins GP, Leung L. A Cost Benefit Analysis of Maize Production and Marketing In Uganda. 2013. Retrieved from

[36] Hanjra MA, Culas RJ. The political economy of maize production and poverty reduction in Zambia: Analysis of the last 50 years. Journal of Asian and African Studies. 2011;**46**(6):546-566

[37] Tibaingana A. Extending the Theory of Storage to a Perishable Commodity in an Underdeveloped Market: A Case of Uganda. University of Pretoria; 2017

[38] Uganda Bureau of Statistics. Ugandan National Agriculture Census of 2008/2009: Agricultural Household and Holding Characteristics Report. 2010. Available from: http://www.fao.org/3/CA0511EN/ca0511en.pdf

[39] Burnard L. What is the Text Encoding Initiative?: How to Add Intelligent Markup to Digital Resources2014. DOI: 10.4000/books.oep.426

[40] Goss JD, Leinbach TR. Focus groups as alternative research practice: experience with transmigrants in Indonesia. Area. 1996;**28**(2):115-123

[41] Fowler SA, Saunders CJ, Hoffman MA. Variation among consent forms for clinical whole exome sequencing. Journal of Genetic Counseling. 2018;**27**(1):104-114

[42] Young JC, Rose DC, Mumby HS, Benitez-Capistros F, Derrick CJ, Finch T, et al. A methodological guide to using and reporting on interviews in conservation science research. Methods in Ecology and Evolution. 2018;**9**(1):10-19

[43] Oliveira M, Bitencourt C, Teixeira E, Santos AC. Thematic content analysis: Is there a difference between the support provided by the MAXQDA® and NVivo® software packages. In: Paper Presented at the 12th European Conference on Research Methods for Business and Management Studies. 2013